Arms Control and Missile Proliferation in the Middle East

This edited volume provides a systematic analysis of the missile threat and proliferation issue in the Middle East region.

The question of how to increase the level of security in the Middle East is not a new one, given the conflict-ridden nature of the region. The solution attempted for this predicament has typically revolved around intense arms build-ups, a strategy which can prove self-defeating due to the subsequent countermeasures employed by neighbouring states.

Arms Control and Missile Proliferation in the Middle East focuses on the strategic proliferation of arms, with a specific emphasis on missiles. This unique emphasis enables the contributors to provide a dynamic new perspective on conceptual and political disarmament efforts, thereby distinguishing this volume from many other related works on the region, which deal mainly with weapons of mass destruction. The book also explores the possibility of a reduction in weapon arsenals, examining a more promising cooperative security concept which includes confidence- and security-building measures.

This book will be of much interest to students of arms control, Middle Eastern politics, Gulf security, war and conflict studies, security studies, and IR.

Bernd W. Kubbig is Project Director at the Peace Research Institute Frankfurt (HSFK/PRIF), Adj. Professor at Goethe University, Frankfurt, and Coordinator of the Academic Peace Orchestra Middle East.

Sven-Eric Fikenscher is a Research Fellow at Goethe University, Frankfurt. He was a Research Assistant at the Peace Research Institute Frankfurt (HSFK/PRIF) from 2006–11 and on the staff of the Academic Peace Orchestra Middle East.

Routledge Global Security Studies
Series Editors: Aaron Karp, Regina Karp, and Terry Teriff

Nuclear Proliferation and International Security
Edited by Morten Bremer Maerli and Sverre Lodgaard

Global Insurgency and the Future of Armed Conflict
Debating fourth-generation warfare
Terry Terriff, Aaron Karp and Regina Karp

Terrorism and Weapons of Mass Destruction
Responding to the challenge
Edited by Ian Bellany

Globalization and WMD Proliferation
Terrorism, transnational networks, and international security
Edited by James A. Russell and Jim J. Wirtz

Power Shifts, Strategy, and War
Declining states and international conflict
Dong Sun Lee

Energy Security and Global Politics
The militarization of resource management
Edited by Daniel Moran and James A. Russell

US Nuclear Weapons Policy after the Cold War
Russians, 'rogues' and domestic division
Nick Ritchie

Security and Post-Conflict Reconstruction
Dealing with fighters in the aftermath of war
Edited Robert Muggah

Network Centric Warfare and Coalition Operations
The new military operating system
Paul T. Mitchell

American Foreign Policy and the Politics of Fear
Threat inflation since 9/11
Edited by A. Trevor Thrall and Jane K. Cramer

Risk, Global Governance and Security
The other war on terror
Yee-Kuang Heng and Kenneth McDonagh

Nuclear Weapons and Cooperative Security in the 21st Century
The new disorder
Stephen J. Cimbala

Political Economy and Grand Strategy
A neoclassical realist view
Mark R. Brawley

Iran and Nuclear Weapons
Protracted conflict and proliferation
Saira Khan

US Strategy in Africa
AFRICOM, terrorism and security challenges
Edited by David J. Francis

Great Powers and Strategic Stability in the 21st Century
Competing visions of world order
Edited by Graeme P. Herd

The Globalisation of NATO
Intervention, security and identity
Veronica M. Kitchen

International Conflict in the Asia-Pacific
Patterns, consequences and management
Jacob Bercovitch and Mikio Oishi

Nuclear Proliferation and International Order
Challenges to the Non-Proliferation Treaty
Edited by Olav Njølstad

Nuclear Disarmament and Non-Proliferation
Towards a nuclear-weapon-free world?
Sverre Lodgaard

Nuclear Energy and Global Governance
Ensuring safety, security and non-proliferation
Trevor Findlay

Unipolarity and World Politics
A theory and its implications
Birthe Hansen

Disarmament Diplomacy and Human Security
Regimes, norms and moral progress in international relations
Denise Garcia

Causes and Consequences of Nuclear Proliferation
Edited by Robert Rauchhaus, Matthew Kroenig and Erik Gartzke

Why Did the United States Invade Iraq?
Edited by Jane K. Cramer and A. Trevor Thrall

Regional Powers and Security Orders
A theoretical framework
Edited by Jane K. Cramer and A. Trevor Thrall

A Perpetual Menace
Nuclear weapons and international order
William Walker

Iran's Nuclear Programme
Strategic implications
Joachim Krause

Arms Control and Missile Proliferation in the Middle East
Edited by Bernd W. Kubbig and Sven-Eric Fikenscher

Arms Control and Missile Proliferation in the Middle East

Edited by Bernd W. Kubbig and Sven-Eric Fikenscher

Routledge
Taylor & Francis Group
LONDON AND NEW YORK

First published 2012
by Routledge
2 Park Square, Milton Park, Abingdon, Oxon OX14 4RN

Simultaneously published in the USA and Canada
by Routledge
711 Third Avenue, New York, NY 10017

Routledge is an imprint of the Taylor & Francis Group, an informa business

First issued in paperback 2013

British Library Cataloguing in Publication Data
A catalogue record for this book is available from the British Library

Library of Congress Cataloging-in-Publication Data
Arms control and missile proliferation in the Middle East / edited by Bernd Kubbig.
p. cm. – (Routledge global security studies ; 29)
Includes bibliographical references and index.
1. Arms control–Middle East. 2. Disarmament–Middle East. 3. Guided missiles–Middle East. 4. Security, International–Middle East. I. Kubbig, Bernd W., 1950–
JZ6009.M628A76 2012
327.1'740956–dc23 2011031885

ISBN: 978-0-415-60111-5 (hbk)
ISBN: 978-0-415-51639-6 (pbk)
ISBN: 978-0-203-13816-8 (ebk)

Typeset in Baskerville
by Wearset Ltd, Boldon, Tyne and Wear

Contents

List of illustrations x
Contributors xi
Acknowledgements xviii
List of abbreviations xix

1 **Introduction: setting the stage: decreasing the security dilemma by gradual missile reductions** 1
BERND W. KUBBIG

PART I
Missiles and their proliferation as a threat 25

2 **Disarmament efforts in the region: lessons from the Arms Control and Regional Security talks** 27
EMILY B. LANDAU AND DALIA DASSA KAYE

3 **Missiles in the Middle East: their destabilizing role** 39
DENNIS M. GORMLEY, COLIN P. CLARKE, AND JÜRGEN ALTMANN

PART II
Conditions for a successful step-by-step approach 63

4 **Taking overall military asymmetries into account: objections from the Arab world and Iran** 65
BERND W. KUBBIG, MOHAMED NOMAN GALAL, MICHAEL HAAS, MURHAF JOUEJATI, SABAHAT KHAN, AHMED SAIF, AND MAHMOOD SARIOLGHALAM

5 **Taking Israel's security interests into account: deterrence policy in a changing strategic environment** 89
URI BAR-JOSEPH

6 **Advancing the control of weapons of mass destruction: an incremental approach** 106
GAWDAT BAHGAT, AVNER COHEN, SVEN-ERIC FIKENSCHER, GIORGIO FRANCESCHINI, BERND W. KUBBIG, AND PATRICIA LEWIS

7 **From (potential) spoilers to guarantors? The role of external powers** 125
SVEN-ERIC FIKENSCHER, SAMEH ABOUL-ENEIN, GÜLDEN AYMAN, ANTON KHLOPKOV, AND HUI ZHANG

8 **The verification challenge: concepts, requirements, and technologies** 149
JÜRGEN SCHEFFRAN, BHARATH GOPALASWAMY, DENNIS M. GORMLEY, BERND W. KUBBIG, UZI RUBIN, AND HARTWIG SPITZER

9 **The prospects for disarmament: the case of Hezbollah** 167
JUDITH PALMER HARIK AND WALID ABU-DALBOUH

10 **Designing disarmament strategies: the case of Hamas** 186
MARGRET JOHANNSEN, GHASSAN KHATIB, AND ANAT KURZ

PART III
Three milestones towards a Missile Free Zone 215

11 **The promise of military transparency: building on East–West experiences and the UN Register of Conventional Arms** 217
SVEN-ERIC FIKENSCHER, PATRICK FLAMM, LINA GOLOB, BERND W. KUBBIG, AND CHRISTIAN WEIDLICH

12 **The Missile Technology Control Regime, the Hague Code of Conduct, and missile proliferation** 234
DINSHAW MISTRY AND MARK SMITH

13 Caps and bans: limiting, reducing, and prohibiting missiles and missile defence 251

MARTIN SENN, JÜRGEN ALTMANN, BERND W. KUBBIG, JÜRGEN SCHEFFRAN, HANS-JOACHIM SCHMIDT, AND OLEG SHULGA

14 Conclusions: the stage reconsidered: findings, concluding remarks, and research perspectives 277

BERND W. KUBBIG

Appendix: the arsenals of actors relevant to a missile free zone in the Middle East/Gulf 303

MICHAEL HAAS AND BERND W. KUBBIG (COMPILERS)

Index 332

Illustrations

Maps

1	The scope of the proposed missile free zone	xxii
2	US military presence in the broader Middle East	132

Tables

13.1	Conceptual dimensions of East–West arms control	256
13.2	Reductions in the START series	257
A.1	Bahrain	304
A.2	Egypt	305–7
A.3	Iran	308–11
A.4	Israel	312–15
A.5	Kuwait	316
A.6	Oman	317
A.7	Qatar	318
A.8	Saudi Arabia	319–20
A.9	Syria	321–3
A.10	United Arab Emirates	324–5
A.11	Yemen	326
A.12	Hamas	327–8
A.13	Hezbollah	329–30

Contributors

Sameh Aboul-Enein is the Egyptian Deputy Ambassador to the United Kingdom and a Visiting Lecturer at the Diplomatic Academy of London. Dr Aboul-Enein was Egypt's Alternate Representative to the Conference on Disarmament and the UN Office at Geneva, an Expert Delegate at the 2005 and 2010 NPT Review Conferences, and Director of the UN Affairs Department at the Egyptian Ministry of Foreign Affairs. His publications focus on disarmament, security, and a Nuclear Weapon Free Zone in the Middle East. His contributions and views are solely in his academic and personal capacity.

Walid Abu-Dalbouh is Assistant Professor and the Director of the American Studies Programme at the Faculty of International Studies, University of Jordan, Amman. He works on security and strategic studies, conflict resolution, democracy and human rights, political economy, and peace studies.

Jürgen Altmann is a physicist and Peace Researcher at the Technical University in Dortmund, Germany. Since 1985 he has focused on the technical problems of disarmament and written on laser weapons, ballistic missile defence, microsystems technology, nanotechnology, non-lethal weapons, and armed uninhabited vehicles.

Gülden Ayman is Associate Professor at the Department of Political Science and International Relations at Marmara University, Istanbul. She has been a Visiting Scholar at the Cesare Alfieri Institute of Political Science, Florence, at the University of Kent, at the Center for Transatlantic Relations of the Free University of Berlin, and she conducted research at Yale University on a Fulbright grant. Her research interest focuses on territorial conflict, conflict resolution, and arms control.

Gawdat Bahgat is a Professor at the National Defense University in Washington, DC. His areas of expertise include energy security, counterterrorism, proliferation of weapons of mass destruction, international political economy, as well as the Middle East, Caspian Sea/Central Asia, and American foreign policy. Dr Bahgat is a prolific writer on these issues.

Uri Bar-Joseph is a Professor at the International Relations Division at Haifa University, Israel. He concentrates on strategic and intelligence studies, especially focusing on the Arab–Israeli conflict and Israeli security policy. In addition to a number of articles in peer reviewed journals, he has written several books.

Colin P. Clarke is a Doctoral Student for foreign policy and international security at Pittsburgh University, PA, as well as a Research Associate at the Research and Development Corporation (RAND). He specializes in counterinsurgency, international terrorism, and irregular warfare, especially in the Middle East and South Asia.

Avner Cohen is a Senior Fellow at Monterey Institute's James Martin Center for Nonproliferation Studies (CNS), Washington, DC. He served as a Public Policy Scholar at the Woodrow Wilson International Center for Scholars (2009/10) following a ten year affiliation with the Center for International and Security Studies at the University of Maryland. Dr Cohen is a two-time winner of prestigious MacArthur Foundation research and writing awards. He is an internationally recognized author and expert on nonproliferation issues in the Middle East and the Israeli nuclear programme.

Sven-Eric Fikenscher is a Research Fellow at Goethe University, Frankfurt. He was a Research Assistant at the Peace Research Institute Frankfurt (HSFK/PRIF) from 2006–11 and on the staff of the Academic Peace Orchestra Middle East. His research interests include US foreign policy, regional security in the Middle East, US–Indian relations, and international terrorism.

Patrick Flamm was a Research Assistant at the Peace Research Institute Frankfurt (HSFK/PRIF) from 2008–10. He studies International Relations, Philosophy of Law, Sociology and Korean Studies at Goethe University in Frankfurt and Korea University in Seoul. His research interests include security studies, civil–military relations and non-Western approaches to international politics and IR theory.

Giorgio Franceschini is a physicist and Research Associate at the Peace Research Institute Frankfurt (HSFK/PRIF) and at Darmstadt Technical University. In 2009, he was a Visiting Fellow at the James Martin Center for Nonproliferation Studies (CNS) in Monterey, CA. His current research focuses on political and technical aspects of nonproliferation, arms control, and nuclear disarmament.

Mohamed Noman Galal, a former ambassador, is currently Advisor to the Foreign Ministry of Bahrain. Previously, he was an Advisor for Strategic Studies and Supervisor of the programme for International Studies and Dialogue among Civilizations at the Bahrain Centre for Studies and Research (BCSR). Dr Galal is a member of the Egyptian Council on Foreign Affairs and a former Assistant Foreign Minister for Policy Planning. As a veteran Egyptian diplomat, he is an expert in Gulf, Arab, Asian, and UN affairs.

Lina Golob was a Research Assistant at the Peace Research Institute Frankfurt (HSFK/PRIF) from 2008–10. She is studying International Relations and Sociology of Law at Goethe University, Frankfurt. In 2010 she worked for the GTZ (German Technical Association) in Damascus, Syria.

Bharath Gopalaswamy, a Mechanical Engineer specializing in Numerical Acoustics, is a Researcher in the Arms Control and Non-proliferation Programme of the Stockholm International Peace Research Institute (SIPRI). Prior to that, he worked at the Indian Space Research Organization's High Altitude Test Facilities and the European Aeronautics Defence and Space Company's Astrium division in Germany. Dr Gopalaswamy's current focus is on space security.

Dennis M. Gormley is Senior Fellow at Monterey Institute's James Martin Center for Nonproliferation Studies, Washington, DC, a faculty member of the Graduate School of Public and International Affairs at the University of Pittsburgh, PA, and a Research Associate at the University's Ridgway Center for International Security Studies. Dr Gormley has served as a Consulting Senior Fellow for Technology and Defence Policy at the International Institute for Strategic Studies in London and has chaired or served on many U.S. Department of Defense advisory committees.

Michael Haas, an Intern at the Peace Research Institute Frankfurt in 2010, studies Political Science at the University of Innsbruck, Austria, where he also works as a Student Assistant. His academic interests are strategic theory, air and sea power, revolution in military affairs, nuclear weapons, and ballistic missile defence.

Margret Johannsen is a Senior Research Fellow at the Institute for Peace Research and Security Policy (IFSH) at the University of Hamburg and Post Graduate Lecturer for the Programme Peace and Security Studies at the University of Hamburg. Her publications focus on arms control, the Israeli–Palestinian conflict, the Middle East peace process, as well as on European and US Middle East Policy. She is the co-editor of the Annual Peace Report (Friedensgutachten).

Murhaf Jouejati is Professor of Middle East Studies at the National Defense University, Lecturer at George Washington University, as well as an Adjunct Scholar at the Near East South Asia Center for Strategic Studies of the National Defense University in Washington, DC. He is a specialist on Middle East affairs and particularly Syrian politics.

Dalia Dassa Kaye is a Senior Political Scientist at the Research and Development Corporation (RAND) in Santa Monica, CA. Beforehand, she was a Council on Foreign Relations International Affairs Fellow and served on the faculty at George Washington University. She is the author of several books and numerous articles related to Middle East security issues.

Sabahat Khan is a Defence Analyst at the Institute for Near East and Gulf Military Analysis (INEGMA) based in Dubai, UAE. He has a special interest in strategic security and military strategy in and around the Arabian Gulf, Iran, and South Asia. His recent work has included studies in Arabian Gulf maritime security, asymmetric warfare, missile proliferation, early warning networks, and missile defence.

Ghassan Khatib is Director of the Palestinian Government Media Center, Ramallah, as well as co-editor of the Palestinian–Israeli political internet-magazine bitterlemons.org. He is Vice-President for Community Outreach at Birzeit University (currently on leave) and a former Palestinian Authority Minister of Planning and Labour. In the years 1991–93 he was a member of the Palestinian delegation for the Madrid Middle East Peace Conference and the subsequent bilateral negotiations in Washington, DC. Additionally, he participated in the Palestinian national reconciliation talks in Cairo.

Anton Khlopkov is Director of the Center for Energy and Security Studies (CENESS), Moscow, an Editor-in-chief of the *Nuclear Club* journal and of the *Nuclear Nonproliferation Encyclopedia* (2009). He was Deputy and Executive Director at the Russian Center for Policy Studies and a member of the Non-proliferation and Export Control Working Group of the Partnership for Peace Consortium. His research focuses on regional aspects of WMD nonproliferation.

Bernd W. Kubbig is Project Director at the Peace Research Institute Frankfurt (HSFK/PRIF) and Adj. Professor (Privatdozent) at Goethe University, Frankfurt. He directs the internet programme on Missile Defense Research International and coordinates the international expert group Academic Peace Orchestra Middle East. He has published extensively on US foreign and security policy, missile defence, and space.

Anat Kurz is Director of Research at the Institute for National Security Studies (INSS), Tel Aviv, and a regular participant in Track II dialogues on Israeli–Palestinian conflict resolution and Middle Eastern regional security. She has published widely on insurgency related issues and substate political organizations. Her current research focuses on the Palestinian national movement and Israeli–Palestinian relations.

Emily B. Landau is Senior Research Associate as well as Director of the Arms Control and Regional Security Project at the Institute for National

Security Studies (INSS), Tel Aviv. She teaches arms control at Tel Aviv University and at the University of Haifa. Her research focuses on regional dynamics and processes in the Middle East as well as global and regional challenges of nuclear proliferation and arms control.

Patricia Lewis is the Deputy Director and Scientist-in-Residence at Monterey Institute's James Martin Center for Nonproliferation Studies, Washington, DC. She served as the Director of the United Nations Institute for Disarmament Research in Geneva, where she also was a member of the United Nations Secretary General's Advisory Board on disarmament matters (1997–2008). Her research focuses on aspects of science, verification, arms control, disarmament, and nonproliferation.

Dinshaw Mistry is Associate Professor of Political Science at the University of Cincinnati, OH. He has been a Fellow at the Woodrow Wilson Center, Washington, DC, at the Belfer Center for Science and International Affairs at Harvard University, and at the Center for International Security and Cooperation, Stanford University, CA. He has written extensively on missile and nuclear proliferation.

Judith Palmer Harik is the President of Matn University, Lebanon, and Associate Professor at the American University of Beirut. Dr Palmer Harik is an expert on Hezbollah, civil-war related issues, and post-war conflicts.

Uzi Rubin is the President of Rubincon Defense Consulting Ltd. He was in the Israeli government service until 2002, where his latest positions included working as Special Assistant for Research and Development Programs in the Office of the Minister of Defense and as Senior Director for Defense Policy. Prior to that, he established and served as head of Israel's Missile Defense Organization, where he oversaw the development of Israel's Arrow anti-missile defence system. He was awarded the Israel Defense Prize in 1996 and 2003. He publishes frequently in the professional and international media on space and missile defence topics.

Ahmed Saif is Director of the Sheba Center for Strategic Studies (SCSS), Sanaa, Associate Dean at Yemen College of Middle Eastern Studies, Associate Professor of Politics at Sanaa University, and a Country Consultant on Parliamentary and Political Development for the United Nations Development Programme. He is also a Member of the Editorial Board of the Journal of Strategic Studies at the Bahrain Centre for Studies and Research (BCSR).

Mahmood Sariolghalam is Professor of International Relations at the Shahid Beheshti (National) University in Tehran. In the 2009/10 academic year, he taught at Kuwait University. He was a doctoral candidate at the University of Southern California and at Ohio State University.

Dr Sariolghalam specializes in Iranian foreign policy and political culture and has made presentations in more than 100 countries.

Jürgen Scheffran is Professor for Climate Change and Security at KlimaCampus and Hamburg University. He was a Senior Research Scientist at the University of Illinois, Urbana-Champaign, and a Visiting Professor at the University of Paris (Pantheon/Sorbonne). He is a co-founder of the International Network of Engineers and Scientists Against Proliferation (INESAP), co-drafter of the Model Nuclear Weapons Convention, and has been an advisor to the UN Department for Disarmament Affairs.

Hans-Joachim Schmidt is a Senior Research Fellow at the Peace Research Institute Frankfurt (HSFK/PRIF), and a member of the Working Group Just Peace of the German Commission Justitia et Pax. He is an expert on conventional arms control and military confidence-building measures in Europe and the Korean peninsula, where he focuses on measures for regulating the North Korean nuclear crisis.

Martin Senn is Assistant Professor at the Department of Political Science at the University of Innsbruck. His dissertation was awarded the 2008 Fulbright Prize in American Studies. His research and teaching interests include nuclear and ballistic missile proliferation, non/counterproliferation, nuclear disarmament, ballistic missile defence, and international relations/foreign policy theory.

Oleg Shulga is President and CEO of Shulga Consulting, providing technical support to the Defense Threat Reduction Agency/Cooperative Threat Reduction Directorate. He interacts with all Russian government agencies and enterprises involved in strategic missile elimination projects in Russia. Mr Shulga holds certificates as Russian Federation INF, START, CFE, and OS Inspector, UN Missile Inspector, international Project Management Professional. His main areas of research are practical disarmament, implementation of arms control, and joint missile defence issues.

Mark Smith is Programme Director and member of Wilton Park Academic Staff. Previously, he was Research Fellow at the Mountbatten Centre for International Studies, Lecturer in International Security at the University of Southampton, and Research Fellow at Swansea University. Dr Smith's special interests include arms trade treaty issues as well as biological, chemical, and nuclear weapons nonproliferation, where he focuses especially on the future of the NPT and the prospects for nuclear reductions and elimination.

Hartwig Spitzer is a Professor Emeritus for Physics at Hamburg University. He has been a Visiting Scholar at Stanford University, CA, and established the Center for Science and International Security at Hamburg

University in 1988. He has focused his research on technical aspects of arms control.

Christian Weidlich is a Research Assistant at the Peace Research Institute Frankfurt (HSFK/PRIF) and on the staff of the Academic Peace Orchestra Middle East. He studied International Relations and Peace and Conflict Studied at Goethe University, Frankfurt. His research interests include autonomous weapon systems with a focus on international humanitarian law, arms control in the Middle East, and civil–military cooperation.

Hui Zhang is a physicist and Senior Research Associate at the Belfer Center for Science and International Affairs at Harvard University. His research includes verification techniques of nuclear arms control, the control of fissile material, nuclear terrorism, China's nuclear policy, nuclear safeguards and nonproliferation, policy of nuclear fuel cycle and reprocessing.

Acknowledgements

The multiple-authorship of several articles in the volume has required a number of meetings to discuss the issues and provide joint drafts. We would like to thank our sponsors for providing the funding to arrange these working groups: the German Federal Ministry for Economic Cooperation and Development, the Protestant Church in Hesse and Nassau, and the Friedrich Ebert Foundation.

We are also grateful to two anonymous reviewers of Routledge for their thoughtful and constructive recommendations regarding the underlying idea of this book and the submission of individual chapters. We would also like to thank the following colleagues worldwide for their helpful comments on earlier drafts of the articles: Sami Alfaraj, Lars Berger, Shlomo Brom, Frieder Bürkle, Shahram Chubin, Justin D. Cook, Michael Elleman, Mark Fitzpatrick, Alexander Frechen, Andrä Gärber, François Géré, Alexander Heerlein, Ole Hexel, Ali Hocaoglu, Ephraim Kam, Isil Kazan, Herbert C. Kelman, Marieke Knußmann, Viktoriya Kokhan, Remke Lohmann, Katharina Maschke, Annika Mildner, William G. Miller, Wencke Müller, Benedikt Münch, Roberta Mulas, Max Markus Mutschler, Spasyana Naydenova, Götz Neuneck, Sebastian Nieke, Rick S. Northrop, Randy Rydell, Sabri Saidam, Mohammad Khair Shiyyab, John Simpson, Etel Solingen, Leila Ullrich, Pieter Wezeman, Hanne Weismann, and Mahjoob Zweiri. Achim Vogt deserves our special thanks for having been the reliable and inspiring link between the meeting of the working groups and the Friedrich Ebert Foundation. We are also very grateful to Dr Michael G. Parker, Frankfurt am Main, who provided detailed editorial advice for the volume, for his competence and commitment. We also thank Cathy Lara, San Francisco, in this respect.

Abbreviations

AA	Anti-Aircraft
AAA	Anti-Aircraft Artillery
ABM	Anti-Ballistic Missile
ACRS	Arms Control and Regional Security
AD	Air Defence
ANZUS	Australia, New Zealand, and United States of America
AP	Additional Protocol
API	Arab Peace Initiative
ASCM	Anti-Ship Cruise Missile
ASW	Anti-Submarine Warfare
ATACMS	Army Tactical Missile System
AWACS	Airborne Warning and Control System
BLU	Bomb Live Unit
BTWC	Biological and Toxin Weapons Convention
B-weapons	Biological Weapons
CBMs	Confidence-Building Measures
CD	Conference on Disarmament
CENTCOM	Central Command
CEP	Circular Error Probable
CFE	Treaty on Conventional Armed Forces in Europe
CSBMs	Confidence- and Security-Building Measures
CSC	Cooperative Security Concept
CSCE	Conference on Security and Co-operation in Europe
CTBT	Comprehensive Test Ban Treaty
CWC	Chemical Weapons Convention
C-weapons	Chemical Weapons
DPRK	Democratic People's Republic of Korea
DVs	Delivery Vehicles
EU	European Union
FMCT	Fissile Material Cut-off Treaty
GBU	Guided Bomb Unit
GCC	Gulf Cooperation Council
GEM	Guidance Enhanced Missile

GLCM	Ground-Launched Cruise Missile
GRIT	Graduated and Reciprocated Initiatives in Tension Reduction
HCOC	Hague Code of Conduct Against Ballistic Missile Proliferation ('Hague Code')
HE	High Explosive
HIMARS	High Mobility Artillery Rocket System
IAEA	International Atomic Energy Agency
IAF	Israeli Air Force
IAI	Israeli Aerospace Industries
ICBM	Intercontinental Ballistic Missile
ICOC	International Code of Conduct Against Ballistic Missile Proliferation
IDF	Israel Defense Forces
I-HAWK	Improved HAWK
INF	Intermediate-Range Nuclear Forces Treaty
INS	Israeli Naval Ship
IR	International Relations
IRBM	Intermediate-Range Ballistic Missile
IRGC	Islamic Revolutionary Guard Corps
LACM	Land-Attack Cruise Missile
MANPADS	Man-Portable Air Defence System
MBFR	Mutual and Balanced Force Reductions
MD	Missile Defence
MFZ	Missile Free Zone
MIRV	Multiple Independently Targetable Re-entry Vehicle
MLRS	Multiple Launch Rocket System
MRBM	Medium-Range Ballistic Missile
MTCR	Missile Technology Control Regime
NATO	North Atlantic Treaty Organization
NPT	Nuclear Non-Proliferation Treaty
NTI	Nuclear Threat Initiative
NTMs	National Technical Means
N-weapons	Nuclear Weapons
NWFZ	Nuclear Weapon Free Zone
OSCE	Organization for Security and Co-operation in Europe
PA	Palestinian Authority
Pac	Patriot Advanced Capability
PLN	Pre-Launch Notification
PLO	Palestine Liberation Organization
PSI	Proliferation Security Initiative
R&D	Research and Development
ROCAME	Register of Conventional Arms in the Middle East
RPV	Remotely Piloted Vehicles
RSAF	Royal Saudi Air Force

RSC	Regional Security Centre
SALT	Strategic Arms Limitation Talks
SAM	Surface-to-Air Missile
SAR	Synthetic Aperture Radar
SEAD	Suppression of Enemy Air Defences
SLBM	Submarine-Launched Ballistic Missile
SLV	Space Launch Vehicle
SNSC	Supreme National Security Council
SORT	Strategic Offensive Reductions Treaty
SPAAGM	Self-Propelled Anti-Aircraft Gun/Missile
SRBM	Short-Range Ballistic Missile
SSM	Surface-to-Surface Missile
START	Strategic Arms Reduction Treaty
TACMS	Tactical Missile System
THAAD	Terminal High-Altitude Area Defense
UAE	United Arab Emirates
UAV	Unmanned Aerial Vehicle
UCAV	Unmanned Combat Aerial Vehicle
UK	United Kingdom
UN	United Nations Organization
UNEF	United Nations Emergency Force
UNIDIR	United Nations Institute for Disarmament Research
UNIFIL	United Nations Interim Force in Lebanon
UNROCA	United Nations Register of Conventional Arms
UNSCR	United Nations Security Council Resolution
US	United States of America
USSR	Union of Soviet Socialist Republics
WHs	Warheads
WMD	Weapons of Mass Destruction
WMDFZ	Weapons of Mass Destruction Free Zone

Map 1 The scope of the proposed missile free zone.

1 Introduction: setting the stage

Decreasing the security dilemma by gradual missile reductions

Bernd W. Kubbig

The question how to increase the level of security in the Middle East/Gulf[1] is not at all new but it is one that, given the conflict-ridden nature of the region, has to be asked – and answered – again and again. The collection of articles in this volume follows a distinctive approach and a central objective, to reduce the area's "particularly pronounced" (Steinberg 1995: 179) security dilemma, focusing on incremental arms control/reduction[2] on the path towards a Missile Free Zone (MFZ).[3] In this context, analysing the security dilemma, characterized by strong mutual threat perceptions, intense arms build-ups, unilateral self-help, and permanent zero-sum thinking, promises to be a viable research focus. As two leading analysts from the conflict region have pointedly noted, the security dilemma reveals the "self-defeating" nature of the attempt to gain security through arms build-ups:

> For almost five decades, states in the Middle East have made every attempt to enhance their security, mostly with little regard to the extent to which these efforts have undermined the security of their neighbors. This was unfortunate because while short-term security requirements could be met in such a fashion, in the longer term this policy proved self-defeating. States whose security was negatively affected by their neighbors' security-enhancing measures took steps to diminish the perceived threat. In turn, these countermeasures sometimes reduced the security of the state whose security-enhancing efforts began the chain reaction.
>
> (Feldman and Toukan 1997: 73)

Reducing and eventually overcoming the current security dilemma requires lowering tensions, increasing arms race and crisis stability, and, as a result, decreasing the likelihood of war. At this writing I have to leave it open whether, and if so to what extent, the unprecedented upheavals in the region expressed in the metaphor 'Arab Spring' will impact the overall political and military relations. However, assuming that an MFZ – given the destabilizing character of missiles as discussed below – can in principle provide greater regional security, we, however, do not present the concept as a solution that is *automatically* feasible. Instead the contributors to this volume start from the premise that the key actors in the region will only participate in an MFZ on the condition that they perceive it as a means for improving their own security. This assumption underlies the research question which guides all articles: under what circumstances are gradual steps towards a Missile Free Zone (technically) feasible and politically acceptable? This question allows us to probe the concept of an MFZ, thereby addressing the region's complex security dilemma in a manageable way, advance the state of the discussion, and offer policy advice to decision-makers. The unique emphasis on missiles enables the assembled authors to provide a fresh look and a new dynamic perspective on conceptual and political disarmament efforts, distinguishing this volume from myriad of other Middle East/Gulf related studies and projects which typically deal with weapons of mass destruction (WMD, i.e. nuclear, biological, and chemical warheads) and neglect consideration of delivery vehicles, especially missiles. By contrast this volume focuses on the broad spectrum of missiles and rockets. Incremental missile reductions leading towards an MFZ are certainly no substitute for, but rather an important complement to, WMD related proposals.

Moreover, our approach is primarily about exploring a perhaps more promising Cooperative Security Concept (CSC) which includes confidence- and security-building measures (CSBMs)[4] as well as the reduction of the weapon arsenals themselves. This implies a progressive and incremental change in the prevailing mind-set in the region and the policies and structures associated with it. In this chapter I argue for establishing the missile problématique as a research and policy field in its own right. Then I make the case for analysing missiles in the context of the particularly intense security dilemma in the Middle East and other regional specifics; however, the importance of external actors will not be neglected. The current security dilemma is not, however, set in stone – it can be changed. In this regard I will appeal to major International Relations (IR) theories. On this basis we have designed three milestones towards missile disarmament as an attempt to reduce the Middle East/Gulf security dilemma. Since all contributors to this volume see this as a classical non-governmental (Track II) effort, they are above all interested in addressing their policy recommendations to (regional) decision-makers. If there is a general consensus among them, it is the importance of revitalizing a version of the modified working group on Arms Control and Regional Security (ACRS) of the 1990s as part of an overall multilateral setting. In fact the Middle East Conference on the establishment of a WMD Free Zone, including its delivery vehicles, which the Nuclear Non-Proliferation Treaty (NPT) Review gathering envisaged in May 2010, presents a new opportunity in this regard. As the focus on missiles and the research design of this project have not been developed in a vacuum, I will use the reports of the UN Panel of Governmental Experts on Missiles as an important point of reference.

'Targeting' missiles: the case for an issue area in its own right

Positioning missiles in the nuclear related literature

The enormous and constantly growing number of studies on a WMD/Nuclear Weapon Free Zone (NWFZ)[5] in the academic literature attests to the theoretical acceptance of zonal approaches (Mützenich 1991: 332) as a legitimate, if controversial confidence-building and arms control/disarmament tool for increasing the level of security in the Middle East/Gulf. However, in the 'world of actual politics' the situation is different. Indeed, there is an enormous and perhaps increasing gap between the relatively optimistic views of leading experts about the imminent possibility of initiating such zones and the sobering effects of political developments in the region. The revival of nuclear power projects in reaction to Iran's ambitious nuclear activities is a case in point. The initially innovative subregional proposal for a Nuclear Weapon Free Zone in the Gulf has been stymied, while older entrenched positions persist and dominate the scene, especially the stalemate between the '(Nuclear) Disarmament First' position, perennially proposed by Egypt, and the 'Peaceful Regional Settlement First' stance, repeatedly called for by Israel. American President Barack Obama's Global Zero vision, which has since revealed the contradictions inherent in his own emphasis on the role of nuclear weapons (N-weapons) on the strategic and operative level, might have had at least one positive effect: Obama seems committed to convening a Middle East Conference which would deal with WMD, including delivery vehicles, and especially nuclear weapons.

The proponents of zonal approaches have identified a number of conceptual issues that have to be resolved before their vision can become reality. Among them are the problems of military asymmetries, especially in the area of delivery vehicles (how to make comparative assessments of missiles against aircraft, for instance), the scope and boundaries of the zone (deciding who is in and who is out), the requirements of an effective verification regime, and the role of extra-regional powers.[6] By considering the conditions for the establishment of an MFZ in a comprehensive and coherent way, we attempt to go beyond the current debate while acknowledging the 'nasty regional realities' impeding progress towards an MFZ. This includes the unprecedented and largely uncontrolled export of delivery vehicles and missile defence (MD) systems by the major weapon suppliers, which is incompatible with missile reduction and eventual elimination. Without being fixated on this final objective, our step-by-step approach is important because it is often missing in discussions on the WMD/NWFZ (by the same token, we do not presume to provide a detailed script of a zonal concept for experienced negotiators). A cautious incremental approach is also suggested to nuclear disarmament. Taking current and possible future 'odd realities' into account, especially the Israeli–Egyptian stalemate, the contributors who outline an incremental approach towards an NWFZ, deliberately stop short of calling for its near- or mid-term realization (see the chapter of Gawdat Bahgat *et al.*). This does not, however, mean abandoning the long-term goal of a Weapons of Mass Destruction Free Zone in the Middle East/Gulf.

Of course, any attempt to establish the missile problématique as a research and policy area in its own right has to acknowledge the potentially disastrous consequences of WMD tipped delivery vehicles, especially nuclear weapons. Because of their dual role, missiles provide a link between purely conventional and non-conventional warheads. If the numbers of missiles were reduced or even eliminated, WMD arsenals would become less threatening and less useful to the regional actors holding or attempting to acquire them (admittedly, alternative means of delivery are also available). The choice between concentrating on missiles or weapons of mass destruction is not mutually exclusive, but needs to be discussed within a common agenda with clear priorities and a timetable. Like the aircraft able to carry WMD, nuclear weapons will have to be taken into account as part of the overall military posture and its built-in asymmetries.

Missiles as a research area

In most of the WMD related literature, missiles are generally regarded as mere add-ons to be factored into a zonal arms control/reduction plan (Prawitz and Leonard 1996). And yet, in recent years there are some clear trends towards endorsing our attempt to treat the means of delivery which are the focus of this book – (un)guided rockets, ballistic and cruise missiles, and unmanned aerial vehicles (UAVs) – as a distinct area. This has always been true for the controversial missile defence systems which have to be included in any analysis. Three authoritative bodies of experts are relevant in this regard. The first is the Weapons of Mass Destruction Commission ('Blix Commission') whose landmark study – the report on "Weapons of Terror" – devotes an entire chapter not only to delivery means, but to missile defence and weapons in space as well (WMDC 2006: 140–9). The same is true for the work of the Geneva-based United Nations Institute for Disarmament Research (UNIDIR); some of its publications deal entirely with the missile issue, although their main emphasis is on WMD related questions, including zones free of weapons of mass destruction (see, for example, UNIDIR 2007).[7]

The third body is the UN Panel of Governmental Experts. Despite its shortcomings the report "The issue of missiles in all its aspects" (UNGA 2002) of 23 July 2002 was the first comprehensive effort by the United Nations to address the subject of missiles.[8] The panel's report of 20 July 2006,[9] the second relevant point of reference,[10] observed:

> There is an emerging tendency to perceive missiles as intrinsically important regardless of how they are armed. From this perspective, a missile can be of great significance even if armed with conventional explosives or not armed at all, relying exclusively on kinetic energy to achieve destruction.
>
> (UNGA 2006: 5)

Moreover, the 2006 report notes that the missile area has increasingly become a distinct field of diplomatic activities. While for decades the UN Security Council and the General Assembly treated missiles merely as means of delivery for nuclear weapons, by the 1990s various categories of missiles had become "a focus of attention in themselves" (ibid.: 6). This inclusive approach, covering most types of missiles whether conventionally or non-conventionally tipped, is evident in the 1992 United Nations Register of Conventional Arms. In chapter VII ("Missiles and missile launchers") the register includes rockets, ballistic missiles, and cruise missiles capable of delivering any kind of warhead with a range of at least 25 km (see the chapter of Sven-Eric Fikenscher *et al.* on Transparency). The following definition of the UN Panel of Governmental Experts should be useful throughout this book when referring to missiles:

> A missile is an unmanned, self-propelled, self-contained, unrecallable, guided or unguided vehicle designed to deliver a weapon or other payload. A ballistic missile is a weapon-delivery vehicle that has a ballistic trajectory over most of its flight path. A cruise missile is an unmanned, self-propelled weapon-delivery vehicle that sustains flight through the use of aerodynamic lift over most of its flight path. Unmanned aerial vehicles (UAV) and remotely piloted vehicles (RPV) share many of the characteristics of cruise missiles and are sometimes undistinguishable from them. Both unmanned aerial and remotely piloted vehicles are intended to be used more than once.
>
> (UNGA 2002: 10–11)

The strongest argument against defining missiles in terms of whether they can carry a nuclear payload is the fact that states and hybrid actors[11] possess and/or threaten to use all the types of conventional missiles available to them, regardless of category – (un)guided rockets, ballistic as well as cruise missiles and UAVs. Missiles of various categories have been used in the Middle East/Gulf wars with lethal and destructive consequences. They have killed soldiers and civilians alike, and they have destroyed military and civilian installations such as airports, hospitals, schools, and UN buildings. While the rockets launched by Hamas and Hezbollah cannot be called weapons of mass destruction, because of their limited 'lethal impact', they are certainly weapons of terror designed to demoralize Israeli citizens, disrupt their lives, as well as to kill and destroy.

Missiles have distinctive and unique characteristics that make a compelling case for treating them as a separate area of research. Unlike manned aircraft, they, once launched, cannot be ordered back. In a crisis situation the fact that they are non-recallable can contribute enormously to instability by fuelling tensions and making accidental wars much more likely. Nevertheless, the capabilities of ballistic missiles have often been exaggerated, as the limited success of less advanced types such as the Scud in the Gulf wars has shown. In contrast, cruise missiles have been underestimated for decades. The time is long past when only ballistic missiles were regarded as a problem, although concern about cruise missiles has yet to translate into reforming the two existing control agreements – the Missile Technology Control Regime (MTCR) and the Hague Code of Conduct Against Ballistic Missile Proliferation (HCOC). The same holds true for unmanned aerial vehicles (see the chapter of Dennis M. Gormley *et al.*). Not only missiles, but also missile defence systems which proponents consider 'purely defensive', can indeed "in the hands of one state … provoke at least uncertainty and possibly real fear in others" as well (Booth and Wheeler 2008: 1). I suggest assessing these threats contextually, in a scenario- and policy-dependent way. Missile defence systems can indeed in principle contribute to regional stability. However, the large-scale importation of MD systems by some Middle East/Gulf countries would violate the 'purely defensive' standard if those systems were to become part of an offensive doctrine, especially in the context of a preemptive strike.[12] Together, these weapons constitute the missile proliferation problem addressed in this volume.

There are additional compelling arguments for focusing on missiles:

- This emphasis permits and requires us to take all relevant conflict formations, major power constellations, and regional dynamics into consideration. In addition to examining conflicts on the state level hybrid actors have to be taken into account as well. By contrast, without disputing the relevance of air power (van Creveld 2011), had we chosen this focus as the central unit of comparison the scope of our analysis would have been much too restrictive.
- By examining the entire range of missiles, we are able to address the multifarious causes contributing to the research and development, procurement and production, flight testing, deployment, stockpiling/storage, and, most importantly, to the use of these weapons. This allows us to consider the constellations of domestic and regional forces that drive missile programmes, as well as the role of external state actors as providers of know-how, technology, and hardware.

In contrast to weapons of mass destruction, especially nuclear ones, there is no equivalent discussion about the threat of missiles, nor are there internationally binding accords with nearly universal membership corresponding to the NPT or regional and bilateral arrangements. The taboo on the use of nuclear weapons has no parallel in the case of missiles (Rosert and Schirmbeck 2007; Tannenwald 2005). Restrictions in the nuclear field contrast sharply with the virtual absence of limits with respect to missiles. The two existing missile regimes, MTCR and HCOC, depend for their enforcement exclusively upon export regulations, voluntary political compliance, and a self-imposed code of conduct, rather than being subject to treaties sanctioned by international law (see the article of Dinshaw Mistry and Mark Smith).

Several factors account for the markedly different developments: from the beginning, the bombing of Hiroshima and Nagasaki – each city being destroyed by a single bomb – demonstrated the horrific potential of the N-weapon. By contrast, the first ballistic missile strikes were the German V-2 attacks on the city of London during the Second World War, resulting in around 7,000 casualties. The rockets were both cause for fear but also fascination. The British scientist Reginald V. Jones (1978: 455) claimed that "our own politicians ... seemed far more frightened by one ton of explosive delivered by rocket than by five tons delivered by aircraft ... no weapon yet produced has a comparable romantic appeal". As Mark Smith (2006: 528) has persuasively observed: "This potent symbolic impact survives to this day."

Nevertheless, in contrast to (nuclear) warheads, missiles are displayed worldwide at military parades as symbols of power and prestige; in addition, their use in the civilian sphere is uncontroversial. The difference in the historical trajectories of the two weapons is symbolized on the personal level by Robert Oppenheimer and many other 'fathers' of the atomic and hydrogen bombs who later regretted their involvement and led the fight for nuclear arms control/disarmament. Wernher von Braun, brainchild of the V-2 rocket, became, by contrast, the representative of the American missile programme in its race against the Soviet Union. Von Braun is remembered as the figure triumphantly carrying the "Explorer" missile above his head with both hands in the famous *Der Spiegel* (1958) cover photo with the ambivalent heading "Germans at the Front".[13] Such pictures became the symbol of the cutting edge of missile technology which was used in the competitive East–West conflict as a suitable means of demonstrating American superiority.

Unlike nuclear warheads, conventionally equipped missiles – perhaps because they have been less fatal – have been built for use rather than only for deterrence. This may explain why they have a secure place in the stockpiles of the armed forces and in the war doctrines of almost all militarily advanced states. The recent use of simple, home-made 'sugar bombs' by Hamas and Hezbollah, however, ridicules the glamour factor associated with missiles on the state level: the garage-made rockets of both actors are technically of great simplicity, are hidden instead of being displayed at parades, and have been used so effectively that Israel's image as the invincible power has been questioned.

Although the missile realm has remained highly controversial among the members of the UN Panel of Governmental Experts (UNGA 2006: 4), it is ironically a specific consensus within the body which creates enormous problems for constructive solutions. To be sure, all parties agree that the "current situation" on missiles is "unsatisfactory", because they "produce significant and, in certain cases, deleterious effects upon the international security environment". However, the panel members do address the central problématique at the heart of the security dilemma:

> [I]t is universally argued that these effects are only negative in the case of another State's missiles. No State regards its own missiles as unduly threatening, and all look upon their missiles as both necessary and legitimate. No State regards its own missiles as part of the problem: on the contrary, missiles are often claimed to be a response to the problem, rather than a part of it.
>
> (ibid.: 12)

Not surprisingly, while there is a consensus about the negative developments in the missile area and about the fact that solutions are not within reach, there is no "shared view on what those solutions might be" (ibid.). In all three of its published reports, the UN Panel of Governmental Experts only vaguely exhorts that it is "essential" to make "continued international efforts" to deal with the issue of missiles (UNGA 2002: 20).

'Regionalizing' missiles: the specific security dilemma and its features in the Middle East/Gulf

> Weapons are the material reality that set up the security dilemma, because they are inherently ambiguous symbols. … Given the stakes involved, the existence of weapons in the hands of one state can provoke at least uncertainty and possibly real fear in others *even when those weapons are not intended to be used except for self-protection (following an attack, or in the event of a threat of an attack).*
>
> (Booth and Wheeler 2008: 1)

Missiles are indeed a part of the "material reality that set up the security dilemma" but the UN Panel of Governmental Experts has chosen to perceive them as a non-specific, universal phenomenon. Yet already the first series of meetings by the panel in 2001 and 2002 suggested that the global approach to coping with "The issue of missiles in all its aspects" was doomed to fail. This is partly due to the fact that no universal norm or treaty governing missile proliferation exists – and that the regional specifics vary enormously. The panel, however, has continued to work under the assumption of universality in its subsequent meetings, because it was bound by the UN General Assembly mandate. The deliberations of the

UN Panel suggest the missile issue should be tackled using mainly (although not exclusively) a regional approach.

While acknowledging the politically important work of the Track I body, this book is to a certain extent the Track II answer to the analysed deficits. As non-governmental experts our proposals aim to mitigate the current deficiencies in missile disarmament politics by 'regionalizing' the missile problématique in the context of the security dilemma and other major defining characteristics of the Middle East/Gulf. However, we do not offer a panacea for removing (all) the above-mentioned deficiencies of the UN Panel reports. The disputes among the panel members are likely to continue at the regional level among the representatives from Egypt, Israel, and Iran. At the same time this is a reliable indicator that the conditions of initiating and implementing an incremental reduction strategy can be found mainly at the regional level.

The "specifically pronounced" (Gerald Steinberg 1995: 179) security dilemma in the Middle East/Gulf with the already mentioned dimensions of strong threat perceptions, intense arms build-ups, unilateral self-help, and permanent zero-sum thinking can best be presented by analysing the overall military asymmetries in the region along with the conflict formations and alliances associated with them. The combination of threat perceptions and arms build-ups (acquisition and development of missiles included) is of special relevance in this regard. Among the relevant features that also have to be considered, the precarious state of the regional order and the problematic relationship among many states loom large, and so does the low degree of cooperation and of institutionalization (see the chapter of Sven-Eric Fikenscher *et al.* on Transparency). As the Middle East is seen despite all specific dynamics as a perforated region, the ambivalent role and interests of the essential extra-regional powers need to be considered as well.

This is complicated by the fact that the United States is not only an external actor (because decisions are made on distant American territory in Washington), but also a regional player (because of its military presence in various forms and degrees[14] in almost all Middle Eastern and Gulf countries as well as in the immediate vicinity, see also the chapter of Sven-Eric Fikenscher *et al.* on External Powers). It will also be important to analyse the consistency of external actors in implementing the current MTCR and HCOC regimes.

Laying the ground: conditions and (cooperative) concepts for reducing the security dilemma

The specific contributions of international relations theories

The following description by Shai Feldman, a leading Israeli expert, contains in a nutshell the central aspects of the Middle East/Gulf predicament as seen from a neo-realist perspective (Waltz 1979):

> Although unaware of the concepts and discourse of international relations theory, by instinct most Israelis [and Arabs as well as Iranians, I assume] are disciples of Kenneth Waltz. First, they regard their regional and international environment as a self-help anarchic system. Indeed, until very recently Israelis viewed their Arab neighbors as extremely hostile and as determined to challenge Israel's very existence. And, they could not identify an external power able and willing to assume responsibility for their security and survival.... As a result, Israelis [and Arabs as well as Iranians] have been conditioned to think about national security, not about mutual, collective or cooperative security. Indeed, until very recently, these latter concepts were foreign to Israeli [and Arabic as well as Iranian] security thinking.
>
> (Feldman 1994: 195–6)

According to this view, the security dilemma has its sources in the anarchic structure of the international system which is regarded as a given and, therefore, as virtually immutable. The role and dynamics of weapons are seen in terms of a traditional and classical action–reaction approach which holds that the arms build-up of a given state (understood as a monolithic entity) results from the military activities of one country or more states in the international environment; the arms dynamics are understood as races between two or more states.[15] This explains why arms build-ups are traditionally understood in terms of military-strategic, geopolitical, threat-based orientations and perceptions,[16] which, in turn, have to be seen as part of (regional) conflicts, military imbalances, and overall military strategies. The inherent ambiguity of weapons as offensive or defensive instruments has, in the literature on deterrence and on the security dilemma, been best summarized by Robert Jervis (1978: 187): if governments were able to distinguish defensive weapons and policies from offensive ones, it would have the impact of reducing the security dilemma, for it would then be possible for a state "to make itself more secure without making others less secure". But Jervis does not specify how governments might overcome the fear that other states could exploit their benign signals.

The adversaries in the East–West conflict showed from the early 1960s on a common interest in restraint which ought to be vital for the Middle East. They signalled a policy of benign intent and rejected an offensive and expansionist policy by accepting and implementing, within a cooperative setting, confidence-building measures as well as steps towards arms control/reduction. This collection of essays, which hopes to bridge the gap between the academic and the policy world, is informed by what is perhaps the most influential policy-oriented study addressing the security dilemma during the Cold War era (Wiseman 2005): the report, entitled "Common Security" was published by the Independent Commission on Disarmament and Security Issues (ICDSI 1982) under the chairmanship of then Swedish Prime Minister Olof Palme. I will build on the common security framework of the Palme Commission, but prior to it analyse the merits of the three mainstream IR theories.

Social constructivists, members of the liberal school of International Relations, and neo-institutionalists/regime theorists[17] agree that the security dilemma is neither an invariant element in an anarchic international system nor an 'objective fact' resulting from sober analysis. *Social constructivists* such as Alexander Wendt have disputed the neo-realist assumption that international relations are based on a fixed, immutable, anarchic structure which they interpret as socially and politically constructed. "Anarchy is what states make of it" (Wendt 1992) – this is the constructivists' programmatic slogan. There is nothing natural or inevitable about states practising self-help and power politics:

> Security dilemmas are not acts of God: they are the effect of practice. This does not mean that once created they can necessarily be escaped (they are, after all, 'dilemmas'), but it puts the causal locus in the right place.
>
> (Wendt 1995: 77)

The same holds true for threat perceptions. According to the constructivist argument, material resources acquire their meaning for human action only through a structure of shared knowledge in which they are embedded. Wendt illustrates this point with an example: the US feels less threatened by 500 British nuclear weapons than five North Korean ones, because its relationship to Great Britain is characterized by friendship, whereas its relationship with North Korea is dominated by hostility. This example "completely eludes" neo-realists' accounts of structure. "Material capabilities as such explain nothing; their effects presuppose structures of shared knowledge, which vary and which are not reducible to capabilities" (ibid.: 73).

Studies informed by the constructivist IR theory see status and prestige as the driving forces behind the acquisition and development of weapons in general and of missiles in particular. If we consider, in addition, the identity of a state/regime and its broader conception of foreign policy, the following question can be raised about the use or threat of use of missiles: is a given government status quo oriented or is it geared towards dominance, e.g. by not accepting established regimes, agreed-upon borders, and by inference in the domestic affairs of other countries? An additional factor is the importance of the past such as the influence of the Holocaust on Israeli military thinking or the Iranian experience of being attacked by chemical weapons during the Iran–Iraq War (see the chapter of Bernd W. Kubbig *et al.*).

For the *liberal IR school*, exemplified by its leading representatives Ernst-Otto Czempiel and Andrew Moravcsik, the security dilemma is a product of domestic, interest-based decision-making processes[18] and seen as the result of deliberate choices by particular governments. Regional uncertainties have often been used to define threats in accord with political, economic, or ideological orientations. From a liberal point of view the security dilemma can be overcome by establishing similar governmental structures among states, the European Union being a role model. As far as the role and build-ups of missiles (as possible barriers for reductions and finally an MFZ) are concerned, the majority of arms dynamic interpretations are rooted in explanations of the domestic politics of a given (democratic) country. These explanations cover only segments of the complex decision-making processes: among them the military-industrial complex (broadly defined as the relationship and interaction between the military and industrial actors with their combined influence on defence policy), the built-in competition among bureaucratic actors, and the momentum of technological development ('Eigendynamik'). As a specific variant of liberal IR thinking, the disciplinary role of the weapon apparatus has also been highlighted for maintaining the order of a state/regime (Senghaas 1972: 325), or for regime survival (Solingen 2007). The control and suppression of internal protests are associated with these kinds of mechanisms.

In addition, the economic system is sometimes seen as the crucial determinant of weapon programmes, albeit with contradictory assessments. For some proponents of this approach, the military sector is a stimulus to economic growth especially in times of crisis (for instance as a job creating measure). Other proponents of the economic explanation, however, emphasize that military expenditures are pursued at the expense of social projects ('guns versus butter') and that excessive military spending may lead to a decreased competitiveness in crucial economic and technological areas. In addition, the importance of the arms sector for the development of the local or regional infrastructure has been underscored, as has the role of the military in initiating and maintaining a national technology policy, whose positive effects for the civilian sector and related technological innovations ('spin-offs') have been called into question by critics. In sum, as important as these internal driving forces are, in the final analysis it is conflict and the way it is perceived by domestic actors that drive the procurement and the development of weapons. A number of empirical studies from a liberal IR perspective corroborate this central finding.[19]

Although based on neo-realist premises, *neo-institutionalists/regime theorists* – and here I am emphasizing the work of Robert O. Keohane – have concluded that cooperation among adversaries is possible. They focus on how nations can establish institutions that capture the potential gains achieved from cooperation, while minimizing the risks of cheating and non-compliance (Keohane 1984; Lipson 1995). The search for international rules and common expectations in inter-state relations is basic to regime studies. Regimes can increase the transparency of accords by establishing mechanisms and creating institutions to provide information which is otherwise either too costly to procure or unavailable. In the military realm, this IR school considers how to minimize the effects of the security dilemma by slowing down or even preventing the arms dynamics or improving crisis stability through risk reduction regarding surprise attacks. Greater transparency and rule enforcement reduces the gains from cheating, although these measures cannot eliminate deception. In the security area, arms control treaties and their implementation have been a major focus of research, whether among superpowers or in the nuclear non-proliferation sector. Regime theorists have summarized obstacles to the creation of security regimes as follows:

> If states view politics as a zero-sum struggle, if they actually desire wars of expansion, if they cannot seek joint gains for domestic political reasons, if they fail to recognize their policy choices are interdependent, if they cannot distinguish each other's offensive and defensive weapons and military deployments, if they are unwilling to reassure other states by permitting adequate verification, then the prospects for security regimes will be poor indeed.
>
> (Lipson 1995: 21)

The Palme Report "Common Security" and the Cooperative Security Concept

The "Common Security" report (ICDSI 1982) not only provides an outline of the normative premises and major substantive elements of cooperative security, but also illustrates how the East–West conflict (both with respect to arms control and the Conference on Security and Co-operation in Europe) contains important points of comparison for analysing Middle East arms control/reduction policies:

> In the nuclear age, nations can no longer hope to protect their citizens through unilateral military measures. All states, even the most powerful, are dependent in the end upon the good sense and restraint of other nations. Even ideological and political opponents have a shared interest in survival. In the long run, no nation can base its security on the insecurity of others. True security requires a cooperative effort, a partnership in the struggle against war which can only be established through dialogue and reconciliation.
>
> All nations of course have the right of self-defense as guaranteed in the Charter of the United Nations. But the pursuit of military superiority is a futile endeavor that can only lead to less security for all. It is evident that most nations have become more powerful militarily over the years, yet it is equally clear that this has not led to a greater sense of security. Common security requires an end to arms competitions through negotiations, national restraint, and a spirit of collective responsibility and mutual confidence.
>
> (Palme Commission on Disarmament and Security Issues 1989: 7)

This programmatic statement can be read as an alternative to neo-realist approaches to the security dilemma. One may of course question whether this cooperative concept applies to a region with one de facto nuclear weapon state (Israel) and one possibly emerging nuclear power (Iran): this will be a topic in several articles of this book. Nevertheless, I take the position that the key elements of such an approach are instrumental for improving the situation in the Middle East/Gulf. In adapting core insights of the Palme Commission for developing a Cooperative Security Concept for the region, one caveat is important: the CSC is not a straightjacket for contributors to this volume, but a vision which embodies a particular ideal and which provides a perspective for the authors as well as for further, more systematic work (see the Conclusions by Bernd W. Kubbig).

A largely regional understanding of the CSC:[20]

- Defines the *problématique* as the effort to increase the security of each single state in the Middle East/Gulf by reducing the security dilemma through a wide range of confidence- and security-building measures, arms control, reduction, and disarmament initiatives in a cooperative way. These steps, which are the military part of a broader culture of restraint, are at the centre of the CSC and so is the specific path towards an MFZ.
- Includes *basic principles:* regional security, which should be based on a mutual interest in avoiding a (nuclear) war, is not a zero-sum game but based on cooperation. Whenever possible, zero-sum approaches and a unilateral self-help attitude should be transformed into win–win thinking. What is needed is a change in the prevailing mind-set towards a cooperative, reciprocal (or mutual) security design. Consensus and compromise are essential principles of this kind of thinking. Regional security can be best achieved in conjunction *with,* rather than *against* an adversary or enemy. It cannot be attained and sustained through military superiority or (nuclear) war-fighting doctrines. On the contrary: fewer weapons can mean more security. Mutual cooperation with adversaries/enemies is the organizing principle that should gradually replace confrontation. Multilateral arrangements may dominate bilateral ones.
- Should in terms of its *scope* be as inclusive as possible. It would engage Iran, Syria, and Lebanon as well as politically central organizations such as Hamas and Hezbollah. By contrast, a more restrictive approach could easily foster the confrontation among alliances which prefer policies of containment, violent confrontation, and even war.
- Favours with respect to the *time frame* pragmatism, flexibility, and incrementalism.[21] An emphasis on short-term policy solutions does not exclude mid- and long-term perspectives. Creating sustainable processes and solutions requires strategies that are effective over the long haul and embody a political vision.

- Acknowledges an *extended notion of security* locating military and arms control related issues in a broader context which includes environmental, resource (water supply), and economic security. This extended notion can become vital, as positioning military/arms control within the overall regional setting ('linkage policy') might under favourable circumstances lead to improvements in arms control/disarmament related measures.
- Favours, regarding the *procedural dimension*, incremental steps such as CSBMs and arms control/reduction initiatives over a sudden leap of confidence which may trigger a new policy (Booth and Wheeler 2008: 234). The more cautious approach taken here to the Middle East problématique is based on the dissatisfying record of success in this policy field. In the context of the poor degree of cooperation and low levels of institutionalization in the region, politically self-binding agreements offer a more realistic chance of success than treaty-based security arrangements. Moreover, *sustainable processes and organizational settings* that gradually instil cooperation, may include forums (formal conferences) as part of a regional peace strategy to deal productively with the security dilemma.

Consequences: the security dilemma can be decreased

The security dilemma is neither a given in an anarchic international system nor an 'objective fact', but the product of political decision-making processes in the domestic sphere. Likewise, apparently natural social structures such as unilateral self-help or intractable political constellations like the Cold War which have developed over time, are subject to change. And cooperation among enemies is in principle possible. If Israelis (and Arabs as well as Iranians, I assume) were to cease to be disciples by instinct of Kenneth Waltz, and instead become deliberate disciples of Alexander Wendt, Ernst-Otto Czempiel, Andrew Moravcsik, or Robert O. Keohane – and even more so of Olof Palme – they could become 'agents of change'. As analytically superior as the constructivist, liberal, and neo-institutionalist IR theory may be over the traditional neo-realist approach, however, these theories have their limits, too. Since the basic theories of arms dynamics were developed in the context the East–West conflict, their applicability to the Middle East/Gulf region remains limited.

Nevertheless, they can provide a checklist for the consideration of major driving forces in the region. For example, one analytical insight, a crucial outcome of arms dynamics research, as well as one remaining challenge will have to be dealt with in this book, even if we provide no in-depth account of how domestic politics influence arms control and missile disarmament. The *insight* – confirmed by recent studies[22] – is that arms build-ups are a multi-causal (redundant or overdetermined) phenomenon. The *crucial result* has already been mentioned: conflict formations are more important for arms dynamics than the variety of internal factors; this informs the conceptualization of arms control/disarmament strategies as well as the assessment of their chances and limitations for reducing the security dilemma. Conflict formations remain paramount – at this point of writing the Syrian regime, so far a pragmatic ally of Iran, would not regard Tehran's missiles as a threat, but on the contrary as an asset, whereas Israel is afraid of Iranian delivery vehicles which can reach its territory and, even more so if they were nuclear tipped. The paramount importance of conflict formations is supported by the relevant research literature on the Middle East, but it departs from the extreme position that regards any arms control effort that does not fundamentally change the grand political landscape as futile. Without contesting the importance of conflict formations, most contributors to this volume affirm the importance of the – albeit limited – role played by arms control (see the chapters of Bernd W. Kubbig *et al.* and Martin Senn *et al.*). But the *basic challenge* remains: the processes of initiating gradual missile reduction by changing the prevailing "self-defeating" (Feldman and Toukan 1997: 73) mind-set will be difficult, cumbersome, and non-linear, i.e. it will not be characterized by uninterrupted progress. Backlashes and recidivism are to be expected.

Concretizing the gradual reduction strategy: the design of the book

Against this background, the development of the research question motivating this book – under what circumstances are gradual steps towards a Missile Free Zone (technically) feasible and politically acceptable? – provides a way of conceptualizing an incremental arms control/disarmament strategy. It combines more modest confidence- and security-building measures and the bolder, far-reaching goal of an MFZ. This concept will be presented by outlining the major divisions of the book and the individual contributions. This study consists of three parts.

Two chapters of Part I lay the groundwork for the succeeding essays. The contribution of Emily B. Landau and Dalia Dassa Kaye ("Disarmament efforts in the region: lessons from the Arms Control and Regional Security talks") presents both the past achievements and the structural barriers to arms control in the region. It examines the relationship between arms control and the political landscape, and draws lessons from the past which can be applied to the current situation. In particular, it emphasizes how CSBMs are a major tool for achieving a win–win situation and for overcoming prevailing zero-sum thinking. The chapter of Dennis M. Gormley, Colin P. Clarke, and Jürgen Altmann ("Missiles in the Middle East: their destabilizing role") makes a compelling case for dealing comprehensively with a broad spectrum of delivery systems, from unguided Qassam rockets to sophisticated ballistic medium-range missiles in the Middle East, including cruise missiles and unmanned aerial vehicles. By focusing on the lethal impact of missiles of all kinds, as well as on the destabilizing technological features of ballistic missiles, the article powerfully argues for establishing delivery systems as a research area in its own right – without overlooking the broader WMD context, since missiles can be used as the delivery vehicles which carry WMD to their targets.

Part II of this volume deals with six conditions for a successful step-by-step approach towards a Missile Free Zone. The chapter by Bernd W. Kubbig, Mohamed Noman Galal, Michael Haas, Murhaf Jouejati, Sabahat Khan, Ahmed Saif, and Mahmood Sariolghalam ("Taking overall military asymmetries into account: objections from the Arab world and Iran") places missiles within the larger military context (see also the table compiled by Michael Haas and Bernd W. Kubbig in the Appendix). The authors, the majority of whom are experts from the region, examine threat perceptions (caused by missiles) both as part of the national military postures and of the broader foreign policy concepts. It is assumed that no country will embark on the proposed strategy of gradual disarmament unless its political leadership sees the move as increasing its own security. It follows that a government will accept a reduction of its missiles (not to mention a 'zero option') only under the condition that the other side – e.g. Israel from the Syrian perspective – is willing to decrease its arsenals equivalently (i.e. in the same *and* in other areas such as air power). The nuclear dimension is a special challenge in this regard. Uri Bar-Joseph ("Taking Israel's security interests into account: deterrence policy in a changing strategic environment") presents the regional security situation from an Israeli perspective. He examines the background of his country's deterrence policy and the importance of nuclear and missile capabilities for protecting the homeland. Given the country's nuclear monopoly and conventional superiority, this article argues that there is leeway for the Jewish state to accept modest compromises in the nuclear field without jeopardizing its security. Such a process would be facilitated, if the Arab states and Iran accepted parallel reductions in the missile area.

Gawdat Bahgat, Avner Cohen, Sven-Eric Fikenscher, Giorgio Franceschini, Bernd W. Kubbig, and Patricia Lewis ("Advancing the control of weapons of mass destruction: an incremental approach") propose concrete, modest steps for overcoming the above-mentioned Egyptian–Israeli gridlock by signing and ratifying the Comprehensive Test Ban Treaty as well as the Additional Protocol, tackling the fissile material cut-off problem, and fuel cycle arrangements in the region. In doing so, this article links up with the discourse on nuclear proliferation and on a Nuclear Weapon Free Zone. The next contribution by Sven-Eric Fikenscher, Sameh Aboul-Enein, Gülden Ayman, Anton Khlopkov, and Hui Zhang ("From (potential) spoilers to guarantors? The role of external powers") lays out the profile and problems of an MFZ as a way of enhancing the security of the states (e.g. where to draw the geographical boundaries and how to gain acceptance for the zone from major extra-regional powers). The experiences of implementing the Tlatelolco and Rarotonga Nuclear Weapon Free Zone agreements provide an important point of reference for answering these questions. Although the authors point to the *non*-proliferation interests of the United States, Russia, and China, the constructive role of these countries in initiating and advancing arms reduction is constrained by their geopolitical and economic interests.

Another key element in any successful arms reduction process and especially the envisaged MFZ is monitoring. Reductions will only be acceptable to the parties in the region if they conclude that the disarmament processes can be adequately verified. Therefore, the authors of the next chapter ("The verification challenge: concepts, requirements, and technologies") – Jürgen Scheffran, Bharath Gopalaswamy, Dennis M. Gormley, Bernd W. Kubbig, Uzi Rubin, and Hartwig Spitzer – follow up on this question by developing clear criteria and considering various technologies for the monitoring of missile related capabilities and activities. They analyse the problems posed for a region which has limited experience with arms control agreements and a mixed compliance record – one dilemma being the trade-off between technical effectiveness and intrusiveness which may politically be hard to accept.

The final challenge for the proposed process of gradual reduction towards an MFZ is to outline strategies for the rocket disarmament of the hybrid actors Hamas and Hezbollah. Judith Palmer Harik and Walid Abu-Dalbouh provide such an account for Hezbollah and Margret Johannsen, Ghassan Khatib, and Anat Kurz do the same for Hamas. In each case, the authors analyse a number of incremental steps for reducing the threat associated with the organizations' rockets and discuss the specific conditions for the options they consider. They range from advancing the norm of 'no use' on the one hand, to a complete disarmament of these actors, on the other. I would like to emphasize that the chapter on Hamas was composed by a trilateral group of authors under the utterly patient and competent leadership of my Hamburg colleague Margret Johannsen, in an effort that can justly be called historic.

Part III contains three milestones on the path towards a Missile Free Zone. These steps are designed to gradually reduce the security dilemma in the region, and we want to take into account realities on the ground by 'picking up the actors where they stand'. Therefore, the chapter of Sven-Eric Fikenscher, Patrick Flamm, Lina Golob, Bernd W. Kubbig, and Christian Weidlich ("The promise of military transparency: building on East–West experiences and the UN Register of Conventional Arms") suggests ways to facilitate dialogue and diplomatic contacts based on the experience of the East–West conflict. As a result, a new framework for negotiations based on the Arab Peace Initiative is outlined. In addition, the authors propose that the UN Register of Conventional Arms be expanded and reformed in order to successfully provide military transparency, especially in the missile realm.

Dinshaw Mistry and Mark Smith ("The Missile Technology Control Regime, the Hague Code of Conduct, and missile proliferation") assess the modest, yet visible impact of these two politically binding missile related regimes on curbing proliferation in the conflict region. The two authors conclude that the MTCR has restrained the proliferation of missiles in the Middle East to some extent, while the Hague Code of Conduct was not very effective. They propose a flight test ban as a major step towards missile control.

As largely confidence-building devices, with the MTCR being an export control regime, the UN Register and the HCOC do not impinge on the current arsenals in quantitative and/or qualitative terms. This claim is the focus of the chapter of Martin Senn, Jürgen Altmann, Bernd W. Kubbig, Jürgen Scheffran, Hans-Joachim Schmidt, and Oleg Shulga, who present the third milestone in the process towards an MFZ ("Caps and bans: limiting, reducing, and prohibiting missiles and missile defence"). In their contribution, which is informed both of East–West and regional arms control experiences, the authors propose three phases for tackling missiles and missile defence systems as part of the overall militarily asymmetric situation. In dealing with the adequate organizational setting for initiating and maintaining such a process, the authors touch upon one of the most relevant and difficult questions: how to create the political will and make a gradual strategy sustainable.

The final chapter by Bernd W. Kubbig ("The stage reconsidered: findings, concluding remarks, and research perspectives") summarizes the central results of the study – including the importance of cooperative arms control/reduction measures – and lists several research perspectives such as broadening the research focus beyond missiles, examining domestic power constellations and conditions for norm-building in greater depth, as well as looking for ways to institutionalize arms control in the region.

Notes

1 In order to avoid the notion 'Near and Middle East/Gulf' the term 'Middle East/Gulf' is used for the entire region.

2 In order to avoid the complete, yet cumbersome term 'arms control/reduction/disarmament', I usually prefer the arms control/disarmament or arms control/reduction combination in this article.

3 The idea of zero ballistic missiles in the Middle East has already been addressed by Lora Lumpe (1993: 16–18). On the concept of regional ballistic missile free zones, see Holton *et al.* (1993).

4 In the process associated with the Conference on Security and Co-operation in Europe, confidence-building measures were defined as "reducing the dangers of armed conflict and of misunderstanding or miscalculation of military activities which could give rise to apprehension, particularly in a situation where states lack clear and timely information" (quoted in Lachowski 2004: 11). The

hardly different term 'confidence- *and security*-building measures' was later added and defined as measures that are politically binding, militarily significant, verifiable when possible, and extending from the Atlantic Ocean to the Ural Mountains (ibid.: 13; see the chapter of Sven-Eric Fikenscher *et al.* on Transparency). If not otherwise indicated both notions are used as synonyms in this book.

5 See for example Mützenich (1991), Prawitz (2009).

6 See for greater detail the corresponding articles in this volume.

7 For one of the rare examples of examining missiles in the context of a universal prohibition and regional bans, see Lumpe (1993). The author addresses in particular the issue of universalizing the Intermediate-Range Nuclear Forces Treaty, a suggestion endorsed by high officials of the Reagan administration such as Kenneth Adelman.

8 The report was prepared by the members of the panel pursuant to UN General Assembly Resolution 55/33 A of 20 November 2000. The process triggered by an Iranian initiative seems to have been painful for a number of panel members.

9 This Track I initiative ("Governmental Experts") has nevertheless distinct Track II features, as qualified non-governmental experts advised the panellists. Moreover, the 2006 publication was prepared by the Geneva-based UNIDIR with the support of additional experts, among them one author of this book.

10 The last publication of 28 July 2008 is a condensed version of the earlier reports without any new insights and with no indication that any new consensus was reached among the panel members who met between June 2007 and June 2008 (UNGA 2008).

11 I refer here especially to Hamas and Hezbollah, which are often called sub-state actors (also by some authors in the volume). While Hezbollah has joined Lebanese governments and Hamas rules the Gaza Strip, 'hybrid' takes their new roles into consideration.

12 To illustrate different scenarios and policies which all in all show that the claim of MD supporters ('purely defensive') is mixed at best: the German Patriot anti-tactical missile system fielded in Israel during the last two Gulf wars, with the Israeli government being passive, supports the MD supporters' claim. The export of the Patriot was seen as unproblematic because it was in fact regarded as 'purely defensive'. When the government in Ankara, however, requested Dutch and German Patriot batteries to defend especially the eastern cities of Diyarbakir and Batman against Iraqi tactical ballistic missiles, the government in Berlin decided in mid February 2003 to deliver 46 Patriot interceptors via the Netherlands to Turkey. Yet this delivery was subject to conditions, as the seemingly defensive Patriots could be used in an offensive scenario. This is shown by Berlin's threat in the early days of the Iraq War to withdraw its Patriots, as well as its Airborne Early Warning and Control Systems, if Turkey became directly involved in the war on Iraqi territory, since the weapons would then become part of an offensive strategy (Kubbig/Nitsche 2005: 524–5).

13 For further information, see Neufeld (2007), Weyer (1999: esp. 120–2).

14 If one wanted to examine the US security relationships with the individual Middle East/Gulf countries systematically (which goes beyond the intention of this book), it would make sense to refer both to hegemonic concepts and to attempts that categorize American security relationships for instance in terms of a continuum ranging from alliance via sphere of influence, and protectorate to (informal) empire – a continuum "defined by the degree of hierarchy between the two parties" and, associated with it, by a different intensity of control (Lake 1999: 26), see also Lake (2009) and Kubbig (2001: 510–15).

15 I am aware of the considerable range of factions among neo-realists, notably among so-called offensive and defensive representatives, but cannot do justice to them simply because of limited space. The same holds true for 'the' constructivists to which I turn later, see Kubbig (2004).
16 But, as explained below, it makes a fundamental difference whether orientations and perceptions are being interpreted within the neo-realist IR theory or within social constructivist, liberal, and neo-institutionalist/regime theoretical approaches, which regard those factors as changeable. At the same time, these aspects show that supporters of neo-realist IR theory have transcended their own central assumption of holding solely the anarchical international system responsible for the security dilemma, see on this Milner (1992: 481, 496).
17 This explains why the notion 'regime' is used with two connotations, the abovementioned one or in terms of a ruling elite.
18 The following paragraphs are based on Czempiel (2002: esp. 15–43); see also Moravcsik (1992).
19 For further information, see Krell (1977), Kubbig (1983, 1988), Müller (1985).
20 On this concept and on the approaches of common, comprehensive, and collective security, how they overlap and where they are distinctive, see Dewitt (1994), Wiseman (1992, 2005). Of special relevance for the Middle East is the milestone study by Jones (1998).
21 These characteristics underscore my preference of the term 'cooperative' (in the sense stated above) to 'common' security – the latter describing a final state in a literal sense when basic differences among the parties involved have already been ironed out.
22 Among them Solingen (2007). Solingen's theory and the necessity of studying proliferation from different angles are highlighted in Potter and Mukhatzhanova (2010a, 2010b).

References

Booth, Ken and Wheeler, Nicholas J. (2008) *The Security Dilemma: Fear, Cooperation and Trust in World Politics*, New York: Palgrave Macmillan.

Czempiel, Ernst-Otto (2002) *Neue Sicherheit in Europa: Eine Kritik an Neorealismus und Realpolitik*, Frankfurt am Main: Campus.

Dewitt, David (1994) 'Common, Comprehensive, and Cooperative Security', *Pacific Review*, 7(1): 1–15.

Feldman, Shai (1994) 'Israel's Changing Environment: Implications for Arms Control', in Shai Feldman (ed.) *Confidence Building and Verification: Prospects in the Middle East*, Tel Aviv: Jaffee Center for Strategic Studies, 195–205.

Feldman, Shai and Toukan, Abdullah (1997) *Bridging the Gap: A Future Security Architecture for the Middle East*, Lanham, MD: Rowman & Littlefield.

Holton, J. Jerome, Lumpe, Lora, and Stone, Jeremy J. (1993) 'Proposal for a Zero Ballistic Missile Regime'. Online, available at: www.fas.org/asmp/library/articles/zerobal93.htm (25 March 2011).

ICDSI/Independent Commission on Disarmament and Security Issues (1982) *Common Security: A Blueprint for Survival*, New York: Simon and Schuster.

Jervis, Robert (1978) 'Cooperation under the Security Dilemma', *World Politics*, 30(2): 167–214.

Jones, Peter (1998) *Towards a Regional Security Regime for the Middle East: Issues and Options*, Stockholm: Stockholm International Peace Research Institute.

Jones, R.V. (1978) *Most Secret War*, London: Hamish Hamilton.

Keohane, Robert O. (1984) *After Hegemony: Cooperation and Discord in the World Political Economy*, Princeton, NJ: Princeton University Press.

Krell, Gert (1977) *Rüstungsdynamik und Rüstungskontrolle: Die gesellschaftlichen Auseinandersetzungen um Salt in den USA 1969–1975*, Frankfurt am Main: Haag+Herchen.

Kubbig, Bernd W. (1983) *Gleichgewicht oder Überlegenheit: Amerikanische Rüstungskontrollpolitik und das Scheitern von SALT II*, Frankfurt am Main: Campus.

Kubbig, Bernd W. (1988) *Amerikanische Rüstungskontrollpolitik: Die innergesellschaftlichen Kräfteverhältnisse in der ersten Amtszeit Reagans (1981–1985)*, Frankfurt am Main: Campus.

Kubbig, Bernd W. (2001) 'Introduction: The US Hegemon in the "American Century." The State of the Art and the German Contributions', in Bernd W. Kubbig (guest ed.), 'Toward a New American Century? The US Hegemon in Motion', *Amerikastudien/American Studies*, Special Volume, 46(4): 495–524.

Kubbig, Bernd W. (2004) *Wissen als Machtfaktor im Kalten Krieg: Naturwissenschaftler und die Raketenabwehr der USA*, Frankfurt am Main: Campus.

Kubbig, Bernd W. and Nitsche, Axel (2005) 'Germany: Selective Security Provider in the Schröder/Fischer Era', in Bernd W. Kubbig (guest ed.) in cooperation with Axel Nitsche, 'The Domestic Politics of Missile Defence', *Contemporary Security Policy*, Special Issue, 26(3): 520–43.

Lachowski, Zdzislaw (2004) *Confidence-and-Security-Building Measures: The New Europe*, Oxford: Stockholm International Peace Research Institute.

Lake, David A. (1999) *Entangling Relations: American Foreign Policy in Its Century*, Princeton, NJ: Princeton University Press.

Lake, David A. (2009) *Hierarchy in International Relations*, Ithaca, NY: Cornell University Press.

Lipson, Charles (1995) 'Are Security Regimes Possible? Historical Cases and Modern Issues', in Efraim Inbar (ed.) *Regional Security Regimes: Israel and Its Neighbors*, Albany, NY: State University of New York Press, 3–32.

Lumpe, Lora (1993) *Zero Ballistic Missiles and the Third World*, College Park, MD: Center for International and Security Studies at Maryland School of Public Affairs, University of Maryland.

Milner, Helen (1992) 'International Theories of Cooperation among Nations: Strengths and Weaknesses', *World Politics*, 44(3): 466–96.

Moravcsik, Andrew (1992) *Liberalism and International Relations Theory*, Cambridge, MA: Harvard University (Typescript).

Müller, Erwin (1985) *Rüstungspolitik und Rüstungsdynamik: Fall USA: Zur Analyse der Rüstungsmotive einer Weltmacht und zur Theorie moderner Rüstungsdynamik*, Baden-Baden: Nomos.

Mützenich, Rolf (1991) *Atomwaffenfreie Zonen und internationale Politik: Historische Erfahrungen, Rahmenbedingungen, Perspektiven*, Frankfurt am Main: Peter Lang.

Neufeld, Michael J. (2007) *Von Braun: Dreamer of Space, Engineer of War*, New York: Alfred A. Knopf.

Palme Commission on Disarmament and Security Issues (1989) *A World at Peace: Common Security in the Twenty-First Century*, Stockholm: The Palme Commission.

Potter, William C. and Mukhatzhanova, Gaukhar (eds) (2010a) *Forecasting Nuclear Proliferation in the 21st Century: The Role of Theory*, Vol. 1, Stanford, CA: Stanford University Press.

Potter, William C. and Mukhatzhanova, Gaukhar (eds) (2010b) *Forecasting Nuclear Proliferation in the 21st Century: A Comparative Perspective*, Vol. 2, Stanford, CA: Stanford University Press.

Prawitz, Jan (2009) 'A Note on the Proposed Zone Free of Weapons of Mass Destruction in the Middle East', updated version of the Paper presented at the Second Conference on the Middle East Weapons of Mass Destruction Free Zone, London (Typescript).

Prawitz, Jan and Leonard, James F. (1996) *A Zone Free of Weapons of Mass Destruction in the Middle East*, Geneva: United Nations Institute for Disarmament Research.

Rosert, Elvira and Schirmbeck, Sonja (2007) 'Zur Erosion internationaler Normen: Folterverbot und nukleares Tabu in der Diskussion', *Zeitschrift für Internationale Beziehungen*, 14(2): 253–87.

Senghaas, Dieter (1972) *Rüstung und Militarismus*, Frankfurt am Main: Suhrkamp.

Smith, Mark (2006) 'Pragmatic Micawberism? Norm Construction on Ballistic Missiles', *Contemporary Security Policy*, 27(3): 526–42.

Solingen, Etel (2007) *Nuclear Logics: Contrasting Paths in East Asia and the Middle East*, Princeton, NJ: Princeton University Press.

Steinberg, Gerald M. (1995) 'Arms Control in the Middle East: Global Regimes vs. Regional Dynamics', in Efraim Inbar (ed.) *Regional Security Regimes: Israel and Its Neighbors*, Albany, NY: State University of New York Press, 175–97.

Tannenwald, Nina (2005) 'Stigmatizing the Bomb: Origins of the Nuclear Taboo', *International Security*, 29(4): 5–49.

UNIDIR/United Nations Institute for Disarmament Research (2007) 'Missile Control?' *Disarmament Forum*, Special Volume, No. 1. Online, available at: www.unidir.org/bdd/fiche-periodique.php?ref_periodique=1020–7287–2007–1-en#contents (25 March 2011).

UNGA/United Nations General Assembly (2002) 'The Issue of Missiles in All Its Aspects', Report of the Secretary-General, A/57/229, New York: United Nations.

UNGA/United Nations General Assembly (2006) 'The Issue of Missiles in All Its Aspects', Report of the Secretary-General, A/61/168, New York: United Nations.

UNGA/United Nations General Assembly (2008) 'The Issue of Missiles in All Its Aspects', Report of the Secretary-General, A/63/176, New York: United Nations.

van Creveld, Martin (2011) *The Age of Airpower*, New York: PublicAffairs.

Waltz, Kenneth N. (1979) *Theory of International Politics*, New York: McGraw-Hill.

Wendt, Alexander (1992) 'Anarchy Is What States Make of It: The Social Construction of Power Politics', *International Organization*, 46(2): 391–425.

Wendt, Alexander (1995) 'Constructing International Politics', *International Security*, 20(1): 71–81.

Weyer, Johannes (1999) *Wernher von Braun*, Reinbek: Rowohlt.

Wiseman, Geoffrey (1992) 'Common Security in the Asia-Pacific Region', *Pacific Review*, 5(1): 42–59.

Wiseman, Geoffrey (2005) 'The Palme Commission: New Thinking about Security', in Ramesh Thakur, Andrew F. Cooper, and John English (eds) *International Commissions and the Power of Ideas*, Tokyo: United Nations University Press, 46–75.

WMDC/Weapons of Mass Destruction Commission (2006) *Weapons of Terror: Freeing the World of Nuclear, Biological and Chemical Arms*, Stockholm: WMDC.

Part I

Missiles and their proliferation as a threat

2 Disarmament efforts in the region

Lessons from the Arms Control and Regional Security talks

Emily B. Landau and Dalia Dassa Kaye

Early efforts to advance weapons of mass destruction arms control

In considering the question of arms control in the Middle East/Gulf, it is crucial to take the lessons from the previous attempts in this regard into account. The earliest efforts from the mid 1970s in this area centred on UN proposals to create a Nuclear Weapon Free Zone (NWFZ) in the region. While in theory the idea had the appeal of pursuing nuclear disarmament in a framework that takes regional realities into account, in practice the UN-based initiatives did not have that effect. When Egypt and Iran first raised the NWFZ idea in 1974, they sought to focus specifically on Israel's assumed nuclear option. Cairo and Tehran hoped for Israel to join the Nuclear Non-Proliferation Treaty (NPT), which would have required it to disclose and eliminate whatever nuclear capability it possessed. Subsequent proposals have been submitted on an annual basis to the UN General Assembly, which has passed resolutions recommending the establishment of an NWFZ in the Middle East (NTI 2008). Since 1980 these have been supported by all states in the Middle East/Gulf, including Israel, but they have nevertheless not proven particularly useful in prompting an effective and practical regional arms control process.

An important juncture was reached in April 1990, when Egypt's President Hosni Mubarak put a new proposal on the table: not an NWFZ, but rather a Weapons of Mass Destruction Free Zone (WMDFZ) for the Middle East. By making this shift in the direction of dealing with *all* categories of non-conventional weapons, the focus also turned away from being exclusively on Israel. Indeed, the immediate impetus for Mubarak's initiative was Iraq, particularly the threat issued by Saddam Hussein on 2 April 1990 to burn half of Israel with binary chemical weapons. Following the issuance of this threat, Mubarak travelled immediately to Baghdad in an attempt to calm the situation and reduce tensions, announcing his new initiative several days later. Still, the new proposal included the demand that Israel join the NPT, which it did not accept as a condition for participating in a regional arms control process.

In the aftermath of the 1991 Gulf War, the Mubarak agenda dovetailed with new thinking in the United States regarding the possibility of initiating a dialogue in the Middle East that would focus specifically on security concerns and threat perceptions, and – together with additional region-wide multilateral discussions – would complement the bilateral peace negotiations to be initiated between Israel and Jordan, Syria as well as the Palestinians in the framework of the Madrid peace conference. A notable parallel UN-based attempt in the early 1990s to flesh out the idea of an NWFZ was undertaken by James Leonard, Jan Prawitz, and Benjamin Sanders (1991) in a paper entitled “Effective and Verifiable Measures which would Facilitate the Establishment of a Nuclear-weapon-free Zone in the Middle East”. While focused on nuclear weapons rather than weapons of mass destruction (WMD), the question of security concerns and the need for effective verification measures were an important aspect of this paper.[1] As such, the study contributed to the consolidation of a new regional perspective on arms control in the Middle East/Gulf that was later incorporated into the Arms Control and Regional Security (ACRS) talks in 1992.[2] ACRS became one of the five working groups that made up the multilateral track of the Madrid peace process.[3]

The ACRS working group was active for four years, from 1992–95. It was noteworthy in that it progressed from the realm of proposals submitted to the UN, to a concrete regional dialogue. It was in the context of ACRS that the concept of arms control discussion in the Middle East was finally developed and tested. Taking its cue from the ideas about regional arms control that were congealing in the late 1980s and early 1990s, the ACRS working group developed into the most important experiment in regional arms control in the non-conventional realm that has taken place in the Middle East/Gulf to date. Its impact went beyond the confines of the talks themselves, and was evident in the Israeli–Jordanian Peace Treaty that was concluded in October 1994. Arms control ideas that were developed in ACRS were incorporated in the text of the treaty – especially regarding the importance of establishing peace and security for disarmament efforts – and the working group itself is mentioned in the accord as a framework for discussing these issues.[4] Due to its importance, ACRS will constitute the major focus of this article. Indeed, it is very likely to return to the spotlight in the next few years in view of the provision included in the Final Document of the 2010 NPT Review Conference that refers to the possible convening of an international conference in 2012 to discuss a WMDFZ in the Middle East (Review Conference 2010).

The achievements of ACRS

In assessing the experience of the ACRS working group, this article begins with the positive side of the talks, in order to better understand the nature and scope of its achievements.[5] There were 14 regional participants: Israel and Arab states from the Maghreb to the Gulf, including Palestinians, beginning in 1993. Iran as well as Iraq were not invited, and Syria as well as Lebanon opted to remain outside the framework. The conveners (or, 'gavel-holders' as they were called) of ACRS were the United States and Russia; in practice, however, the US became the primary external party navigating the discussions. Washington was guided by the assumption that the quality of the security relations that prevail among regional states is a key factor to take into account when devising regional arms control initiatives. In other words, improving relations among Middle Eastern states must be at the heart of arms control efforts.

The seminar framework chosen for the talks was thus purposely designed to encourage dialogue among the participants, enhance their appreciation of mutual threat perceptions and concerns, and create the basis for building confidence and stability. The goal was to reach agreement on a shared understanding of arms control that made sense for the participants and for the region as a whole. In the words of then-U.S. Secretary of State James Baker (1992), the agenda would be to consider

> a set of modest confidence-building or transparency measures covering notifications of selected military-related activities and crisis-prevention communications. The purpose would be to lessen the prospects for incidents and miscalculation that could lead to heightened competition or even conflict.

This arms control logic – drawn directly from the superpower experience of the Cold War – was designed to improve understanding and mutual confidence and create stability in a step-by-step manner, beginning with confidence- and security-building measures (CSBMs). CSBMs have been defined as any militarily significant measure that is mutually agreed upon among states, and viewed as contributing to their common sense of security, without impinging upon their core security interests and concerns. They are the embodiment of a win–win approach to international security, through a dynamic step-by-step inter-state process (Levite and Landau 1997). Once confidence was established – and with parallel movement on improving inter-state relations more generally through peace negotiations – it was believed that the actors would be able to move to the more difficult stage of arms control, when attention would turn to the control of WMD arsenals themselves.

Therefore, while the motivating factor for convening ACRS was no doubt the commonly perceived interest to mitigate the dangers associated with WMD in the region, the talks themselves dealt mainly with CSBMs that were not directly related to WMD. It was with regard to these CSBMs that the initial successes of the ACRS working group were achieved, before the discussions were put on hold indefinitely in 1995. Dialogue in ACRS began with two plenary sessions that were conducted basically as educational seminars; they provided the setting for the Americans and Russians to present their own experiences with arms control from the 1970s and 1980s to the regional participants. From about the third plenary, work began on defining a specific agenda for the Middle East, and inter-sessional discussions were initiated on conceptual and operational topics. These meetings continued through 1995, although the last plenary (which had the authority to approve decisions) took place in Tunis, in December 1994.

From 1993 to 1995, discussion in the operational basket of ACRS focused on four categories of CSBMs, and at the sixth and final plenary significant progress was achieved with regard to all four:

1 *Maritime issues.* Two documents were endorsed on search-and-rescue and incidents-at-sea, and Tunisia agreed in principle to host an additional exercise at sea as well as a meeting of senior naval officers from the region.
2 *Pre-notification and military exchange.* An agreement on prior notification of military exercises was concluded, and the parties agreed to exchange information regarding military personnel, unclassified military documents, and military training and education.
3 *Regional communications network.* Six parties (Israel, Jordan, the Palestinians, Tunisia, Oman, and Egypt) agreed in principle to participate in a temporary network set up in The Hague, and Egypt also offered to later host the permanent hub in Cairo.
4 Establishment of three *Regional Security Centres.* A decision was taken to set up a primary centre in Jordan, with secondary ones in Qatar and Tunisia. Their objective was defined as crisis prevention, management, and resolution (Landau 2006: 42–7).

Although the talks were put on hold before these CSBMs could be formally adopted and put into practice, clear agreement on them was secured within the framework of ACRS – which was no small achievement. They thus created a tangible outcome which proved that such agreements are possible, and they are sure to be an important point of reference for any future regional security dialogue in the Middle East/Gulf. The primary success of the working group – reflected in the discussions over, and initial agreement on CSBMs – was that it followed the logic of *placing weapons in context.* This meant attributing importance and granting priority to the reality of security concerns and threat perceptions among states in the region. The goal was to initially lower tensions rather than to place exclusive emphasis on the elimination of weapons per se, particularly at early stages of the process. ACRS thus placed emphasis on *states* and the *region,* rather than on *treaties* and the *global* disarmament context and experience. And it took relations in all seriousness, instead of just focusing on arsenals.

Why were the talks put on hold indefinitely?

Despite some significant and surprising progress in the ACRS working group, this experience also highlighted a number of barriers to security cooperation and arms control in the Middle East. Such obstacles are rooted not only in the difficult strategic conditions of the region, but also in political realities that will pose challenges to any future arms control effort. While such strategic and political impediments are not insurmountable, they are considerable, and a better understanding of them is critical to formulating more successful initiatives in the future. As such, understanding the failures of ACRS is at least as instructive for future arms control scenarios as understanding its achievements.

Strategic barriers

Two fundamental strategic barriers presented challenges to the process and are inherent in any comparable effort in this region: asymmetries of power (and perceptions of power imbalances), and classic security dilemmas. At present – and certainly at the outset of the ACRS talks in 1992 – Israel is assumed to be the only nuclear power in the Middle East/Gulf.[6] It has also long pursued a policy of maintaining a qualitative edge in its military capabilities to offset more numerous regional forces arrayed against it.[7] But Israel's military prowess has also created incentives for its adversaries to develop asymmetric capabilities, including non-conventional weapons. While WMD development among Arab states and Iran was not necessarily pursued to counter superior Israeli conventional capabilities – for example, Tehran and Baghdad worked on such capabilities in the context of Gulf security dynamics – the development of WMD across the Middle East has contributed to an extremely unstable regional security environment. Under these conditions arms control is particularly difficult because different actors will want to focus on areas where the adversary is strongest.

So in the ACRS case, the Arab states, particularly Egypt, consistently tried to steer the course of the discussions into the nuclear arena, whereas Israel preferred to maintain a broader definition of arms control, to include conventional capabilities, chemical and biological weapons, and long-range missiles. As we discuss later, Cairo's motivation for doing so related more to political considerations than to strategic ones. But the fundamental dilemma with promoting arms control in a region with significant military disparities was nonetheless a serious impediment to progress in the ACRS working group. The fact that key strategic players in the Middle East/Gulf with significant non-conventional capabilities – Iran, Iraq, and Syria – were absent from the ACRS process further complicated the situation. This limited regional participation made it more difficult to address the security concerns of those states that did take part. Without the inclusion of major regional powers like Iran and Iraq, for example, the process could not adequately address the security concerns of Israel and Arab states regarding these states' WMD development.

Such power asymmetries and the exclusion of key strategic players are especially problematic for arms control processes in regions where the security dilemma and balance of power thinking are acute. Indeed, the Middle East is in many ways the archetype of zero-sum politics. Regional arms races – surpassing other regions in both quantities and quality – feed into the security dilemma, whereby measures taken by one state to bolster its security are inevitably viewed by others as threatening theirs, leading to continued escalation in armaments and capabilities (Grimmett 2009).[8] This is not just the case between Israel and its neighbours, but also among Arab states and between them and Iran, with actors like Hezbollah and Hamas playing a significant role in affecting perceptions of power balances. Because of this, no player in the region is comfortable with any other actor gaining preeminence.

Despite such structural impediments, ACRS and subsequent Track II security dialogues have played an important role in shaping regional elites' views of arms control in less threatening ways by underscoring the advantages, not just risks, to engaging in such processes. The CSBMs within the ACRS working group created the basis for conducting a win–win dialogue, tension reduction, increased confidence, and improved relationships among adversaries. The demise of a major arms patron (the Soviet Union) and the end of Cold War rivalry in the early 1990s also presented an opportunity to reshape regional views towards security cooperation and arms control, and in fact helped foster the considerable progress that did emerge in ACRS.

Nevertheless, the dominant view among the security elites in the Middle East – who have real command over arms control policies – remains strongly rooted in balance of power thinking, largely sceptical and sometimes hostile to regional arms control efforts. In fact, this is perhaps one issue on which Israelis, Iranians, and Arabs can all agree. Authoritarian and relatively closed political and economic systems throughout the region further contribute to resistance towards security cooperation and arms control.[9] Such a strategic context would present formidable enough challenges. But, in fact, the political impediments to arms control posed an even greater challenge than these strategic calculations.

Regional political barriers

Domestic pressures facing Arab regimes – particularly anti-Israeli sentiment among publics and professional associations – posed a problem for all the multilateral working groups, including the one on ACRS. Arab leaders were reluctant to move ahead of public sentiment on normalization with Israel absent significant progress in the Israeli–Palestinian bilateral negotiating track. This explains why all the multilateral working groups received a significant boost after the Oslo Declaration in September 1993. Likewise, the assassination of Yitzhak Rabin and the overall demise of the peace process – starting in 1996 and concluding with the beginning of the second *Intifada* in the autumn of 2000 – led to the freezing of most formal regional cooperation efforts. That said, ACRS actually broke down *before* the general demise of Israeli–Palestinian relations, suggesting that while domestic and regional political factors certainly affected the climate of these talks, they alone cannot explain its ultimate collapse.

Instead, regional political dynamics and rivalries – in particular Egypt's view of itself in the region – played a major role in the deterioration of ACRS. From the outset of the talks, Cairo consistently raised the nuclear issue, fearing that the Israelis were trying to ignore it by emphasizing security cooperation more than arms control (the Israelis, for example, would often refer to ACRS as the Regional Security Group, dropping or downplaying the arms control aspect). While ACRS was still able to make progress from 1992–94, the international conference in April 1995 to review and extend the NPT indefinitely provided the context for Egypt to push this issue more aggressively, demanding that Israel join the Nuclear Non-Proliferation Treaty and adhere to inspections of its nuclear facilities by the International Atomic Energy Agency.

With Israel's expected refusal to do so, the ACRS working group sessions after this dispute became increasingly contentious, and ultimately the participants stopped meeting formally after September 1995. At these last meetings, Cairo blocked a number of initiatives, including the establishment of a Regional Security Centre in Amman, which Jordan was very much interested in setting up. Egypt refused to continue with ACRS in the aftermath of the NPT dispute unless the issue of a WMDFZ was placed on the agenda, and this essentially put a halt to the group's work. Given Cairo's critical regional position and the fact that decisions could only be made by consensus, the United States viewed Egyptian objections and threats as sufficient to freeze the process, especially due to the possible adverse effects on Egyptian–Israeli bilateral relations.

Why was Cairo so reluctant to continue to move ahead with regional arms control? To be sure, the Egyptians genuinely believed the process was unbalanced from the start in that its content favoured the Israeli preference for incremental CSBMs over disarmament goals, particularly towards an NWFZ that Cairo had been advocating for years. But the Egyptians also viewed ACRS as a problem because of political concerns, particularly the challenge that the talks posed to Cairo's sense of itself as a regional leader, especially on issues related to arms control. Any diminishing of the nuclear issue in the process also suggested a weakening of Egypt's status, as its championing of this issue contributed to its bona fide claims to Arab leadership. Moreover, a multilateral process boosted the importance of smaller Arab states like Jordan and Qatar to the great annoyance of Cairo. These countries played important and active roles in ACRS, leading Egypt to perceive them as a challenge to its rightful position as regional leader on arms control issues. And Egypt worried that greater Israeli interaction with the broader region would enhance Israeli dominance, further challenging Egypt's central role and influence in regional affairs.[10]

Finally, although regional political impediments proved most detrimental to ACRS's continued progress, the lack of sustained American and international commitment presented another challenge. All multilateral working groups that emerged from Madrid were not a high priority for top US policy-makers, and always took a back seat to the more high-profile bilateral peace tracks. In fact, the multilaterals were created somewhat as an afterthought – as a device to improve the regional climate for bilateral peace-making, with very little attention devoted to the content of the discussions themselves at their outset, particularly among high-level American officials.[11] Very quickly the ACRS forum and the other groups were relegated to competent but working-level officials in various State Department bureaus. While these officials very rapidly constructed an impressive and substantive agenda, ACRS did not receive significant high-level attention or support from top policy-makers after its inaugural meeting in Moscow in 1992 co-hosted by U.S. Secretary of State James Baker. Future efforts would benefit from higher-level support and investment in such processes from the United States, Russia, and other international partners, including China (see the chapter of Sven-Eric Fikenscher *et al.* on External Powers). A greater focus on the substance and merits of the discussions themselves – as opposed to the role they play in facilitating Arab–Israeli peace – is another lesson we can learn from the short-lived ACRS experiment.

Disarmament depends on the state of political and security relationships

Effective arms control usually depends on progress in political and security relationships, and ACRS was no exception. The expectation that technical cooperation could in and of itself overcome and help solve political challenges was unrealistic, particularly in a region where any issue has political ramifications. The experience of the ACRS working group teaches us that we should not expect arms control to bring about peace in the Middle East; rather, addressing the political and security dimensions of regional politics can help facilitate arms control over the long run. This does not suggest that addressing arms control and disarmament await the resolution of all political conflicts, just as demanding disarmament without addressing fundamental political disputes is unrealistic. Such strict sequencing in either direction is not likely to lead to a productive arms control and security process in the future. Instead, it is important to pursue both political and security confidence-building measures, while at the same time establishing a process that allows for discussion of disarmament aims. In ACRS, the more incremental elements focused on limited CSBMs among regional parties made significant progress. At the same time, an arms control process that begins with confidence-building and tension-reduction measures can help to move problematic regional relations in a more positive direction and pave the way for the discussion of more serious arms reduction measures.

Moreover, the long-term goal of a region free of WMD, including ballistic missiles, is critical to maintain and sustain more incremental efforts and to ensure the support of key Arab states like Egypt. But ACRS underscores that arms control in the Middle East/Gulf will have to focus on changing political and security relationships and the regional psychology towards security more generally if such processes hope to successfully tackle large-scale disarmament.

Conclusions

The explanation for the demise of the ACRS working group offered here significantly expands upon the commonly cited reason for the failure of the talks – namely, the insolvable zero-sum positions on the nuclear issue held by Egypt and Israel. It develops an approach that places this bilateral dynamic in a broader regional framework; in this regard it evaluates the impact of regionwide dynamics, especially as these relate to inter-Arab politics and Egypt's role within them. We noted that the major success of ACRS was the fact that WMD were considered within the context of regional inter-state relations; not surprisingly, perhaps, the explanation for the final breakdown of the talks is to be found in the effects of this context as well. The lesson is that arms control efforts must take the underlying context of threat perceptions and security interests in all seriousness, both for enhancing prospects of success, as well as minimizing the risks of failure.

The regional focus of the working group indicated recognition of the importance of the Middle Eastern security context to disarmament efforts. Moreover, the priority accorded to CSBMs was a clear indication that this process took the security context seriously. However, such an approach meant that arms control efforts would necessarily take *time* and would need to recognize political concerns of participating states, particularly regional leaders like Egypt. It would take years to tease out the different threat perceptions and find common ground for building up the basis for cooperation. Cairo in particular began to get restless, and pressed for a move to more focused discussion of concrete arms control (as opposed to confidence-building). Moreover, it became apparent that once the regional frame became the primary reference point of the talks, other salient Middle Eastern/Gulf dynamics also surfaced – specifically, Egypt's leadership interest.

Developments regarding Iran's nuclear ambitions over the past decade have moved the arms control focus in the Middle East back to the international sphere, and in the direction of an NPT-based approach. This has had the effect of pushing the regional logic applied to ACRS to the sidelines of the Middle East/Gulf arms control debate. However, in May 2010, the idea of a WMDFZ reemerged in the context of the NPT Review Conference's Final Document, and could again become a focus of regional arms control efforts. In light of this recent development, it is more important than ever to recognize that what failed in ACRS was not the logic of the regional approach, but rather its implementation, and especially its lack of attention to broader regional dynamics such as Cairo's leadership agenda vis-à-vis the other Arab states. The insights gained in the 1990s as to the importance of easing tensions as an essential first step for discussing disarmament issues should not be discarded, or allowed to disappear from the agenda. Moreover, in any future endeavour, all regional states must be present at the table (see the chapter of Sven-Eric Fikenscher *et al.* on Transparency). These insights have important implications as far as Iran's participation in any regional dynamic that is initiated is concerned.

Unfortunately, Tehran's current nuclear advances, especially in the context of internal political turmoil and repression in the aftermath of the June 2009 presidential election, render the situation today highly unfavourable for regional efforts of any kind, not least in the realm of arms control. If Iran becomes a nuclear weapon state, attention will be directed to the issue of deterrence, and how it might be expected to play out (see the chapters of Bernd W. Kubbig *et al.* on Military Asymmetries and of Uri Bar-Joseph). This will be a serious step backwards for arms control efforts and may have the added negative effect of spurring additional nuclear proliferation in the region and reinforcing the zero-sum security outlook of neighbouring states.

Such a potential future underscores that the need for a serious regional arms control process drawing on the successes of ACRS and incorporating the lessons of its failures may be more urgent than ever. What this implies is an incremental, step-by-step process that should relate to a broad category of WMD and missiles, taking all the relevant threat perceptions of the regional actors into account. Whether the new idea for a conference on a WMDFZ in the Middle East in 2012 will incorporate these insights and prove conducive to moving regional dynamics in a positive direction remains to be seen.

Notes

1 For an earlier study on the concept, see Karem (1988). For an Israeli perspective on regional conditions for an NWFZ, see Steinberg (1996).

2 For more information on the new arms control thinking for the Middle East that developed in the early 1990s, see Kemp (1991), Steinberg (1993).

3 For a discussion of how the multilateral working groups emerged from the Madrid peace process, see Kaye (2001).
4 The wording of Article 4, section 7 of the peace treaty is most significant:

> The Parties undertake to work as a matter of priority, and as soon as possible, in the context of the Multilateral Working Group on Arms Control and Regional Security, and jointly, towards the following: A. The creation in the Middle East of a region free from hostile alliances and coalitions; B. The creation of a Middle East free from weapons of mass destruction, both conventional and non-conventional, in the context of a comprehensive, lasting and stable peace, characterized by the renunciation of the use of force, and by reconciliation and good will.
>
> (Hashemite Kingdom of Jordan and State of Israel 1994)

5 This section draws on the writings of Landau (2006, 2008).
6 For a history of Israel's stance on nuclear weapons, see Cohen (1998).
7 For current assessments of the balance of military power, see Cordesman and Nerguizian (2010), IISS (2009), INSS (2009).
8 For a conceptualization of the security dilemma, see Herz (1950) and the Introduction to this volume by Bernd W. Kubbig.
9 On the role of domestic political coalitions and prevailing economic practices in affecting dispositions towards nuclear proliferation, see Solingen (2007).
10 For further elaboration of this argument, see Jentleson and Kaye (1998). For a comprehensive explanation of the normative context within which Egypt's leadership role evolved in the Middle East, and its ensuing effects in ACRS, see Landau (2006).
11 See Kaye (2001), particularly chapter three on the origins of the multilateral talks.

References

Baker, James A. III (1992) 'Remarks by Secretary of State James A. Baker, III before the Organizational Meeting for Multilateral Negotiations on the Middle East', House of Unions, 28 January. Online, available at: www.mfa.gov.il/MFA/Archive/Peace%20Process/1992/SECRETARY%20OF%20STATE%20BAKER-%20ORGANIZATIONAL%20MEETING%20F (25 March 2011).

Cohen, Avner (1998) *Israel and the Bomb*, New York: Columbia University Press.

Cordesman, Anthony H. and Nerguizian, Aram (2010) *The Arab-Israeli Military Balance: Conventional Realities and Asymmetric Challenges*, Washington, DC: Center for Strategic and International Studies.

Grimmett, Richard F. (2009) *Conventional Arms Transfers to Developing Nations: 2001–2008*, Washington, DC: Congressional Research Service. Online, available at: www.fas.org/sgp/crs/weapons/R40796.pdf (25 March 2011).

Hashemite Kingdom of Jordan and State of Israel (1994) 'Treaty of Peace between the Hashemite Kingdom of Jordan and the State of Israel', 26 October. Online, available at: www.kinghussein.gov.jo/peacetreaty.html (25 March 2011).

Herz, John H. (1950) 'Idealist Internationalism and the Security Dilemma', *World Politics*, 2(2): 157–80.

IISS/International Institute for Strategic Studies (2009) *The Military Balance 2009*, London: IISS.

INSS/Institute for National Security Studies (2009) *Middle East Military Balance Database*, Tel Aviv: INSS. Online, available at: www.inss.org.il/programs.php?cat=37 (25 March 2011).

Jentleson, Bruce W. and Kaye, Dalia Dassa (1998) 'Security Status: Explaining Regional Security Cooperation and Its Limits in the Middle East', *Security Studies*, 8(1): 204–38.

Karem, Mahmoud (1988) *A Nuclear-Weapon-Free Zone in the Middle East: Problems and Prospects*, New York: Greenwood Press.

Kaye, Dalia Dassa (2001) *Beyond the Handshake: Multilateral Cooperation in the Arab–Israeli Peace Process 1991–1996*, New York: Columbia University Press.

Kemp, Geoffrey (1991) 'A New Arms Race or Arms Control?' in Geoffrey Kemp (ed.) with the assistance of Shelley A. Stahl *The Control of the Middle East Arms Race*, New York: Carnegie Endowment for International Peace, 1–14.

Landau, Emily B. (2006) *Arms Control in the Middle East: Cooperative Security Dialogue and Regional Constraints*, Eastbourne: Sussex Academic Press.

Landau, Emily B. (2008) 'ACRS: What Worked, What Didn't, and What Could Be Relevant for the Region Today', *Disarmament Forum*, No. 2: 13–20.

Leonard, James, Prawitz, Jan, and Sanders, Benjamin (1991) 'Effective and Verifiable Measures Which Would Facilitate the Establishment of a Nuclear-Weapon-Free Zone in the Middle East', Report of the Secretary-General, A/45/435, New York: United Nations Department for Disarmament Affairs.

Levite, Ariel and Landau, Emily (1997) 'Confidence and Security Building Measures in the Middle East', *Journal of Strategic Studies*, 20(1): 143–71.

NTI/Nuclear Threat Initiative (2008) 'Chronology of Events Related to the Creation of Nuclear Weapon Free Zones'. Online, available at: www.nti.org/h_learnmore/nwfztutorial/chronology.html (25 March 2011).

Review Conference of the Parties to the Treaty on the Non-Proliferation of Nuclear Weapons (2010) 'Final Document', Vol. 1. Online, available at: www.un.org/ga/search/view_doc.asp?symbol=NPT/CONF.2010/50%20%28VOL.I%29 (25 March 2011).

Solingen, Etel (2007) *Nuclear Logics: Contrasting Paths in East Asia and the Middle East*, Princeton, NJ: Princeton University Press.

Steinberg, Gerald M. (1993) 'The History of Arms Control in the Middle East', in Avi Beker (ed.) *Arms Control without Glasnost: Building Confidence in the Middle East*, Jerusalem: Israel Council on Foreign Relations, 7–27.

Steinberg, Gerald M. (1996) 'Peace Security and Deterrence in the Middle East: The Obstacles to a Nuclear Weapons Free Zone', Paper prepared for presentation at a Workshop on Nuclear-Weapons-Free Zones, Australian National University Peace Research Centre, 28 November. Online, available at: http://faculty.biu.ac.il/~steing/arms/austral.htm (25 March 2011).

3 Missiles in the Middle East

Their destabilizing role

Dennis M. Gormley, Colin P. Clarke, and Jürgen Altmann[1]

The mounting missile threat

Meeting before more than 300 defence contractors at the Seventh Annual U.S. Missile Defense Conference in Washington, DC, on 23 March 2009, the Vice Chairman of the U.S. Joint Chiefs of Staff, General James Cartwright, stunned his audience when he asserted: "Ballistic missiles are about as passé as sea mail. Nobody does it anymore.... [E]ven countries we consider 'Third World' have gone beyond that" (quoted in Carden 2009). Although the General's blunt comments contained a kernel of truth – namely, the ballistic missile threat to US territory has not matured as missile defence (MD) advocates have predicted – they nonetheless reflect a virtually singular focus on the American homeland while belying proliferation realities facing countries in the Middle East/Gulf as well as in South and Northeast Asia. Seen from these comparatively confined geographic regions of the world, not only have arsenals of ballistic missiles increased, but so too has the spread of cruise missiles and, notably in the Middle East, conventional rockets (see the Table compiled by Michael Haas and Bernd W. Kubbig in the Appendix). And because the toxic combination of ballistic and cruise missile proliferation dramatically increases the costs of MD, a growing number of regional states facing these threats have become enamoured with preemptive strike doctrines either instead of acquiring missile defences or as a complement to them. The adverse consequences of these developments for regional stability over the next decade could be profound.

This article addresses the destabilizing role of missiles in the Middle East/Gulf, a region of the world where missile use stands out: nearly 91 per cent of the 5,025 ballistic and cruise missiles fired in combat since the end of the Second World War occurred in the Middle East (Gormley 2008: 48). We first tackle the question of precisely what in the character of missiles (including ballistic and cruise) as well as unmanned aerial vehicles (UAVs) make them so profoundly destabilizing as instruments of warfare. We then examine lessons drawn from the use of these missiles in past wars in the Middle East, before turning to the contemporary setting of their proliferation within the region. Finally, the article addresses the relatively new phenomenon of the frequent use of rockets, with ranges from a few to over 200 km, by actors like Hamas and Hezbollah. We conclude with an argument for addressing missiles in their own right as sufficiently destabilizing means of warfare – no matter whether they carry nuclear or non-nuclear payloads.

The destabilizing characteristics of missiles, rockets, and UAVs

Mark Smith (2006) has argued wisely that we should avoid exaggerating the link between missiles and weapons of mass destruction (WMD) as a foundation for controlling the spread of delivery vehicles. Instead, he has suggested focusing on the unique characteristics that distinguish missiles from other delivery systems. These would include, in Smith's view, range, speed, and low-susceptibility to missile defences, with improvements in accuracy adding additional strategic weight. Although Smith's list had ballistic missiles exclusively in mind in the context of norm construction, Dennis Gormley (2008: 161–2) has argued elsewhere that land-attack cruise missiles (LACMs) meet each of these unique characteristics while also having the advantage of low cost compared with ballistic missiles – making saturation attacks more probable.

Range and speed are prominently associated with ballistic missiles, especially to the extent that they can – from great distances today – impart shock effect, paralysis, and even lasting psychological damage on the targeted nation. The psychological effect first became evident during the Nazi German use of primitive V-1 cruise missiles in June 1944 followed by V-2 ballistic missiles three months later. Although many more V-1s were fired towards Allied cities than V-2s, the latter missiles immediately eclipsed V-1s as the more fearful Nazi 'vengeance' weapons. Of course, in neither case were these arms particularly effective. Steve Fetter's (1991: 12) analysis showed that each V-2's 1 ton warhead landing in London produced, on average, five deaths and 13 injuries, while damaging 40 buildings. V-1 cruise missiles produced just over two deaths and six injuries per attack, largely because the V-1's pulsejet engine was so noisy and its slow speed provided more warning for civilians to take cover compared with the fast-travelling V-2 ballistic missile.

Despite the fact that the V-1 and V-2 proved so relatively ineffective, the latter engendered a palpable sense of defencelessness due to the inability of Allied forces to stop them. During the last four weeks of V-1 cruise missile attacks, British air defences intercepted, respectively, 24 and 79 per cent of incoming V-1s. The absence of any defences against fast-moving V-2s (flying at more than four-and-a-half times the speed of sound, or mach 4.6) proved psychologically demoralizing. According to Allied Supreme Commander Dwight D. Eisenhower, had the Germans perfected their V-weapons six months earlier, the 'D-Day' invasion of Europe might not have been possible (Betts 1981: 5). Since their first use in 1944, the capacity of ballistic missiles to produce shock effect and a feeling of defencelessness, regardless of poor accuracy, made them increasingly appealing weapons of choice for a variety of reasons. According to Aaron Karp (2005: 9), ballistic missiles possess extraordinary symbolic value; hence, their featured role in military parades offering an iconic image of the wherewithal to render harm from great distances. As Karp has written, ballistic missiles "have come to symbolize both the erratic threats of paranoid dictators and the final defense of insecure nations".

Two recent developments, however, have altered the perceived value of ballistic missiles, one potentially negatively, the other positively. In the first regard, the vulnerability and defencelessness that ballistic missiles have always produced have begun to recede somewhat due to improvements in MD. Between 1962 and 1996, despite America's investment of roughly $100 billion in anti-ballistic missile defences, the US population remained vulnerable to nuclear attacks (Schwartz 1998: 284–98). Since 1996, at least another $50 billion has been expended with the only notable improvements occurring in defences against short-range capabilities. Where Patriot systems performed poorly against Iraq's ballistic missiles during the 1991 Gulf War, greatly improved Patriots intercepted all nine of the ballistic missiles that threatened US forces in the 2003 War with Baghdad. Yet, their performance against the regime's surprise use of primitive cruise missiles contributed to precisely the opposite effect: cruise missile defences were broadly seen to be grossly inferior to MD (Gormley 2003/4). Indeed, in the immediate aftermath of the 2003 War, the widely perceived shortcomings of American cruise missile defences quickly became a central narrative element in increasing state interest in acquiring land-attack cruise missiles (Gormley 2008: 107–22). It is rapidly becoming apparent that the best way to cope with missile defences is to challenge them with *both* ballistic and cruise missiles.

Improvement in ballistic missile accuracy is also beginning to manifest itself, most notably in the case of China. Beijing has benefited greatly in this respect from the technological and intellectual support of Russia. Its large arsenal of over 1,000 short-range ballistic missiles facing Taiwan would not be nearly as effective as conventional delivery systems without Moscow's guidance and navigation assistance beginning in the early 1990s. The People's Liberation Army borrowed heavily from late Soviet-era writings about how conventionally armed ballistic missiles can leverage the effectiveness of one's attack aircraft. Chinese strategists now view their growing arsenal of conventional ballistic and cruise missiles as a means of trumping Taiwan's superior air force. The sheer shock value of these missiles, not to speak of the difficulty of defending against large-volume attacks, is also seen as conveying coercive value short of actual use (ibid.: 19–26).[2] And, like their Soviet-era counterparts, Beijing's military strategists devote considerable space in their writings to the importance of seizing the initiative from the very outset of combat. The goal of such preemptive action, in which missiles play a featured role, is not to win in one blow, but to adjust the initial conditions so as to predetermine the favourable outcome.

Land-attack cruise missiles compare favourably with ballistic missiles in that they both possess unique characteristics that confer a strategic advantage. Cruise missile range is currently ample and growing in every regional conflict area, including the Middle East/Gulf. Although the maximum range of these missiles today stands at around 3,000 km, cruise missiles can readily achieve strategic range by means of their multiple-launch possibilities – equally from land, air, and sea means. Whereas US space-based sensors can detect ballistic missile launches, making them more susceptible to missile defences, cruise missile launches are much more difficult to detect from afar. As cruise missiles can fly low and alter their flight paths unpredictably into the target area, or incorporate stealth materials to reduce their detection by radar, they too can achieve shock effect, if not total surprise, independent of their speed. Although some observers might argue that modern cruise missiles are weapons of discrimination rather than mass destruction, it is also true that they are fully capable of delivering mass destruction payloads significantly better than ballistic missiles. For instance, Pakistan's new land-attack cruise missile, the Babur, is designed to carry either a nuclear warhead or a conventional payload for precision delivery. Finally, the cost advantages cruise missiles have over ballistic ones not only give them added appeal but also make the prospect of effective MD increasingly more problematic.

Stability in the Middle East is not only threatened by missiles but also by rockets launched by actors such as Hamas and Hezbollah. Despite Israel's status as the most advanced military force and only nuclear power in the entire region, the country is particularly vulnerable to rocket attacks because of its geography, surrounded on several sides by territories inhabited by belligerent groups. Hezbollah's use of rockets has provoked military campaigns of its southern neighbour into Lebanon and in July 2006 this strategy has been used to respond to Israeli attacks. Moreover, Hamas' rocket fire was cited as the reason for Israel's latest foray into the Gaza Strip. The ensuing battles have resulted in severe loss of life and destroyed critical infrastructure.

The use of rockets by groups like Hezbollah and Hamas should not be dismissed as a mere nuisance. Hardly any training is needed to employ these weapons. Furthermore, compared to other arms they are both cheap and expendable. If a rocket team fires a dozen Qassams with no results, nothing is lost. However, if only a single rocket causes damage or death, the attack is considered a success. Even for the battle-hardened Israelis, the rockets have a psychological impact well beyond the physical toll they exact. The costs of rocket use are felt most widely in physical and economic terms, as well as social, political, and diplomatic effects. The following broad points capture these costs. Because the rockets used by these groups are highly mobile weapons, they can be assembled in a matter of minutes and fired quickly, allowing the terrorists to blend surreptitiously into a civilian backdrop. Since these rockets can, and sometimes do, result in casualties, the Israel Defense Forces (IDF) are always going to seek out both the production facilities and launching sites. Detecting and targeting mobile launchers requires rapid response and, thus, incomplete decision-making based on what can at times be ambiguous intelligence. Mitigating the rocket threat requires aggressive intelligence work, likely including penetration of surrounding states as well as fast-move and helicopter patrolling of the border areas for detection.

Unlike a missile launched from Iran, a rocket shot from Gaza or the West Bank does not leave Israel enough time to detect the threat and to deploy an adequate defence. Satellites can detect Scud launchers, but Qassam rockets can be assembled quickly, fired, then folded up and concealed (Karon 2002). Due largely to the difficulties inherent in identifying these missiles, prompt excessive measures such as attacking entire villages to locate one truck loaded with Katyushas might be seen as an option. This reality puts civilians increasingly at risk. In turn, civilian deaths on both sides put pressure on political leaders to be decisive and sometimes influence them to take a course of action that they otherwise would have eschewed. This is just one way that 'tit for tat' violence can escalate into a more prolonged and deadly conflict. In addition to the physical toll exacted on the warring parties, the diplomatic and political costs of rocket attacks have arguably been more disastrous. While some critics disregard such strikes as minor irritants, their potential to destabilize the region must be recognized for what it is – a serious threat to the prospects for peace. The more effective Hezbollah and Hamas perceive rocket use to be, the more they will seek to acquire stockpiles, and the deeper the relationships between these groups and Iran will become. One of the most critical outcomes of Israel's July 2006 War was the marked gain in prestige and credibility garnered by Hezbollah throughout the Arab and Muslim world. The organization was widely recognized as a group that has both the will *and* the capacity to stand up to the Israeli Goliath. Throughout the Middle East, legions of citizens from various Arab nations waved pictures of Hezbollah Secretary General Hasan Nasrallah whose popularity seemed to eclipse even that of former Egyptian President Gamal Abdel Nasser, the last Arab leader to stand up to Israel (MacFarquhar 2006).

The Gaza War, too, had repercussions beyond what many Israelis expected. According to Anthony Cordesman (2009: ii), the

> impact of the 'Gaza War' on the Arab world and Israel's neighbors is far clearer. The IDF's success may have enhanced some aspects of Israel's military 'edge' and ability to deter, but it also did much to

> provoke. Reactions built on the anger caused by both the steadily deteriorating situation of the Palestinians and the impact of civilian casualties and collateral damage – not only in the fighting in Gaza but in Lebanon in 2006.

Though the numbers are still contested by both sides, estimates show a nearly 100:1 ratio of Palestinians killed compared to Israelis.[3] Meanwhile, each time the IDF undertakes such an operation public opinion in their own country is often divided and they fail to achieve their stated goal of making the Jewish state more secure. In fact, it may be just the opposite. The real question is the following: are these actions, sometimes contrived to provide short-term benefit to politicians, going to make Israel any safer in the long run? Hamas may be notorious for its suicide bombing campaign, but its rockets also have a psychological effect. Since the erection of the security barrier, suicide bombings have decreased drastically. Still, the rockets have grown in sophistication and could pose an even greater threat in the future. The actual number of deaths is low, but the potential for instability is high.

Unmanned aerial vehicles, often called drones, have been used by armed forces of highly industrialized countries for decades, mostly in reconnaissance roles. In recent years, however, there have been two problematic trends. One is proliferation to many more countries – by now, more than 60 countries have UAV activities, more than 50 develop, and over 20 export them (Jane's 2008: 23–8; van Blyenburgh 2010). In the Middle East region Bahrain, Kuwait, Qatar, and Saudi Arabia operate UAVs; Egypt, Jordan, as well as the United Arab Emirates (UAE) manufacture or develop them; and Iran as well as Israel export them. The latter two countries have the greatest number of different types: Israel 80 and Iran 36. Nevertheless, most of these drones, which range in wing span from less than 1 m to more than 25 m, are used for observation with visible or infrared light cameras or radar.

The second problematic trend, which began with the United States and Israel but is now being taken up by other countries, is putting weapons on UAVs. Since 2002 the United States has routinely attacked targets in Afghanistan, Iraq, and Pakistan with missiles launched from Predator and Reaper UAVs (wing spans of 17 and 20 m, payloads of 340 and 1,700 kg, and ranges of 740 and 3,100 km, respectively). These attacks are remotely controlled from the United States. In Pakistan alone more than 1,000 militants and hundreds of civilians have been killed using UAVs (New America Foundation 2010). Israel is much less forthcoming about its armed drones, but reportedly it has used the Hermes 450 (10.5 m wing span, 150 kg payload, 200 km range) for attacks in Lebanon and the Gaza Strip. The Eitan (or Heron TP, 26 m wing span, above 1,000 kg payload, more than 1,000 km range) has been discussed for long-range strategic attack missions and missile defence; according to one report it was involved in the strike against an Iranian arms convoy in Sudan in 2009 (Mahnaimi 2009). Iran has had the Ababil-T (3.3m wing span, 45kg explosive warhead, above 150km range) for many years and in 2010 announced that it possessed the Raad[4] and the Nazir, two long-range UAVs equipped with bombs. To counter Iranian threats, the UAE has asked the United States to supply them with Reapers (Rawnsley 2010); however, the UAE is unlikely to be able to import them due to restrictions imposed by the 1987 Missile Technology Control Regime (MTCR).

Several industrialized countries are developing and testing prototypes of pilotless combat aircraft, called unmanned combat aerial vehicles, which should be able to execute all the functions of traditional piloted combat aircraft in 10–20 years. Absent arms control restrictions many more countries are likely to follow this trend. A special problem with armed drones is their relationship to cruise missiles. Whereas the U.S. Department of Defense (DoD 2007) explicitly excludes the latter from its definition of an U(A)V, cruise missiles are clearly unmanned aircraft. There are some differences between (ground-launched) cruise missiles and (new types of) drones: the former are fired from a launcher and travel one way along a preprogrammed path, destroying themselves at the target, while many of the latter take off and land on runways, can be controlled in flight, discharge a weapon at the target, and return after the operation. However, there are similarities – some UAVs need a launcher, some armed types are expendable – and these similarities will increase in the future, as cruise missiles are equipped with an abort or mission change capability. Thus, many arguments about the destabilizing role of cruise missiles made in this chapter apply to armed drones too. In addition, armed UAVs above 500km range could undermine the Intermediate-Range Nuclear Forces Treaty since they are not classified under the definition of a ground-launched cruise missile, but can execute similar missions.[5]

Armed UAVs can be more destabilizing than piloted aircraft in several respects (Altmann 2009). Some will be difficult to detect at low altitudes, allowing them to penetrate deeply into enemy territory for precision surprise attacks. Without a crew on board to be endangered or captured, UAVs could be sent more easily and on more dangerous missions. They could also carry weapons of mass destruction, which need not be recognizable from outside. But even with conventional armament, they can be destabilizing in a crisis. If two such fleets were to fly at short distances from each other (along a border, in international territory), watching for indications of attack, misinterpreting each other's intentions could lead to war. Even unarmed drones could bring instability if they operate along controversial borders and are perceived as intruding in foreign airspace. For the time being, weapon release from UAVs is done under remote control by a human operator (who in the US case sits at a console, for example, in Nevada, observing events through the UAV camera picture and controlling the drone, for instance in the Middle East, all by satellite communication).

There are several factors that are likely to drive the military towards UAV autonomy both in flight and in weapon use including technical advances, cost savings, increased reaction times, and the chance of communication failure. This raises a fundamental ethical question: should an inanimate machine ever be allowed to 'decide' when to kill a human being? Roboticists warn that assessing a complex battlefield situation as required for compliance with the international law of armed conflict will remain beyond the capabilities of 'artificial intelligence' for a long time. Proliferation of drones will likely increase instability in the Middle East. From 2004–06 Hezbollah sent Iranian Ababil-T UAVs into Israeli airspace. It was only in 2006 that manned Israeli combat aircraft managed to shoot down two such UAVs which turned out as armed with explosives (Gormley 2008: 65; Jane's 2007: 76). These incidents indicate the possible uses to which armed UAVs could be put in the future. Small drones (payload tens of kilogrammes), accessible to militias, could be effective against high-value targets if the targeting was sufficiently precise. Hence they have an advantage over unguided rockets, and their range is much greater as well. If they were deployed in great numbers, small UAVs could overwhelm efforts to intercept them. Defence by a fleet of similarly small 'fighter UAVs' may be feasible, but this would accelerate the arms race and the associated increase in offensive capacity could also increase crisis instability.

Very small drones (wing span, payload, and range around or below 1 m, 1 kg and a few kilometres, respectively) could be effective tools for terrorists, for example to distribute biological agents or to assassinate very important persons.[6] At present, Israel is in a technologically and numerically superior position as compared to other regional powers with respect to armed and unarmed UAVs. However, this could change as others acquire similar capabilities through imports or indigenous development. In the long run, Israel may become more threatened by armed drones of close or less close neighbours than vice versa. In any case, the use of armed drones will grow and thereby add another layer to military instability in the Middle East/Gulf. Because armed drones could provide similar capabilities for surprise attack as cruise missiles proper, efforts to limit and ban the latter in the Middle East should include UAVs that could carry a significant weapon payload (say, above 500 kg) over significant ranges (say, above 50 or 200 km; the MTCR range threshold is 300 km). However, because piloted aircraft could fulfil similar functions, quantitative limits on such aircraft are recommended in parallel.

The past as prologue?

Ballistic and cruise missiles were used in five different conflicts in the Middle East, including the Egyptian and Syrian missile attacks on Israel in the 1973 Yom Kippur War, the 1985–88 'War of the Cities' between Iraq and Iran, the 1991 Gulf War, the Yemen Civil War of 1994, and the 2003

US-led invasion of Iraq. The most notable were the 'War of the Cities' and the two Gulf Wars of 1991 and 2003. In the Iraq–Iran War, both sides sparsely used ballistic missiles until 1985, whereupon Baghdad began to attack Tehran with its new longer-range weapon, the al-Hussein. This campaign culminated in a period of six weeks during 1988 in which about 190 Iraqi missiles killed some 2,000 Iranians. More broadly, these attacks terrorized the population of the Islamic Republic. Tehran's responsive attacks on Baghdad, involving about 90 Scud missiles, by contrast, seemed to have little effect on Iraqi morale (Karp 1996: 45).[7] There appears little doubt that Saddam Hussein's war-fighting strategy of employing ballistic missiles as terror weapons powerfully affected Iran's decision to seek a ceasefire and eventual conclusion of the war in 1988. According to some specialists, Iraq's threat of attacking the Islamic Republic with chemically armed missiles shaped the latter's decision to accept a less than balanced outcome to the war (GlobalSecurity.org 2005a). Baghdad had repeatedly employed blister and nerve agents against military targets with some notable tactical success, but possible strategic use of such potent weapons on population centres was apparently ruled out.

Missile use in the 'War of the Cities' also influenced Saudi Arabia to obtain, from China, CSS-2 ballistic missiles with a range of 2,800 km – a deal struck reportedly a few months after Iran's Scud attacks on Iraq in March 1985 (Carus 1990: 4–5). Clearly, ballistic missile use in the war left a substantial and unstable legacy in the Middle East/Gulf. Despite the fact that these weapons were not especially effective in delivering conventional weapons, both Baghdad and Tehran accorded ballistic missiles a much higher status in the war's aftermath, which led to significant missile modernization efforts that in the case of the Islamic Republic are only now showing evidence of truly worrisome capabilities (Rubin 2006; see also the chapter of Bernd W. Kubbig *et al.*). For Iran, one key consequence of the war with Iraq was to make ballistic missiles the preferred alternative to combat aircraft, since they could achieve long-range effects, however inconsequential, without having to train highly skilled pilots or logistical support units (Nolan 1991: 85). And because the clergy-led government mistrusted their professional air force, missiles under the control of the Revolutionary Guards took on added salience (Zaloga 1988: 1,423, quoted in Nolan 1991: 85).

In the 1991 Gulf War, Saddam Hussein threatened to use Scud and al-Hussein ballistic missiles as both a means of deterrence and escalation control (Karp 1996: 36). Iraq's missile threat failed to deter the US-led United Nations coalition from commencing "Operation Desert Storm" to retake Kuwait. Once active combat began in January 1991, however, Baghdad used its ballistic missile force – successfully launching 81 weapons altogether, 31 of which were fired at Israel – in part to escalate the conflict by disrupting coalition operations and drawing Israel into the war, an objective the regime nearly achieved (Pedatzur 2005: 624). Because

US-deployed Patriot systems could not successfully intercept most of Saddam Hussein's ballistic missile attacks on Israel and Saudi Arabia, and American war planes were unable to find and destroy Iraq's mobile missile launchers, Israel came perilously close to intervening, which risked splitting the fragile UN coalition at a decisive point.

However, although many Patriots failed to perform as desired, they did seem to play a mitigating role in allaying Israeli concerns about Baghdad's attacks. The strategic lessons of Iraq's ballistic missile attacks for both Israel and the United States became clear in the war's aftermath. The Jewish state's previous caution about investing in costly ballistic missile defences was no longer politically viable. After the war, with American financial support, Israel would embark on developing the Arrow MD system. America, too, had entered the war unprepared to deal with Baghdad's missile threat. The U.S. General Accountability Office (GAO 1994: 36) determined that at most only 9 per cent of Iraq's missiles were successfully engaged.[8] An Iraqi Scud hit an American military barracks at Dhahran, Saudi Arabia at the war's close, killing 28 US soldiers and wounding another 99. This was enough to justify what would become more than a $3 billion investment in improving Patriot missile defences in the decade after the 1991 War.

The Gulf War in 2003 at once legitimated American investments in upgrading the system while also demonstrating its continuing weaknesses. Intelligence assessments, which proved wrong in the end, suggested that Iraq might have possessed a small number of longer-range al-Hussein ballistic missiles, capable of delivering a primitively armed chemical or biological payload to a range of 650km. Iraq had admitted to the UN's Special Commission that before the 1991 War it had filled 25 warheads with anthrax, botulinum toxin, and aflatoxin, and 50 others with the mixed chemical agents sarin and cyclosarin. Also of concern was knowledge that before the Gulf War, Saddam Hussein had handed predelegated authority to his commanders to launch biologically and chemically equipped delivery vehicles if nuclear weapons were used against his regime or Baghdad was threatened. With regime change a declared US goal this time around, the fear was that Saddam Hussein might be more inclined to use WMD armed missiles both against civilian and military targets. But instead, Iraq used al-Samoud and al-Ababil ballistic missiles, which, due to their shorter ranges and slower speeds, proved easier to intercept. In all, of the 19 ballistic missiles Iraq launched at coalition targets in Kuwait and Iraq, only nine threatened potential targets (the others landed in deserted areas). All nine were successfully intercepted and destroyed, vastly exceeding Patriot's performance record in 1991.

America's missile defences were not without flaws, however. While the MD system batted 1,000 against ballistic missiles, it struck out entirely when Patriot batteries and missile warning systems, such as the U.S. Air Force's Airborne Warning and Control System, called AWACS, failed to detect or intercept any of the five primitive Iraqi cruise missiles launched against coalition targets. One came close to hitting a Marine command post on the first day of the war, while the others produced only minor damage to civilian targets or none at all. More ominous for the future, however, was the fact that Iraq's use of both ballistic and cruise missiles forced Patriot radars to search for high- and low-angle missiles, thereby probably contributing to the unfortunate series of friendly-fire incidents, including the loss of two aircraft, and, worse, the deaths of three crew members. None of this should have surprised military planners. Over the preceding decade of the 1990s, the Pentagon had detailed in depth the abysmal state of cruise MD and steps needed to rectify shortcomings (Gormley 2001: 66–76). What did surprise American war planners was the unexpected Iraqi use of both ballistic and cruise missiles, which portended more demanding MD challenges to come.

Ballistic and cruise missile as well as UAV proliferation trends

Ballistic missiles have become a prized commodity among several belligerent states in the Middle East/Gulf faced with Israeli technological and military superiority. These actors include Syria, Iran, and Saudi Arabia, although Riyadh's acquisition of Chinese-supplied DF-3 medium-range ballistic missiles in the 1980s is of far less military significance than Damascus' and Tehran's growing missile holdings. At present, neither Syria nor Iran has the capacity to undertake military attacks with conventionally armed ballistic missiles aimed at significantly disrupting Israel's decisive offensive air superiority over these two adversaries. Both Damascus and Tehran, whose air forces do not remotely measure up to Israel's, seem more inclined at present towards developing the capacity to threaten Israel with chemical, biological, and, eventually, in the Islamic Republic's case, nuclear weapons.[9]

Syria's missile arsenal is largely a Scud-based one, the only exception being the SS-21 (ca. 100 km). Since the 1960s, Damascus has depended on foreign assistance, first from the former Soviet Union, and more recently from North Korea, China, and Iran (NTI 2010). Pyongyang is believed to have transferred the Scud-D, which is thought to possess a separating warhead and course correction capabilities that could conceivably improve its accuracy. While US intelligence does not provide estimates of the number of Scuds or SS-21s that Damascus has produced or acquired, it does report "fewer than 50" launchers available for every system save for the longest-range missile, Scud-D (NASIC 2006: 4; see also the article of Dinshaw Mistry and Mark Smith). With what could be as many as 100 launchers facing Israeli targets, there is more than an ample potential, in theory, for generating a militarily significant volume of fire.

In light of the cramped geography of the region (the two nations' capitals are less than 100 km from their common borders), all of Israel is easily within range of Syrian missiles. However, any short-term effects from conventional missile attacks – most likely SS-21s, due to their greater accuracy – would probably remain unexploited by Syrian air forces, which may be large but are marked by poor pilot training and low operational readiness. Moreover, orchestrating an effective combined arms operation would require that Damascus' air and missile forces be more skilled in coordinating and executing closely timed missile and air strikes than they appear to be. Israeli missile defences may further devalue the threat of Syria's conventionally armed ballistic missiles.[10] Tehran presents a decidedly different kind of ballistic missile threat from that of Damascus. Iran has comparative geographic advantages. With roughly nine times the land mass of Syria, the Islamic Republic is slightly larger than Alaska. Its huge area furnishes valuable strategic depth, allowing it to deploy its fixed and mobile missile launchers deep within the country's interior.

The distance from Tel Aviv to Tehran (1,598 km) is almost 13.5 times the distance to Damascus (214 km). And there is nearly another 900 km of strategic deployment depth to the east of Tehran, where missiles could be based. That said, strategic depth cuts more than one way. While it affords operational breathing room, it also means that short-range ballistic missiles are unlikely to satisfy one's military needs. After first purchasing Scud-B missiles from Libya in 1985, the Islamic Republic quickly turned to providing cash to help finance North Korea's missile development in exchange for substantial hard and soft technology transfers and, essentially, the right to buy whatever Pyongyang produced. Initially, Iran acquired over 200 Scuds, but what Tehran really coveted was the capacity to produce ballistic missiles on its own. This relationship has permitted the Islamic Republic to expand its holdings of Scud-Bs and -Ds, as well as develop the Shahab-3, which can reach Israel. Most of Tehran's ballistic missiles are liquid-fuelled. But both solid and possibly storable liquid systems are either deployed or on track for deployment.[11] In the case of solid systems, Iran obtained the 150 km range CSS-8 solid fuel ballistic missile from China and is suspected of attempting to purchase M-11 solid fuel missiles, although evidence is thin.

The United States did later impose sanctions on China; however, Beijing's assistance in not only solid-propellant production but also in guidance and missile testing has continued (Feickert 2004). Such support probably aided the Islamic Republic in its development of the Fateh-110, a 200 km range solid-propellant missile first tested in 2002 and now deployed widely as an Iranian battlefield delivery system. Of much more significance is Tehran's surprise success in developing and apparently successfully testing (in November 2008 and May 2009) a two stage solid fuel missile, called Sajjil-2, which has a range of approximately 2,200 km (IISS 2010: 58). Rather than a small incremental jump from the 2 ton Zelzal-2 solid rocket to something slightly heavier, which is the normal developmental pattern, the Sajjil-2's first stage used a 10 ton solid motor. Still, to move from this point to a satisfactory mastery of large solid rocket motor production will probably require another three to five years of intensive development (IISS 2009, 2010: 133).

Iran also faces stiff challenges in improving its indigenous capacity to produce better navigation and guidance systems that increase these missiles' military value. The Islamic Republic has saved years of development in the past by turning to China and Russia.[12] Besides its vastly superior conventional capability, Israel possesses its own force of nuclear armed ballistic missiles that would allow it to attack any regional target of interest. The heart of that force consists of 50 single stage Jericho-1 solid fuel ballistic missiles deployed on mobile launchers, 100 two stage Jericho-2s on underground wheeled launchers or railroad flat cars, and an uncertain number of three stage Jericho-3s, which reportedly became operational in 2005. They can deliver payloads ranging between 500 and 1,000 kg. Were Israel in need of achieving even longer ranges, it could readily convert its space launcher, the Shavit, into a missile capable of delivering a nuclear weapon to a range over 7,000 km.

Israel also dominates development and acquisition of cruise missiles and UAVs in the region. The Jewish state is a major developer of reconnaissance drones and the air-launched Popeye LACM, an advanced version of which is suspected of having been adapted for launch from Israel's German-furnished Dolphin class submarines to a range of 1,500 km, carrying a small nuclear warhead. Although the latter possibility is just that, there is little doubt that the Popeye class missiles could be readily upgraded to achieve ranges much greater than the reported 250 km (FAS 2000). Israel originally developed the Popeye as a short-range (less than 100 km) stand-off missile powered by a solid-propellant rocket motor. It morphed into several variants. Israel has also turned to UAVs to perform counterforce missions against enemy missile launchers. While the Jewish state struggled to find and attack Hezbollah's ubiquitous short-range Katyusha rockets, which can be emplaced by one man in seconds, air and ground forces flooded the skies with UAVs networked together to provide loitering aircraft with precise targeting coordinates of detected medium- and long-range Hezbollah rockets deployed on mobile launchers. Reportedly, Israel destroyed between 80 and 90 per cent of these launchers (around 125), all within 45 seconds to one minute between detection and attack (*Defense News*, 2 October 2006: 1). The Jewish state has also investigated using UAVs to carry interceptors in order to attack missiles during their boost phase (within 90 seconds after launch).

Iran's increasing interest in cruise missiles and UAVs may well be a response to Israel's substantial efforts at deploying ballistic missile defences. Besides providing Hezbollah with its own indigenously produced UAVs, Tehran has obtained cruise missile systems and technology from Beijing to support its new anti-ship cruise missile, called Nur, which is essentially the Chinese C-802 missile produced locally in the Islamic

Republic (NTI 2007). Iran also has reportedly converted 300 HY-2s by means of fitting them with a turbojet engine and new guidance systems. Its search for a suitable turbojet engine for converted HY-2 anti-ship missiles may also explain how Tehran's new Raad cruise missile has achieved a range of 350 km, and why it may appear as both an anti-ship and land-attack cruise missile (Ibrügger 2004). Ambitions for even substantially longer-range land-attack cruise missiles appear evident in the Islamic Republic's 2001 illegal acquisition of six Russian/Ukrainian Kh-55 cruise missiles, using dealers from both countries, for nearly $50 million (*Washington Times*, 6 April 2005). The theft of these 3,500 km range cruise missiles likely means that if Iran can muster sufficient foreign technical assistance, it is probably on a path to developing a long-range LACM.[13] Confronted with Israel's substantial investment in ballistic missile defences, Tehran appears to view a cruise missile arsenal as an efficient way to increase the return on investment in its ballistic missile programme (IranFocus.com 2004).

The special case of rockets

Rockets are the perfect weapon for terrorists, insurgents, and militias as they are cheap, easy to use, and can inflict significant psychological damage on one's enemy. These weapons are also highly effective because they are light and can be fired from a makeshift stand or mortar, making them a popular choice for groups concerned with mobility, concealment, and sustained fire.[14]

Past use of rockets

While the use of rockets by actors such as Hamas and Hezbollah in the Middle East is a relatively new phenomenon, the origin of the Katyusha rocket, favoured by Hezbollah, dates back nearly 80 years. The Soviet Air Force developed this weapon, or "little Katie" in Russian, in the 1930s (Aerospaceweb.org 2006).[15] The Katyusha gained notoriety during the Second World War when the Soviets used it against the Germans. More recently, it has become synonymous with Hezbollah, the Shiite actor based in Lebanon and backed by Iran and Syria (see the article of Judith Palmer Harik and Walid Abu-Dalbouh). Katyushas are 122 mm surface-to-surface rockets and can be either short-range (up to 25 km) or extended range (25–35 km). The Palestine Liberation Organization, and several smaller militias first experimented with the use of Katyushas throughout the course of the Lebanese Civil War (1975–90), during which these groups fired them from southern Lebanon into northern Israel. The Israelis responded by invading the country in 1982 and stationed a robust military force there for the next 18 years, until withdrawing in June 2000.

During its July 2006 War with Israel, Hezbollah maintained a consistent barrage of rocket fire into towns and villages along the border. According to Andrew Exum (2006: 12), during the first ten days of the conflict, Hezbollah fired an average of 150 to 180 rockets per day, with a single day high of 350, on 18 July. While the average eventually dipped to 100 per day towards the end of the fighting, in a last-ditch display of potency the group managed to fire 250 rockets into Israel one day prior to the cease-fire. The Palestinian actor Hamas, currently responsible for governing the territory of Gaza, has adopted the use of the Qassam rocket as a means of conducting asymmetric warfare against the much more powerful Israeli military (see the article of Margret Johannsen *et al.*). The Qassam rocket is manufactured and used almost exclusively in Gaza, although the Israelis have seized the weapon in various parts of the West Bank as well. This 60–170 mm home-made iron rocket is cylindrical with a small warhead at the end, four stabilizing wings, an engine, and a detonating fuse.

Construction of the Qassams began during the outbreak of the Al-Aqsa *Intifada* and made their first appearance in March 2002 when the Izz ad-Din Brigades used these crude devices. Between 2002 and 2004 Hamas fired roughly 450 Qassam rockets into Israel, with the city of Sderot and smaller towns and villages adjacent to the Gaza Strip bearing the brunt of the attacks (GlobalSecurity.org 2009). In 2006, the rockets, which in addition to Qassams also included Katyushas, reached the city of Ashkelon. Following the expiry of a six month truce, sporadic rocket fire into Israel emanating from Gaza began again in November 2008. The Israelis responded with "Operation Cast Lead", one of several goals of which was terminating the shelling into southern Israel. The incursion also aimed at stymieing the smuggling of arms in Gaza from tunnels that connect the territory with Egypt. Through these underground conduits Hamas was able to smuggle 81/82 mm and 120 mm mortars. What has become known as the Gaza War lasted from 27 December 2008 until 18 January 2009. Over the course of the three week battle Hamas fired approximately 600 rockets into Israeli territory (Cordesman 2009: 7).

Rocket holdings

In 2004, the head of the Israeli Military Intelligence reported that Hezbollah's arsenal consisted of approximately 13,000 rockets, the majority of which had a 25 km range (GlobalSecurity.org 2006). Furthermore, the group was thought to possess an additional 500 rockets with a range of 45 km (Fajr-3) to 75 km (Fajr-5) and several dozen 115 km range rockets (Zelzal-2).[16] Following its July 2006 War with Israel, a Congressional Research Service report estimated that Hezbollah still maintained an arsenal of 12,000 "munitions", the vast majority of which were Katyushas (Sharp 2006: 8). Hamas' rocket arsenal was estimated at approximately 3,000 at the beginning of "Operation Cast Lead". In addition to the 600 rockets fired by the group, the IDF destroyed an additional 1,200. This would leave the actor with 1,200, although almost immediately following the ceasefire it is believed that Iran began resupplying Hamas to replenish its inventory (Cordesman 2009: 58). The majority of this arsenal is made up of Qassam rockets, although it also includes other types of arms obtained from Iran, Syria, and Hezbollah. Three versions of the Qassam have been identified, simply named the Qassam-1, -2, and -3. The original version was 60mm, carried a payload of 5.5kg, and had a range of 3–4.5km while the improved Qassam-2 was 180cm long, weighed between 5 and 9kg, and had a maximum range of nearly 10km (*Jane's Defence Weekly*, 28 January 2009: 7). The latest iteration, Qassam-3, was developed in 2005 and can weigh as much as 20kg while travelling as far as 12km, thus extending the weapon's reach farther into Israel's territory.

Actively defending against rockets

The discussion of Hamas' and Hezbollah's rocket arsenals and their accompanying capabilities to destabilize the region would not be complete without an assessment of Israeli MD capabilities. The two systems that were being considered as tools against Qassam and Katyusha rockets are the Israeli-made Iron Dome and the American-made Skyguard, formerly known as Nautilus. Much of the debate inside the Jewish state has focused on the decision-making process within Israel's highest reaches of government – including disagreements between policy-makers and the military. Still, scant attention has been devoted to the actual merits and shortcomings of both systems, and whether or not missile defence systems are a worthwhile investment and an effective deterrent. Iron Dome, made by Rafael Defense Systems, employs an advanced radar that locates and tracks rockets, which can then be destroyed by a kinetic missile interceptor (Katz 2009b). The IDF plans to establish a battalion in its anti-aircraft division to operate the system. The main criticism levied against Iron Dome is that it is ineffective against close-range rockets and that, as such, many homes in Sderot, near the border with Gaza, will remain vulnerable to rocket and mortar fire (Pedatzur 2008). Moreover, it should be noted that there is an immense difference in cost between rockets employed by organizations such as Hamas and Israeli MD programmes.

However, such criticisms may be overblown with the successful tests of the system in summer 2010.[17] Furthermore, opponents of missile defence argue that robust MD would merely encourage Israel's enemies in the region to seek more sophisticated rockets with an extended range. This argument lacks evidence and fails to consider that Hamas and Hezbollah will likely pursue such ends regardless of the countermeasures. Certainly, though, one possible consequence of a staunch MD system would be that Hamas and Hezbollah resort back to less sophisticated but equally deadly methods of attack, including ambushes, sniper shootings, kidnapping, and suicide bombing. Nevertheless, in July 2009 the Iron Dome system was hailed as a success following tests against three 122 mm Grad type weapons (Spacewar.com 2009). But this still begs the question of the criteria used to measure success when the Iron Dome testing went through dozens of simulations on live data from the thousands of rockets launched into Israel over the past several years. One cannot simply eschew the immense psychological impact that these rockets inflict on the Israeli public.

Critics come from both sides of the aisle, and recently the state comptroller chastised the Defense Ministry's mismanagement of the Iron Dome and warned that slipshod development, poor coordination between military users, and an overall dearth of oversight could doom the project before it has a chance to become effective. The Skyguard system, made by Northrop Grumman, uses a laser beam for interceptions, and according to some Israeli commentators, is cheaper, ready to deploy, and more effective against the threat of rockets, although opinions vary widely on this issue. Some speculate that the Israelis favoured Iron Dome over Skyguard in which they are no longer participating because the former is Israeli-made while the latter is not. Another possible alternative to Iron Dome was the Raytheon-made Phalanx, a heavy calibre automated machine gun whose bullets destroy incoming shells. Finally, the Stunner/David's Sling is a static and towed surface-to-air missile system which will be incorporated into the Magic Wand project and is meant to defend Israel against rockets with a range of 70–250 km. This would directly affect some of Hezbollah's rockets. The Stunner system is a component of the medium- and short-range missile defence programme and is regarded both by the US and by Israel to fulfil an indispensable and urgent need for a medium-range, highly efficient, and cost effective weapon.

The system is expected to be available in 2012 with full operational capability the following year (Susser 2009). Hamas and Hezbollah will undoubtedly devise countermeasures in an attempt to overcome the implementation of Iron Dome. Adaptability and innovation are the hallmarks of such actors. This does not mean that missile defence systems are not worthwhile. In fact, as Uzi Rubin commented regarding the Arrow (already in operation) and the Iron Dome, "[Israel] has something to stabilize the situation: the knowledge that an attack will fail" (quoted in Schneider 2009). In most nations that can afford them – particularly so in the Jewish state – the political obligation to protect the nation's citizenry creates enormous pressure on politicians to deploy missile defences, the level of effectiveness and costs notwithstanding. Metrics such as 'cost effectiveness at the margin' – Paul Nitze's Cold War notion that it ought to be less costly to deploy MD than for one's adversary to add more offensive missiles – generally buckle under such political pressure.

Final observations

Because of their inherent capacity to produce shock effects with little warning, military strategists have always favoured assigning offensive missiles important roles to accomplish preemptive or preventive results. This preference is by no means receding. In fact, there is a truly global contagion underway in which regional states are acquiring new missiles, ballistic and cruise alike, and linking them to preemptive or preventive strike doctrines (Gormley 2008: 123–45). The Middle East/Gulf is certainly no exception to these global trends. Preemption has been an option in Israeli military decision-making since its first application in the Six Day War of 1967. The confined geographic circumstances facing both the Jewish state and Syria surely may increase incentives to strike first during intense crises. Indeed, were Damascus to obtain from Moscow significant quantities of Iskander-E ballistic or Iskander-K cruise missiles – arguably, which furnished Damascus with even a temporary preemptive advantage – crisis stability would only suffer. For Israel's part, the development of multi-tier MD should assuage its traditional predilection towards preemptive action. Yet, some Israeli strategists, while championing missile defences, still strongly affirm the need to act preemptively, especially in light of emerging nuclear threats (Beres 2007: 37–54). The increase of armed UAVs on the part of many actors in the Middle East will tend to exacerbate preemption fears and pressures.

Senior leaders of Iran's Islamic Republican Guards Corps (IRGC) have studied closely US military actions in Afghanistan and Iraq to draw lessons for restructuring the IRGC primarily to cope with American rather than neighbouring threats. Not surprisingly, the conclusion is that the best defence against preemption is rapid deployment of asymmetrical forces together with increasing reliance on missiles and real-time intelligence activity. Accordingly, the chief of the Guards Corps, Mohammad Ali Ja'fari, who spent two years at the IRGC Strategy Center preparing for the restructuring, intends to incorporate the lessons he learned from the war with Iraq about the use of surprise attacks into a new IRGC structure emphasizing lighter combat units and missile use. Ja'fari believes Iranian-backed rocket attacks by Hezbollah played a critically important role in the war with Israel in southern Lebanon (Gormley 2008: 138). However global the trend might be in favour of linking missiles and rockets to preemptive action, the Middle East stands out as unique compared with other regional circumstances. The increasing use of short-range rockets only exacerbates the toxic mix of historical, social, cultural, political, and geographic circumstances in the region. And no matter whether their range is a few kilometres or ten or more times that distance, they have killed or wounded an unusually large number of people in their own right, while precipitating substantially larger military operations that produce even more suffering and loss of lives.

Notes

1 Author of the UAV section.
2 The best western source on China's growing dependence on missiles is Stokes (2005).
3 In Gaza, Palestinian sources report (PCHR 2009) a total of 1,417 killed and 5,303 wounded, while Israeli sources claim that 1,166 were killed on the Palestinian side, without reference to the wounded. A total of 13 Israelis were killed in 2008, including three civilians.
4 Raad means 'thunder' and is commonly applied to a variety of weapons in Iranian press statements. This explains why another type of Raad is denoted as anti-ship and land-attack cruise missile of 350 km range in the text.
5 The Treaty definition in Art. II is fairly general:

> 2. The term 'cruise missile' means an unmanned, self-propelled vehicle that sustains flight through the use of aerodynamic lift over most of its flight path. The term 'ground-launched cruise missile (GLCM)' means a ground-launched cruise missile that is a weapon-delivery vehicle.

If taken verbatim this definition would include armed UAVs. But only one existing type of GLCM is named in Art. III, and the United States maintains that cruise missiles are different from UAVs.
6 For terrorist attacks using larger UAVs, see Gormley (2005).
7 According to Karp (1996: 45) the Iraqi missile attacks during the 'War of the Cities' in 1988 "killed some 2,000 people, drove half the people of Tehran out of the city and damaged public morale".
8 This figure is contested though. George N. Lewis and Theodore A. Postol (2000: 315) have argued that "the actual success rate for Patriot against the Iraqi missiles was very low, and most likely was zero".
9 Syria's chemical weapons reportedly include between 50 and 100 missile warheads, while the Syrians have worked on mustard gas, sarin, tabun, and VX (GlobalSecurity.org 2005b). Although Iran signed and ratified the Chemical Weapons Convention and declared and eliminated two chemical weapons production facilities, it remains uncertain whether or not the Islamic Republic has completely eliminated its arsenal. US sources claim that Iran used chemical weapons in retaliation to Iraq's use in the 'War of the Cities' (GlobalSecurity.org 2008).
10 In addition to the Arrow-2 missile defence system, Israel also deploys US Patriot missile interceptors. In May 2009, the United States informed Israel that it would fully fund the Arrow-3 missile defence programme, which aims to deal with longer-range ballistic missiles coming in from higher altitudes. The US also co-funds the Israeli programme called David's Sling, which aims to defeat missiles with ranges between 70 and 250 km (Katz 2009a). Without success thus far, Syria has sought to obtain from Russia the solid fuel Iskander-E, or SS-26, with a range of 280 km. Its terminal velocity and purported manoeuvrability could possibly stress Israeli missile defences.
11 On the possibility of North Korea's provision of 18 storable liquid fuel missiles with a range of 3,500 km to Iran in 2006, see Rubin (2006: 29–30).
12 According to US intelligence, Russia is believed to have saved Iran years on the Shahab-3 programme, while China, besides its solid motor support, supplied Iran with missile guidance systems and computerized production tools, see IranWatch.org (2010). See also the chapter of Sven-Eric Fikenscher *et al.* on External Powers, the chapter of Bernd W. Kubbig *et al.*, as well as the contribution of Dinshaw Mistry and Mark Smith.
13 Uzi Rubin (2006: 28) believes that China and North Korea are likely collaborators in assisting Iran.

14 A standard rocket used for military purposes is a cylindrical tube containing an explosive warhead and powered by a basic propulsion system, usually a solid rocket motor. Rockets, unlike missiles, are not equipped with guidance systems and are therefore much more erratic and difficult to use effectively against fixed or moving targets.

15 The Katyusha had two variants, a 82 mm rocket equipped with a fragmentation warhead used by fighter aircraft and a 132 mm rocket used by bombers.

16 The Fajr-3 is a 240 calibre (mm) rocket with 12 barrels and used with a truck-mounted launcher. The Fajr-5 is a 333 mm rocket with four barrels and used with a truck-mounted launcher. The Zelzal-2 is a 610 mm rocket with a maximum range of 210 km, see Exum (2006: 6).

17 However, the Israeli Air Force cancelled a final test at the end of 2010 after encountering a number of technical malfunctions (*Jane's Defence Weekly*, 12 January 2011: 6).

References

Aerospaceweb.org (2006) 'Weapons Questions, Katyusha and Qassam Rockets'. Online, available at: www.aerospaceweb.org/question/weapons/q0279.shtml (25 March 2011).

Altmann, Jürgen (2009) 'Preventive Arms Control for Uninhabited Military Vehicles', in Rafael Capurro and Michael Nagenborg (eds) *Ethics and Robotics*, Heidelberg: Akademische Verlagsgesellschaft, 69–82.

Beres, Louis R. (2007) 'Israel's Uncertain Strategic Future', *Parameters*, 37(1): 37–54.

Betts, Richard K. (1981) 'Innovation, Assessment, and Decision', in Richard K. Betts (ed.) *Cruise Missiles: Technology, Strategy, Politics*, Washington, DC: Brookings Institution, 1–30.

Carden, Michael J. (2009) *Missile Defense Requires New Focus, Vice Chairman Says*, Washington, DC: U.S. Department of Defense. Online, available at: www.defenselink.mil/news/newsarticle.aspx?id=53605 (25 March 2011).

Carus, W. Seth (1990) *Ballistic Missiles in the Third World: Threat and Response*, Washington, DC: Center for Strategic and International Studies.

Cordesman, Anthony H. (2009) *The "Gaza War": A Strategic Analysis*, Washington, DC: Center for Strategic and International Studies. Online, available at: http://csis.org/files/media/csis/pubs/090202_gaza_war.pdf (25 March 2011).

DoD/U.S. Department of Defense (2007) *Unmanned Systems Roadmap 2007–2032*, Washington, DC: DoD.

Exum, Andrew (2006) 'Hizballah at War: A Military Assessment', *Policy Focus*, No. 63, Washington, DC: Washington Institute for Near East Policy.

FAS/Federation of American Scientists (2000) *Popeye Turbo*, Washington, DC: FAS. Online, available at: www.fas.org/nuke/guide/israel/missile/popeye-t.htm (25 March 2011).

Feickert, Andrew (2004) *Iran's Ballistic Missile Capabilities*, Washington, DC: Congressional Research Service.

Fetter, Steve (1991) 'Ballistic Missiles and Weapons of Mass Destruction: What Is the Threat? What Should Be Done?' *International Security*, 16(1): 5–41.

GAO/U.S. General Accounting Office (1994) *Desert Shield and Desert Storm Reports and Testimonies: 1991–93*, Washington, DC: GAO. Online, available at: www.gao.gov/new.items/ns94134w.pdf (25 March 2011).

GlobalSecurity.org (2005a) 'Iran–Iraq War (1980–1988)'. Online, available at: www.globalsecurity.org/military/world/war/iran-iraq.htm (25 March 2011).

GlobalSecurity.org (2005b) 'Syria: Weapons of Mass Destruction (WMD), Chemical Weapons'. Online, available at: www.globalsecurity.org/wmd/world/syria/cw.htm (25 March 2011).

GlobalSecurity.org (2006) 'Hizballah Rockets'. Online, available at: www.globalsecurity.org/military/world/para/hizballah-rockets.htm (25 March 2011).

GlobalSecurity.org (2008) 'Iran: Weapons of Mass Destruction (WMD), Chemical Weapons'. Online, available at: www.globalsecurity.org/wmd/world/iran/cw.htm (25 March 2011).

GlobalSecurity.org (2009) 'Hamas Rockets'. Online, available at: www.globalsecurity.org/military/world/para/hamas-qassam.htm (25 March 2011).

Gormley, Dennis M. (2001) *Dealing with the Threat of Cruise Missiles*, Oxford: Oxford University Press.

Gormley, Dennis M. (2003/4) 'Missile Defence Myopia: Lessons from the Iraq War', *Survival*, 45(4): 61–86.

Gormley, Dennis M. (2005) 'Unmanned Air Vehicles as Terror Weapons: Real or Imagined?' Online, available at: www.nti.org/e_research/e3_68a.html (25 March 2011).

Gormley, Dennis M. (2008) *Missile Contagion: Cruise Missile Proliferation and the Threat to International Security*, Westport, CT: Praeger Security International.

Ibrügger, Lothar (2004) *Missile Defences and Weapons in Space*, Brussels: NATO Parliamentary Assembly. Online, available at: http://natopa.ibicenter.net/default.asp?SHORTCUT=497 (25 March 2011).

IranFocus.com (2004) 'Iran Seeks Cruise Missile to Support Shihab'. Online, available at: www.iranfocus.com/en/index.php?option=com_content&view=article&id=12:iran-seeks-cruise-missile-to-support-shihab&catid=9:terrorism&Itemid=31 (25 March 2011).

IranWatch.org (2010) *Iran's Ballistic Missile Program*, Washington, DC: Wisconsin Project on Nuclear Arms Control. Online, available at: www.iranwatch.org/wmd/wmd-iranmissileessay.htm (25 March 2011).

IISS/International Institute for Strategic Studies (2009) 'Iran's Missile Development: Further Tests needed to Cement Recent Advance', *IISS Strategic Comments*, 15(1). Online, available at: www.iiss.org/EasySiteWeb/getresource.axd?AssetID=23857&type=full&servicetype=Attachment (25 March 2011).

IISS/International Institute for Strategic Studies (2010) *Iran's Ballistic Missile Capabilities: A Net Assessment*, London: IISS.

Jane's (2007) *Jane's Unmanned Vehicles and Aerial Targets*, No. 29, Bracknell: IHS Global Limited.

Jane's (2008) *Jane's Unmanned Vehicles and Aerial Targets*, No. 31, Bracknell: IHS Global Limited.

Karon, Tony (2002) 'The Homemade Rocket That Could Change the Mideast', *Time*, 11 February. Online, available at: www.time.com/time/world/article/0,8599,202159,00.html (25 March 2011).

Karp, Aaron (1996) *Ballistic Missile Proliferation: The Politics and Technics*, Oxford: Stockholm International Peace Research Institute.

Karp, Aaron (2005) 'Going Ballistic? Reversing Missile Proliferation', *Arms Control Today*, 35(5). Online, available at: www.armscontrol.org/act/2005_06/Karp (25 March 2011).

Katz, Yaakov (2009a) 'US to Fully Fund Arrow 3 Missile System', *Jerusalem Post*, 20 May. Online, available at: www.jpost.com/Israel/Article.aspx?id=142935 (25 March 2011).

Katz, Yaakov (2009b) 'Iron Dome System Successful in Tests: Israeli-Developed Missile Defense System Intercepts Rockets Mimicking Kassam, Grad-Model Katyushas', *Jerusalem Post*, 15 July. Online, available at: www.jpost.com/Home/Article.aspx?id=148792 (25 March 2011).

Lewis, George N. and Postol, Theodore A. (2000) 'Technical Debate over Patriot Performance in the Gulf War: American Physical Society Panel Correctly Rejects Criticisms of Analysis Showing Patriot Failed to Destroy Scud Warheads', *Science and Global Security*, 8: 315–56.

MacFarquhar, Neil (2006) 'Tide of Arab Opinion Turns to Support for Hezbollah', *New York Times*, 28 July. Online, available at: www.nytimes.com/2006/07/28/world/middleeast/28arabs.html?_r=1 (25 March 2011).

Mahnaimi, Uzi (2009) 'Israeli Drones Destroy Rocket-Smuggling Convoys in Sudan', *Sunday Times*, 29 March.

NASIC/National Air and Space Intelligence Center (2006) *Ballistic and Cruise Missile Threat*, Wright-Patterson Air Force Base, OH: NASIC.

New America Foundation (2010) 'The Year of the Drone: An Analysis of U.S. Drone Strikes in Pakistan, 2004–2010'. Online, available at: http://counterterrorism.newamerica.net/drones (25 March 2011).

Nolan, Janne (1991) *Trappings of Power: Ballistic Missiles in the Third World*, Washington, DC: Brookings Institution.

NTI/Nuclear Threat Initiative (2007) 'Iran Missile Chronology: 2005'. Online, available at: www.nti.org/e_research/profiles/Iran/Missile/chronology_2005.html (25 March 2011).

NTI/Nuclear Threat Initiative (2010) 'Syria Missile Overview'. Online, available at: www.nuclearthreatinitiative.org/e_research/profiles/Syria/Missile/index.html (25 March 2011).

PCHR/Palestinian Centre for Human Rights (2009) 'Confirmed Figures Reveal the True Extent of the Destruction Inflicted upon the Gaza Strip: Israel's Offensive Resulted in 1,417 Dead, Including 926 Civilians, 255 Police Officers, and 236 Fighters', Gaza: PCHR. Online, available at: www.pchrgaza.org/files/PressR/English/2008/36–2009.html (25 March 2011).

Pedatzur, Reuven (2005) 'Israel: Coping with a Proliferated Region', in Bernd W. Kubbig (guest ed.) in cooperation with Axel Nitsche, 'The Domestic Politics of Missile Defence', *Contemporary Security Policy*, Special Issue, 26(3): 621–38.

Pedatzur, Reuven and *Haaretz* Correspondent (2008) 'Iron Dome System Found to Be Helpless against Qassams', *Haaretz*, 22 February. Online, available at: www.haaretz.com/hasen/spages/956859.html (25 March 2011).

Rawnsley, Adam (2010) 'WikiLeaks Reveals Everybody's Christmas List: The World Wants Drones'. Online, available at: www.wired.com/dangerroom/2010/11/wikileaks-reveals-everybodys-christmas-list-the-world-wants-drones (15 February 2011).

Rubin, Uzi (2006) *The Global Reach of Iran's Ballistic Missiles*, Tel Aviv: Institute for National Security Studies. Online, available at: www.inss.org.il/upload/%28FILE%291188302022.pdf (25 March 2011).

Schneider, Howard (2009) 'Israel Finds Strength in Its Missile Defenses: Advanced System Could Alter Strategic Decisions in Region', *Washington Post*, 19 September. Online, available at: www.washingtonpost.com/wp-dyn/content/article/2009/09/18/AR2009091801787.html (25 March 2011).

Schwartz, Stephen I. (ed.) (1998) *Atomic Audit: The Costs and Consequences of U.S. Nuclear Weapons since 1940*, Washington, DC: Brookings Institution.

Sharp, Jeremy M. (2006) *Lebanon: The Israel-Hamas-Hezbollah Conflict*, Washington, DC: Congressional Research Service.

Smith, Mark (2006) 'Pragmatic Micawberism? Norm Construction on Ballistic Missiles', *Contemporary Security Policy*, 27(3): 526–42.

Spacewar.com (2009) 'Israel Assembles Iron Dome Unit'. Online, available at: www.spacewar.com/reports/Israel_assembles_Iron_Dome_unit_999.html (25 March 2011).

Stokes, Mark A. (2005) 'The Chinese Joint Aerospace Campaign: Strategy, Doctrine, and Force Modernization', in James Mulvenon and David M. Finkelstein (eds) *China's Revolution in Doctrinal Affairs: Emerging Trends in the Operational Art of the Chinese People's Liberation Army*, Alexandria, VA: CNA Cooperation, 221–305.

Susser, Leslie (2009) 'The Shield of David: Israel is Putting Together Multi-Layered Missile Defenses to Deter Enemies from Starting a War', *Jerusalem Post*, 25 May. Online, available at: www.jpost.com/JerusalemReport/Article.aspx?id=142018 (25 March 2011).

van Blyenburgh, Peter (ed.) (2010) *Unmanned Aircraft Systems: The Global Perspective 2010/2011*, Paris: Blyenburgh & Co.

Zaloga, Steven (1988) 'Ballistic Missiles in the Third World: Scud and Beyond', *International Defense Review*, 11: 1423–7.

Part II

Conditions for a successful step-by-step approach

4 Taking overall military asymmetries into account

Objections from the Arab world and Iran

Bernd W. Kubbig, Mohamed Noman Galal, Michael Haas, Murhaf Jouejati, Sabahat Khan, Ahmed Saif, and Mahmood Sariolghalam

Developing a framework for asymmetries

The central interest of this article is to find out why missiles have been imported and indigenously developed by the Arab countries and Iran and what can be done about reducing or even eliminating them. No government with an inventory of these weapons will accept drastic measures if it concludes that even gradual steps towards a Missile Free Zone (MFZ) jeopardize its country's security. Examples of such delivery vehicles include the missiles procured by Syria to counter superior Israeli fighter jets and its unique nuclear arsenal, or Saudi Arabian imports of both aircraft and missile defence (MD) systems as a response to Iranian missiles, nuclear activities, and Tehran's quest for regional supremacy. The challenge, therefore, is not only to assess and compare missiles against missiles, but also missiles against other weaponry such as combat aircraft. Static comparisons by category can give a misleading picture, if a numerical balance with respect to missiles ignores large disparities in other sectors. In addition, while soldiers and weapons may easily be compared quantitatively, perhaps the more relevant factors such as the motivation, loyalty, skills, and training of the fighters as well as command, control, and communication capabilities are hardly quantifiable. The same holds true for technical and organizational quality as well as for the availability and reliability of warheads and/or delivery vehicles. The picture becomes even more complex if one includes operative, tactical/strategic dimensions, combined with geographical, logistical, and coalition related factors.[1]

The attempt to assess these asymmetries is complicated by (potential) weapons of mass destruction (WMD) capabilities in the region, notably Israel's nuclear arsenal. Unlike missiles the existence of biological and chemical weapons remains unclear, since precise and reliable data from independent sources are not available.[2] This lack of information coupled with the problem of how to distinguish and compare nuclear, biological, and chemical warheads (all three categories constituting WMD), as well as how to weigh them against conventional arsenals, is a challenge we have to leave to the official negotiators. We take WMD related problems into account insofar as we focus on missiles with a range over 70 km and on fighter jets, which both can be equipped with nuclear, biological, and chemical warheads; they account for a large part of the asymmetries. As for nuclear weapons specifically, we refer to a separate contribution in this volume which copes conceptually in a compromise-oriented way with this aspect (see the chapter of Gawdat Bahgat *et al.*). We will show that static comparisons of weapon categories are 'dynamized' by the paramount action–reaction history of the procurement both of missiles and aircraft. We will simplify the asymmetries problem by ignoring Hamas and Hezbollah as part of the overall equation at this point (see the articles of Margret Johannsen *et al.* as well as the one by Judith Palmer Harik and Walid Abu-Dalbouh). Missile defence systems, however, will be taken into account, since they are increasingly becoming an important part of the regional arms postures.

This article is not about weapon arsenals as such, not to mention about bean counting, but rather focuses on the constellation of regional conflicts and (volatile) alliances. Here we follow a core insight of constructivist International Relations theory that weapons are embedded in a social – domestic and regional – context. (Threat) perceptions matter in state relationships, and not the presence of weapons per se (Wendt 1992; see also the Introduction by Bernd W. Kubbig). Germany for example feared Soviet arsenals, but no longer fears them as Russian weapons, provided that they are kept safe and not launched accidentally. Our approach puts two states – Israel and Iran – as the major reason for the Arab countries to procure missiles and missile defence systems on the spot: their traditional fears regard Israel as a de facto nuclear weapon state and, more recently (with the exception of Syria), a possibly emerging nuclear Iran. The Islamic Republic is in an ambivalent position, because it is not only the focus of fears, above all from most Gulf states, but also has concerns itself regarding Israel and even more so with respect to the military potential of the United States which in fact has become a regional power.

This nation- and coalition-centred article with its emphasis on perceptions of external threats is concerned not only with weapon capabilities,[3] but also with the associated military strategies/doctrines, whether predominantly offensive or defensive, and political intentions, expressed especially in strategy papers and in the rhetoric of official representatives. At issue is the basic foreign policy orientation of a country, as well as its actual behaviour. Is it status quo oriented? Does it have aspirations for regional hegemony? Is it expansionist, seizing opportunities, and filling political vacuums? Or is it imperialist? The military and foreign policy dimensions are linked, and, in fact, mutually enforcing: a militarily stronger state may assert itself more vigorously in the foreign policy domain. A 'menace' will be defined broadly: first, if a state aims at destabilizing other countries and their regimes by interfering in their domestic spheres or even threatening them with invasion; and, second, if it impedes a regional order that takes the security needs of all states in the Middle East/Gulf into account.

The multifaceted threats reflect the basic elements of security which all states seek: domestic stability, especially regime maintenance, and territorial integrity. The military factor needs to be taken into account regarding the overall structure of conflict: the activities (missile related ones included) in this area seek to protect the state against its adversaries without recognizing the counterproductive effect on its neighbours. If the latter do feel threatened and start an arms build-up, it can lead to an interactive dynamic of escalation, reinforcing the already existing specifically pronounced security dilemma in the Middle East/Gulf. As we shall see, the extent of asymmetries and interactive arms race dynamics vary in the adversarial relationships. Indigenous factors motivating the build-up or procurement of missiles will be included in our analysis whenever possible. These include historical experiences, military-industrial-bureaucratic interests, and, broadly with regard to Iran, domestic power constellations and foreign policy orientations. Instead of choosing between considering asymmetries either in narrowly military terms or exclusively in terms of geographical, political, economic, and cultural factors, we have tried to steer a middle course between these two extremes.

The article begins with an analysis of the perception of exclusively external threats in the regional environment; next, it examines the process of missile procurement and production in the context of arms asymmetries with varying interactive dynamics. We conclude by summarizing the countries' demands for arms reduction leading to an MFZ. We will pursue the idea that slowing the arms race is necessary for reducing the security dilemma; otherwise the attempt by states to acquire a comparative military advantage (superiority) increases the temptation to use this position to enlarge their sphere of influence, making war more likely.

The GCC members and Yemen: military measures in the shadow of a possibly emerging nuclear Iran

Threat perceptions and the regional environment

Some of the six members of the Gulf Cooperation Council (GCC) – Saudi Arabia, the United Arab Emirates (UAE), Kuwait, Qatar, Oman, and Bahrain – have ongoing disputes amongst themselves about territorial, economic, and political issues. Although these conflicts are an impediment to achieving deeper cooperation, they are not a threat to cooperation overall. These disputes have their origins in the region's colonial heritage, the nature of the ruling regimes, and the Sunni–Shia divide. The different foreign policy profiles are also due to the hierarchical structure of the GCC; Saudi Arabia is the only dominant power in the group with regional ambitions. The competing interests explain the slow pace of cooperation among the Council's members and their bilateral 'hub-and-spokes' cooperation profile with the United States, though a long-term commitment to the GCC charter remains strongly in place. At this time of writing (mid 2011) it is not clear what this year's developments, called metaphorically 'Arab Spring', mean for the internal situation of each GCC member and its foreign policy and whether (and if so to what degree) they impact the missile issue (with the exception of Bahrain, all other GCC countries have not experienced uprisings).

Despite the traditionally caustic rhetoric of all six Gulf states concerning Israel's policy towards the Palestinians, Israel is not regarded as the preeminent national security threat despite its long-standing adversarial role and its position as the sole nuclear power in the region. Although Israel is still seen as a major menace to regional security, most GCC states would identify the most imminent threat as emanating from the Islamic Republic of Iran. This is in part due to a new sub-regional approach to security questions in the Middle East. In addition, progress has been made on the question of Israel's right to exist; the Arab Peace Initiative of 2002 (renewed 2007) initiated by Saudi Arabia and accepted by all 22 Arab League members, is the case in point. In contrast to Iran, which borders on many GCC states, Israel is relatively far away. It seems that the traditional Gulf states' list of concerns towards Israel has lost some of its significance. Nevertheless they are still worried by the quantitative and qualitative military parameters that demonstrate Israel's superiority: these parameters are force multipliers that include highly sophisticated command, control, communication, and intelligence systems ('Revolution in Military Affairs') which compensate for the lacking *hinterland*; Israel's forces (standing versus reserve manpower) are larger than any of the Arab armed forces; and Israel has a superior capacity to mobilize reserve forces, within 72 hours if necessary (Feldman 1997: 270; Shapir and Brom 2008: 44).

The GCC has announced the initiative to declare a Gulf Weapons of Mass Destruction Free Zone (Alani 2008: 359–60). In addition to including its member countries, the plan calls for an incremental integration of the other three Gulf states, Iran, Iraq, and Yemen, and finally the entire Near and Middle East/Gulf, including Israel. While the Israelis have balked at the concept, overcoming their objections is not regarded as "insoluble" at least according to Mustafa Alani (ibid.: 360), the major Track II initiator of the Gulf Weapons of Mass Destruction Free Zone. Like similar diplomatic activities of the GCC states since, this initiative is directed towards Tehran's nuclear and missile ambitions. However, there is no common threat assessment. The reason is that Iran's nuclear aspirations and its military activities are seen in a different way as part of its regional foreign policy. The different experiences from history and recent events have affected the overall relationship of the GCC states towards the Islamic Republic.

Oman and Qatar, whose foreign policy profiles are characterized generally by a high degree of pragmatism, have managed to maintain relatively good relations with Tehran and Washington simultaneously. In the 1980s, for instance, Muscat supported UN Security Council resolutions calling for an end to Iraq's war with Iran. But Oman was also the first state in the Gulf to allow American forces access to its military facilities after the Islamic Revolution. Likewise, Qatar, which owns the third-largest natural gas reserves in the world and shares the North Field/South Pars Field off its coast with Iran, is an important partner of the US. The country hosts the Air Operations Center for the Middle East (the biggest American air base in the region), and also the United States Central Command (CENTCOM) Forward Headquarters. In March 2007, the government publicly declared, however, that it will not allow its territory to be used by US forces for an attack on Iranian nuclear facilities (IISS 2008: 50–2).

The other four Gulf states have increasingly expressed fears about their Iranian neighbour for three reasons: first, Tehran's hegemonic aspirations in the region; second, military and environmental threats associated with Iran's nuclear programme; and third, the menace the Islamic Republic poses to the regional order.

Tehran's hegemonic aspirations – meddling in the domestic politics of the Gulf States

The four GCC countries have expressed fears about Iran primarily because of its confrontational foreign policy. Especially since the election of President Mahmoud Ahmadinejad (June 2005) Tehran is no longer perceived as pursuing a pragmatic policy or even rapprochement such as in the Rafsanjani and Khatami years (Alshayji 2002: 226). Rather the leadership in Tehran is regarded as a revisionist regime, resuming the earlier practice of exporting the revolution abroad: from the Bahraini perspective the current regime interferes in the domestic affairs especially in the Gulf countries and spreads radical Shiism in some Sunni states, thereby attempting to destabilize them and to increase its own influence in the entire region (Al-Ghatam *et al.* 2010: 95). Saudi Arabia has had its own experiences in this respect and its relations with Iran are basically competitive and factional due to Riyadh's traditional quest for hegemony in the Gulf region. Saudi Arabia is the dominant GCC member country; because of its vast territory, its enormous oil resources, and its unique role as the custodian of the two holy cities of Mecca and Medina, the kingdom sees itself as the most eminent Islamic state (Al-Mani 1995: 86). Riyadh pursues its policies through an enormous arms build-up and through attempts to secure advantages in proxy wars, for instance in Yemen. Saudi Arabia also fears Iranian meddling in Yemeni domestic affairs, particularly in the northern part of the country. Iran is accused of exploiting the rebellion led by the Houthis against Yemen's central government, supported by the Saudis, in order to destabilize the country (Döring 2010: 17–19).

The greatest bone of contention between the United Arab Emirates and Iran is the fate of the three Persian Gulf islands Abu Musa, and Greater and Lesser Tunbs. Although the UAE leadership has become more rhetorically assertive towards Iran, it basically pursues a low-key policy of engagement because Tehran is one of its most important trading partners. The two major power centres in the Emirates, Abu Dhabi and Dubai, do not even share a common view of the Iranian threat and how to deal with it. For Dubai doing business is more important than security concerns, whereas the government in Abu Dhabi, in part under US pressure, is much more concerned about security threats; Washington has obviously used Dubai's financial crisis by demanding a more restrictive policy also of this emirate (*Arab News*, 28 June 2010).

According to Abdullah Alshayji (2002: 225), Kuwait "bore the brunt of Iranian wrath more than any GCC state", many inhabitants "remember vividly" the bombing of the oil refineries, direct threats against the country, and sabotage operations committed by some Kuwaitis of Iranian extraction. Bahrainis are utterly sensitive to Iranian meddling in their internal affairs by exploiting the schism between the elite and the Sunni royal family on the one hand and the majority of the population which is Shii on the other (Al-Ghatam *et al.* 2010: 93–4).

Military and environmental threats associated with Iran's (potential) nuclear status – menaces to the regional order

Saudi Arabia, the UAE, Kuwait, and Bahrain fear the Islamic Republic because they are unclear about the "real objective" of its nuclear programme (ibid.: 90). In their view, a (near) nuclear Iran might feel less vulnerable and therefore more likely to project its power and influence abroad. Moreover, Tehran's nuclear ambitions are seen as further threatening strategic stability by intensifying the continuing arms race. A preemptive attack on its nuclear facilities by Israeli or joint Israeli–American forces, however, is portrayed by some Gulf state leaders as "the most fearsome threat to the region" (Al Khalifa 2009). The question, however, is whether elites in all four GCC states would not in fact tolerate or even prefer such a strike.[4] In 2010, the UAE Ambassador to the United States, in an unusually blunt statement publicly endorsed the use of the military option for countering Iran's nuclear project, if sanctions fail (*Haaretz*, 27 July 2010). More importantly, King Abdullah of Saudi Arabia has repeatedly urged the US to attack Iran and to destroy its nuclear programme.[5] In addition, Gulf states stress the fear of the environmental impact should there be an accident at the Bushehr nuclear power plant, located directly on the Persian Gulf coast. The four Gulf states are also concerned that the Obama administration might establish a rapprochement with Tehran thereby recognizing its regional role at the expense of the interest of the GCC members (Al-Ghatam *et al.* 2010: 80; Al Shayji

2010). For Saudi Arabia this would mean that the United States, even if it did not coalesce with Riyadh's rival, might adopt the entirely new role of balancing the two regional powers with hegemonic aspirations. The four Gulf states have made clear that securing the regional order is at the heart of their concerns and that they are not willing to take any steps to include an intransigent and politically ambitious Iran into any formal cooperative framework which, from a Saudi perspective, might pave the way for "a Greater Iran, so that [this country] can become a leading power" (MEMRI 2010).

The motives for and strategies of procuring missiles and missile defence systems – and the problem of military asymmetries

All GCC states have had experience with missiles in the past, particularly in the 1991 Second Gulf War, in which Saudi Arabia was persistently targeted by Iraqi Scuds. For the Saudis, that experience starkly illustrated both the growing threat of missiles and the emerging relevance of missile defence. Even prior to the war, the Saudi military had begun to understand the gravity of the rapidly evolving air threat. In 1981, the Royal Saudi Air Force (RSAF) was established as an independent branch of the armed forces and was quickly equipped through the US$5.6 billion Peace Shield programme. As their shopping lists have revealed in recent years, GCC country procurements are increasingly driven by the perceived threat posed by Iran. However, Oman is not part of the American weapon packages of 2007 and of 2010, missile defence systems included. Qatar which operates several old short-range surface-to-air missile systems only received some logistical equipment. As far as Saudi Arabia is concerned, its hegemonic aspirations are part of the overall multifaceted rivalry with Tehran and an additional driving force behind its weapon procurement. Although Saudi Arabia has some military-bureaucratic elements and the UAE has strengthened its own defence industry (*Jane's Defence Weekly*, 11 February 2009: 64–8) a military-industrial complex in the traditional sense does not exist in these countries. They, too, remain highly dependent on US equipment, although they show their independence by occasionally buying non-American equipment. The big weapon deals of 2007 and 2010 follow this pattern, and include a "glitter factor" (Cordesman and Obaid 2005: 137), i.e. an interest in importing the most recent technology.

As far as Yemen is concerned, while most of its weaponry – including air defence capabilities – is imported from Moscow, Beijing is the second major source of delivery. Meanwhile, however, the US has provided equipment to the Republican Guards and one can expect a shift towards American technology in the future given the airbase in Hodiedah and the naval base likely to be built in Socotra (see the Table compiled by Michael Haas and Bernd W. Kubbig in the Appendix). As for Saudi Arabia, the imports of 2007 and 2010 will add significantly to its existing capabilities in the combat aircraft and missile (defence) areas. Riyadh reportedly procured the CSS-2/DF-3 intermediate-range ballistic missile of some 2,800 km range from China in the 1980s. Also, the RSAF operates 16 batteries of Pac-2 MD systems and possesses a total of some 2,000 surface-to-air missiles of various types, including the R460 SCIA/Shahine and MIM-23B I-Hawk that can be launched from fixed and mobile platforms. Over the next few years Saudi Arabia might integrate a higher altitude interceptor system into its missile defence shield as well as continue to procure advanced air-to-surface missiles that would enable it to destroy adversarial missile launch sites in case of an attack. With the $60 billion deal of 2010, Riyadh easily surpassed Abu Dhabi which had the lion's share in the 2007 package, amounting to about $20 billion, including some $16 billion for various missile defence systems. Kuwait's purchases in the missile sector exceed $1.3 billion, and include 80 Pac-3 missiles as well as technology to upgrade the older 60 Pac-2 missiles and radar systems (see the chapter of Sven-Eric Fikenscher *et al.* on External Powers). These recent purchases show a strong interest in MD technology and reflect Kuwait's fear of a major regional conflict.

Bahrain, the smallest member of the GCC, has not been a recipient of modern missile defence systems in recent years. Being host to the headquarters of the U.S. Fifth Fleet and the naval component of CENTCOM, Manama benefits from one of the strongest missile defence umbrellas anywhere in the world. Despite their massive military spending, the GCC members – Saudi Arabia is no exception (Al-Mani 1995: 91) – still rely upon the United States to protect their territory. In fact, the distinctive technological bilateralism of the US (Davis *et al.* 1999) prolongs the asymmetric dependency and impedes processes within the GCC integrated defence planning; any progress in this area has been "largely cosmetic" (Cordesman and Obaid 2005: 138). What does this all mean for the issue of military asymmetries and arms dynamics? First, taken individually the GCC countries are, with the exception of Saudi Arabia, militarily inferior to Tehran. Second, Saudi Arabia as the most powerful GCC member, however, complicates this picture; especially through its most recent interest in purchasing high-quality combat aircraft and missile defence systems, it seeks to expand its advantage over Tehran in this area (Cordesman and Al-Rodhan 2007: 30–1). At the time of this writing it is not clear whether US exports of conventional weapons to the Saudis will constrain Riyadh from going nuclear in the long run. The GCC states have commissioned a study to establish a joint programme in the civilian nuclear energy area, signalling their willingness to limit the nuclear option for the moment. However, Iran and Israel have both been alarmed by Riyadh's interest in unprecedented arms purchases from the United States as a central part of Washington's containment strategy towards Tehran (*Jane's Defence Weekly*, 8 September 2010: 18).

Third, one has to factor in the United States as the long-standing and reliable partner of all GCC states in the overall regional picture. While it makes sense at the negotiating table to juxtapose Saudi combat aircraft and missile defence with Iranian missiles at the regional level, in fact, one has to include on Riyadh's side of the ledger the US arsenal which is by all standards militarily superior to the Islamic Republic of Iran. Its bilateral agreements with and its presence in the GCC states constitute the above-mentioned umbrella thus demonstrating the asymmetric and bilateral 'hub-and-spokes' relationship with all six members. Fourth, in combat scenarios, being regarded by Tehran as part of a superior coalition might turn out to be a liability for GCC countries. This would be the case for instance if Washington decides to strike at Iran's nuclear facilities which may ironically put the missile defence systems delivered by the US in action; for it seems likely that in return Tehran would retaliate against American facilities in various Gulf countries.

Egypt: coping militarily and diplomatically with the superior Israeli neighbour and distant provocative Iran

Threat perceptions and the regional environment

Despite the normal and peaceful relations between Egypt and Israel codified in the Camp David Peace Treaty (1979), their bilateral relationship has also been a remarkably precarious one. Israel has invaded Egyptian territories more than once and remains Cairo's main adversary. The peace treaty, which has been domestically controversial from the beginning, was defended by representatives of the ruling elite in the context of its thirtieth anniversary as "a plus for Egypt". Its territory was returned, freeing up assets "to stimulate economic growth and promote development domestically".[6] Moreover, Egyptian and Jordanian territories have been "excluded" from "the process of Israeli expansion and entrenchment" (Said 2009). Unlike the geographically distant Gulf states, Cairo considers the Israeli missiles, nuclear weapons, as well as their superior combat aircraft to be a primary (potential) threat to Egypt's status and role as a regional power.

As for Cairo's relations with Tehran, 1979 also marks a milestone, yet a negative one. The Islamic Revolution led one year later to the severing of diplomatic relations, which were resumed in 1991, but only on the level of interest offices. Although there have been several high-level contacts between both governments in recent years (Bahgat 2009), there are a number of contentious bilateral issues and similar fears to those between the four concerned Gulf countries and Tehran, albeit with country-specific dimensions. One of the bilateral issues is Iranian interference in Palestinian affairs and its accusation that Egypt blocked support for Hamas during the Gaza War 2008/09 (*Daily News – Egypt*, 21–22 March 2009). One incident was the

> arrest of an armed Hezbollah cell in the Egyptian province of Sinai, and the escalation of mutual discourse between Iran and Egypt in this

> regard. According to comments made by Egypt's Foreign Minister Ahmed Aboul Gheit, in a television interview with the BBC on the eve of the US President's visit to Egypt in June 2009, the dispute boils down to the fact that Iran's escalation of tension in the region is an attempt to involve Egypt in a war with Israel.
>
> (Al-Ghatam *et al.* 2010: 95)

In addition, Cairo fears Iran's strengthening of Hamas, which is strongly supported by the Muslim Brotherhood; this actor is becoming increasingly influential in Egypt in the wake of the 'Arab Spring'. Cairo, too, has been concerned about how to position Iran in the regional order (MEMRI 2010). But for the government – and here it differs from the Gulf regimes for which Israel is geographically distant – Iran's nuclear activities put Cairo's neighbour Israel, as the unacceptable de facto nuclear weapon state on the spot. Egypt has developed a culture of sensitivity towards the double standard by which Israel and the non-nuclear Arab states are treated in the region. Egypt shares the assessment of the Arab world that Israel is to date the only existing nuclear menace in the entire region. But a nuclear Iran would be an additional threat to the status and role of Egypt in the region – and it is this prestige related factor (Jentleson and Kaye 1998: 229–35; Landau 2006: 141–3), which cannot be overrated in explaining Cairo's foreign policy behaviour. However, it remains to be seen how the post-Mubarak Egypt will position itself in the region.

The motives for and strategies of procuring missiles in the context of military asymmetries

Egypt's efforts from the late 1950s to import and to produce missiles of various ranges indigenously or in cooperation with the Soviet Union, North Korea, and Argentina/Iraq (Condor-2) were not only driven by the prestige factor, but also by Israel's military activities, and the quest for regional leadership (Mistry 2003: 75–6, 100–3). Today, despite "forty years of missile efforts, Cairo has essentially only upgraded its Scuds but had not built a more advanced missile although it received foreign technical assistance and has a modest defense industry" (ibid.: 101) which builds jet engines, rockets, and other military equipment. Very little is reliably known about the country's missile and rocket forces.[7] It currently fields the Scud-B short-range ballistic missile (300 km) and operates 9–24 Scud-B transporter-erector-launcher vehicles. The number of missiles in its inventory is generally estimated at 100. The Scud-B, which can carry a single, 1,000 kg conventional high-explosive or chemical warhead, is an obsolete weapon system of doubtful value in a conventional role, due to its long launch sequence (60–90 minutes) and low accuracy. Some or most of its Scud-B missiles might be of the North Korean Hwasong-5 variant, which features minor improvements over the original model. The development of an indigenous derivative of the Scud-B with an estimated range of 450 km and a payload of 1,000 kg, aided by Pyongyang and code-named Project T, has been reported (Cirincione *et al.* 2005: 106).

It is unknown whether Egypt has ever deployed the Scud-C or its Hwasong-6 derivative from North Korea, with an increased range of 600 km and a smaller warhead. However, it is known that it has received Hwasong-6 components from Pyongyang during the 1990s. In addition, Cairo continues to deploy the Frog-7 artillery rocket with a range of 70 km. The original Frog-7A can deliver both nuclear and chemical warheads, while the Frog-7B was used to deliver a cargo warhead similar to the indigenously produced Sakr-80 (a large 80 km artillery rocket). The importance of missile and rocket forces in Egypt's overall defence posture is limited. The army's centre of gravity is its armoured and mechanized formations, with supporting artillery, making for a very conventional posture. The modernization of the air force is also given high priority because recent wars have highlighted its importance. In all likelihood, missiles will not, for the time being, constitute a vital part of Egypt's military posture. The missions of the missile and rocket brigades are most probably limited both by capabilities and doctrine. The short-range ballistic missile force might be part of a limited deterrent vis-à-vis Israel that is centred on Egypt's possible possession of chemical weapons and/or conventional 'punishment' attacks against Israeli cities. The rocket force is probably trained for conventional, battlefield missions only.

The two major reasons for Cairo's limited advancement in the missile area are the Missile Technology Control Regime and US pressure. The United States is aware of the prestige factor surrounding the missile issue and sensitive to the overall disparity in Israel's favour. Both countries pursue a sensitive policy in which 'red lines' are not crossed: while Washington has not pressed Cairo to curb its short-range activities, Egypt has not crossed much beyond the Scud threshold (Mistry 2003: 101–3). Much more importantly, it has refrained from becoming a nuclear weapon state (Solingen 2007: 229–45), although in 2006, the Mubarak government joined the Gulf countries in announcing a civilian nuclear energy programme.

Syria: missiles against the superior adversary Israel – pragmatic ally of Iran

Threat perceptions and the regional environment

For Syria, Israel, located only 40 miles southwest of Damascus, represents the greatest threat to its security; it is seen as the scion of imperialism, an aggressive, expansionist, and settler-colonial state. Syria, along with other Arab states, went to war against Israel in 1948, 1956, 1967, and 1973. Since the 1967 War, Israel has occupied Syria's Golan Heights. In 1982, the two rivals confronted each other again, this time in Lebanon. Despite the continuity in the Arab–Israeli conflict, a major shift in Syria's security alliance took place after the collapse of its superpower patron, the Soviet Union. The Asad regime turned to Iran and since then Damascus has remained Tehran's pragmatic ally.

Syria also feels vulnerable because of the US-dominated regional security environment, although there have been some recent positive developments: its traditionally tense relations with its northern neighbour Turkey, a powerful American partner and NATO member, have improved dramatically in recent years. Also, Damascus benefited from the Israeli–Turkish crisis in 2010, when Israeli naval commandoes mounted a deadly raid on one of several vessels sent by a Turkish organization to break the Israeli blockade on the Gaza Strip. East of Syria, in Iraq, almost all US troops – regarded by the Asad regime as a threat to its security – will have left at the end of 2011, while the influence of its Iranian ally is increasing. To its southern neighbour, Jordan, Syria maintains an uneasy relationship in large part because the Jordanian elite is among America's closest Arab partners. Finally, off the western coastline, the U.S. Sixth Fleet is firmly anchored in the Mediterranean. During the George W. Bush administration American relations with Damascus were mostly strained, i.e. Syria withdrew its troops in 2005 under American pressure from Lebanon. The cautious rapprochement under Obama, finding expression in the opening of embassies in Washington and Damascus, came to an end due to the regime's decision to use military force against the opposition in early 2011. Suddenly the Syrian government had to cope with unprecedented internal challenges.

The motives for and strategies of procuring missiles in the context of military asymmetries

Syria's missile procurement and development is a direct response to Israel's massive accumulation of power. Damascus has always been militarily inferior to its adversary: from 1948 through the 1980s, when Hafiz al-Asad attempted to bolster the armed forces (Cobban 1991: 50–8) until today, Syrian capabilities are no match for the technologically superior Israel Defense Forces, thanks to large-scale US military assistance. By contrast, its arsenal is "the impoverished stepchild of the region" (Cordesman *et al.* 2008: 11). It has become "something of a military museum" (ibid.: 12). Nowhere is the asymmetry more glaring than in air power: "Syria may have the largest numbers, but [it] has one of the least capable air forces" (ibid.: 40). Israel's air superiority was evident in 1982 during its invasion of Lebanon, when Syria lost 80 combat aircraft in two days, whereas Israel lost only a single fighter jet.

Furthermore, Israel has a "distinct lead in its ability to use these systems and its tankers in long-range strike missions" (ibid.: 41). Given Israel's obvious air superiority, the Syrian High Command opted to decrease its investment in the aircraft sector and instead concentrate on the acquisition and development of a credible missile force. According to Israeli press reports, Syria now has hundreds of 300 km range Scud-Bs, 600 km range Scud-Cs, 700 km range Scud-Ds, and highly accurate 75 km range SS-21 missiles (Hilal *et al.* 2007). Damascus has also aimed at strengthening its defensive capabilities against Israel's superior offensive ones by purchasing above all air defence systems and C-802 anti-ship cruise missiles from Iran. At the time of this writing, Damascus and Moscow have agreed on a deal involving the transfer of the Yakhont anti-ship cruise missiles, as part of Syria's efforts to modernize its naval and costal defence capabilities. None of these efforts can match Israeli naval superiority (*Jane's Defence Weekly*, 8 September 2010: 24), and the same disparity can be said about Damascus' capabilities to counter Israel's air power.

In retrospect, Syria's decision to switch its focus from air power to surface-to-surface missiles – a direct result of its defeat by the Israeli Air Force in Lebanon's skies in 1982 – was a sound one. This decision was corroborated by the successes Hezbollah scored against Israel in the 2006 War, thanks partly to Damascus-supplied rockets and anti-tank missiles. Therefore, the Asad regime has adopted a doctrine of asymmetrical warfare in which missiles have above all two important functions. First, preemption: the Scud-D surface-to-surface variant for instance, puts most of its adversary within range, enabling Syria to disrupt Israel's mobilization centres before it can carry out a large ground offensive against Damascus' forward positions. Second, deterrence: missiles, which are far less costly than combat aircraft, can be armed with chemical warheads. In recent years Damascus has achieved the capacity to test-launch a large number of medium- and long-range rockets. This provides Syria with a deterrent capability, which may cause a number of casualties beyond the threshold of Israel's tolerance. In early 2007, Damascus reportedly deployed great numbers of short- and medium-range artillery rockets to the Israeli border, many of them in fixed underground positions (Shapir and Brom 2008: 43–4). All this indicates that the contribution of missiles to Syria's overall military power cannot be overestimated.

Iran: missiles as the core element in a Tehran-averse Middle East/Gulf

Threat perceptions, the domestic, and regional environments[8]

Like all centralized regimes (with some democratic elements), Iranian foreign policy is predominantly determined and conducted by the composition of its political leadership. Two factions have dominated: the internationalist and the revolutionary. Members of the first group attach importance to the country's wealth and global standing, and to the economic well-being of the average citizen. In contrast, the revolutionary school is committed to unerring infallible Islamic rule, the struggles of the Muslim world, and the fight against what it perceives to be Western imperialism. More recently, contrary of depicting the country's elite as a 'clerical regime', it is the military-industrial complex, especially the Pasdaran, that has become the dominant force. The regime has kept, despite massive opposition, its revolutionary order – institutions, ideological atmosphere, and ideas – intact. The most powerful idea shaping the revolution was that of putting an end to foreign intervention. The year 1979 changed the geopolitical orientation of the country overnight from being one of the closest strategic allies of the United States to being one of its most vehement opponents. Iran's threat perception and foreign policy priorities changed with respect both to its immediate environment and to the larger surroundings, the increased US presence in the region posing the major problem: the buffer or physical space between Iran and the United States disappeared, so that they are now literally neighbours in the Gulf.

The Iranian elite perceives American opposition to the Islamic Republic as the central external threat and this shapes Tehran's security doctrine. This has not changed under the Obama administration. The Iranian leadership has not reacted positively to attempts by Washington to reach out to Tehran (its offer came from the regime's perspective in view of the mass protest under extremely unfavourable circumstances). At the same time, the leadership has pointed out that US operative 'policies on the ground' have worsened. The Obama administration has indeed been the major driving force behind a tougher sanctions policy both on the bilateral level (pressing its allies to pursue stricter measures) and in the multilateral context – the UN Security Council continues to insist that Tehran forego its enrichment activities as the major condition for economic and technological assistance.

Security issues remain at the core of the American–Iranian confrontation. Tehran fears the talk of regime change and attacks against its nuclear installations, either by the United States (in general perceived as the primary threat) or Israel (considered as the secondary menace), or some combination of the two. While regime change by a military invasion is not on the agenda of the Obama administration, the US war against Baghdad that toppled Saddam Hussein nevertheless remains vivid in the minds of Tehran's elite. The current US executive branch does not in principle rule out military attacks against the Iranian nuclear and missile related infrastructure. Washington will continue a policy of isolating Tehran if it does not stop its enrichment activities (President of the United States 2010: 26), by containing it through sanctions and substantially increasing its exports, including missile defence systems (DoD 2010: 33–4), to the Gulf states.

Iran's motives for and strategies of acquiring missiles in an overall asymmetrical situation

Tehran's domestic and regional environments have shaped the build-up of its missile arsenal. Although procurement efforts date back to the 1970s, it was primarily the missile use during the Iraq–Iran War (1980–88) and Saddam Hussein's chemical weapon attacks against the Iranian population that exerted the strongest influences on both the public and the leadership. Iran felt isolated in its war with Iraq since the latter had the full support of relevant countries of the Arab and Western world, which either backed or ignored the use of chemical weapons by Iraq that violated international law. In view of the Islamic Republic's deficient air power, its limited anti-ship and anti-missile capabilities,[9] missiles became the centrepiece of Tehran's military strategy. Iran started its production and acquisition activities in the mid 1980s, expanding the military-industrial base established under the Shah and engaging in cooperative efforts with China, North Korea, Libya, and Syria. In the second half of the 1980s Pyongyang became its principal partner for supplying 300 km range Scud-Bs (Shahab-1), 600 km range Scud-Cs (Shahab-2), and the longer-range (about 1,000 km) Nodong (Shahab-3) missile.

The Islamic Republic continued its rapid build-up after the Second Gulf War. Revelations about advances made by Baghdad on its missile and WMD projects came as a shock to the Iranian leadership and triggered another big push. In addition, the war revealed that – at that time – only ballistic missiles were able to penetrate the American Patriot defence system. While after the Second Gulf War the elites in Tehran continued to see Baghdad as the paramount threat, it was the demise of the Soviet Union that considerably changed their threat perceptions:[10] the United States, as the 'sole superpower' now became the primary source of the Islamic Republic's fears. Ballistic missiles were the core element of its asymmetrical warfare strategy to counter Washington's containment approach, and to deter any hopes of regime change after the US had toppled Saddam Hussein in early 2003 (Taremi 2005: 101–2). Tehran's missile activities are primarily a response to US capabilities that are superior across the board by all standards. Another factor is Israel's conventional offence and defence capabilities, its nuclear weapons, and its preemptive strike doctrine. As outlined above, the capabilities and doctrine have to be related to the not unlikely scenario of American, Israeli, or to joint strikes, with which Iran would have to cope on its own. The Islamic Republic has not attained and cannot reach an all-encompassing security alliance with any other country except for Syria, since its distinctive revolutionary credentials disallow any such partnership.

Iran's strategy and military doctrine has been portrayed as "primarily defensive" (Knepper 2008: 461) in the sense that – the support for hybrid actors notwithstanding – the main focus is on deterrence which has even been acknowledged in the Israeli defence literature (Rubin 2006: 14). In the event that deterrence fails, Iran's missiles will be a vital part of its asymmetric war-fighting efforts. They may be launched against US bases, facilities, and soldiers in the region, and they are likely to be fired against targets in Israel. The missile activities of the Islamic Republic stand for technological prestige and national pride. They signal the leadership's determination to stand its ground and contribute to a greater international standing. These aspects point to the indigenous importance of Tehran's missile programmes which have become entrenched in a techno-cultural network: universities, institutes, companies, facilities, a myriad of engineers, and their supporters in the politico-military system are all involved, although publicly accessible reliable facts and figures on this network do not exist. This explains the broad spectrum and the speedy development of the country's missile capabilities. The Shahab-3, which originates from the Soviet Union and constitutes the core element of Iran's missile programmes, with its range of some 1,000 km, is capable of reaching Israel. It is estimated that the Islamic Republic has between 200 and 300 Shahab-1s and -2s. In addition, Tehran is developing a new medium-range, solid-propellant delivery vehicle (Sajjil-2) with a range up to approximately 2,200 km; the leadership hopes that it will carry Iran's first nuclear warhead.

There is no doubt that the Islamic Republic has invested enormously in its indigenous missile projects; its progress and accomplishments have been assessed as impressive. This includes the redesign of the Shahab-3, which led to the longer-range Ghadr-1; in 2009 Tehran succeeded in integrating a second stage on to a modified Ghadr-1 airframe. Thus the Safir space launch vehicle which took a small satellite into low-earth orbit was created. In February 2010, the Iranians surprised the military world by the unveiling of a two stage, liquid-fuelled Simorgh launch vehicle with a take-off weight double that of the Safir booster. In August 2010, national television broadcast images of what they claim was the first test of the Qiam-1, which is probably an upgraded Shahab-2. Yet the Islamic Republic still faces considerable obstacles to its missile programme because it continues to be dependent on foreign imports and assistance to a considerable extent. Its programmes require extensive, year-long testing, thus providing its adversaries with considerable warning time. The deployment of a three stage intermediate-range solid-propellant Sajjil-2, which would be capable of delivering a 1 ton warhead to 3,700 km or even to 4,000–5,000 km, is a number of years away (IISS 2010: 143). The view from a leading Israeli rocket expert that there is "some doubt as to the quality and precision of the more advanced missiles" (Rubin 2006: 8) has been corroborated by the substantive IISS report (2010: 139). It states that military utility of the Iranian missiles is "severely limited".[11] The delivery vehicles equipped with B- or C-weapons would largely remain weapons of terror against regional adversaries, resulting probably in less than a few hundred casualties.[12]

The prospects of an arms reduction process towards an MFZ: country-specific demands

The members of the GCC and Yemen

The GCC states do not see Israel as the paramount threat and their subregional proposal of a WMD Free Zone in the Gulf is proof of this. This will not prevent the GCC members from pursuing their traditional demand for a Nuclear Weapon Free Zone, which in the final analysis would require Israel to become a party to the Nuclear Non-Proliferation Treaty (NPT). The question is, however, whether the Arab states are prepared to give their long-time adversary leeway in terms of time and substance to respond to such a position. Concessions in this area will probably depend upon Israel adopting a more forthcoming position on the Palestinian issue. As regards Iran, the fears of the Gulf states that we have canvassed suggest that Tehran will need to discontinue or at least slow down its nuclear and missile programmes and reduce or terminate its connections with the Shii groups within Arab states. To be sure, the military imports of the five small Gulf states are a reaction to the efforts of the Islamic Republic. The situation is more complex in the case of Saudi Arabia because its own hegemonic aspirations and its unprecedented and ongoing arms build-up frighten Iran. The Saudi Kingdom, and also the UAE, are becoming more inclined towards acquiring the military-industrial capabilities to produce missiles domestically. At present, this does not constitute the major obstacle towards a reduction process with the MFZ as the ultimate objective. Yet the trend towards intensified arms dynamics in the region clearly does *not* lead in the direction of an MFZ.

Egypt

From a purely military-strategic and technological standpoint, abandoning its ballistic missile force, which is so far oriented entirely towards Israel, would weaken Cairo's defences in a rather limited way, suggesting that it might be supportive of establishing such a zone in the region, if Israel – based on the principle of reciprocity – was prepared to do the same. But this optimistic conclusion needs to be modified, because the topic of missiles is taboo in the Egyptian strategic discourse, whether inside or outside the government. It is a delicate issue, and given the current asymmetries (including WMD) is frequently regarded as a way of distracting attention away from Israel's nuclear arsenal.[13] This generalization holds for the position Cairo has taken within the multilateral Arms Control and Regional Security working group; it applies with regard to its innovative and constructive proposals since 1974 of a Nuclear Weapon Free Zone, and, since 1990, of a WMD Free Zone in the Middle East. The formidable success of the Egyptian delegation at the NPT Review Conference in 2010 and its input in shaping the envisaged Middle East Conference for 2012 also have to be seen in this light. Israel's nuclear arsenals are the focus of Cairo's concern. Egypt's perception of the Israeli threat is not only military, but also a question of status, and its quest for Israeli nuclear disarmament is part of its interest in exercising regional leadership. Again, it remains to be seen which policies will be pursued by the post-Mubarak administrations.

Syria

In contrast to Egypt, Syria's missile arsenal forms the most formidable part of its armed forces and its doctrine. It would only agree to an MFZ if Israel renounced its air superiority and if the US, Israel's largest supplier of military equipment, altered its policy of promoting its partner's supremacy. Absent a peace agreement between Syria and Israel, there is very little likelihood that Damascus will voluntarily participate in an arms control regime that strips it of its most potent weapon category. Nor is the Asad dynasty likely to engage in arms control talks with its adversary before Israel commits itself to UN 'land for peace' principles, namely its withdrawal from the Golan Heights to the lines of 4 June 1967.

Iran

We have analysed several factors behind Tehran's missile activities which are obstacles for an arms reduction process towards the final objective of an MFZ: Iranian experiences in the First and Second Gulf Wars; the overwhelming US military presence after the end of the East–West conflict combined with Israel's offensive capabilities, strategy, and rhetoric; the technological and domestic entrenchment of the missile programme in the infrastructure of the Islamic Republic; and the prestige associated with missiles – all these factors have convinced Tehran's military planners that this weapon category is important for the country's defence, both to deter external attempts to overthrow the regime, as well as to defend it against attacks of its nuclear and missile related facilities. Their motives are not only reactive, however, but also part of its foreign policy strategy of increasing regional influence.

Conclusions

By focusing on the perception of threats we have been able to provide a new perspective on the military asymmetries, which are an obstacle to serious disarmament, in a way that goes beyond quantitative comparisons. In fact, this approach has brought to light the interests and fears motivating various missile programmes, and it has provided us with insights into conflict structures and foreign policies which are only in part commensurable.

The following 'hot spots' stand out: Israel's adversarial relationship with its neighbours Egypt and Syria; Tehran's confrontational behaviour towards the Arab countries and its distinct rivalry with Riyadh for regional hegemony; finally, of paramount importance, the problematic relationship between Iran and the US, and, secondary, that between the Islamic Republic and Israel. Without ignoring the importance of Egypt and Syria, the handling of the relations between Iran, the United States, Saudi Arabia, and Israel is of utmost relevance. This relational rectangle must be improved incrementally through compromise-oriented and balanced measures at the regional level as a modus vivendi on the way to an overall framework for reducing the current security dilemma. A serious disarmament strategy in the missile sector can be seen as part of such a concept. In brief the cornerstones of this framework follow from the analysis of our article: Iranian caution in its foreign and military policy towards other actors and a security guarantee for Tehran by the United States combined with Israeli military restraint.

To be more specific: asking the Islamic Republic to stop or dismantle its missile programmes would simply not work. But Tehran may be persuaded to terminate developing longer-range missiles as a part of a comprehensive solution to problems with the United States. Washington cannot hope to dissuade its adversary from crossing the nuclear threshold as long as the military option is still on the table. Likewise, for the purpose of easing security tensions, the Islamic Republic cannot expect to hold a fruitful dialogue with the US without altering its attitude on contentious issues, above all with respect to the Israeli–Palestinian conflict. It may be possible to persuade the Iranian leadership to deploy the Shahab-3 at a location from which it cannot reach sensitive areas in Europe and Israel. A verifiable regime could be agreed upon to check this. This would be an important confidence-building measure. At the same time the leadership in Tehran would certainly ask for a quid pro quo from Israel. Its demands will encompass military assets, perhaps limiting the range of some of the Israeli missiles, as well as a less offensive strategy and rhetoric (see the article of Martin Senn *et al.*). Here the major problem is how to bring both sides together.

Of course the concerns of Iran's Arab neighbours have to be addressed in a satisfactorily way, too. The Islamic Republic may be willing in this respect to work towards building trust and creating a more cooperative security environment. The first step down this path should be based on serious dialogue. In fact, Tehran should make all efforts to develop cooperative relations with all its neighbours, although to pursue strategic relations with many of them may not be an easy objective in the near future. Trade, cultural exchange, and elite communication could be part of the agenda for gradual cooperation. Our analysis has revealed two sensitive factors in this respect. First, Iran is not a member of any military or security alliance. Second, especially Riyadh fears a rapprochement between

Tehran and Washington. Here, the US would play a central role. It would ideally encourage the Arab countries in the Gulf to engage the Islamic Republic in discussions with the ultimate aim of establishing a mutually satisfactory regional security arrangement.

Notes

1 See Brom 2003, Forndran and Krell 1984, Kubbig 1983, 1985, 1988, Shapir and Brom 2008.

2 Egypt and Israel are suspected to have important biological research and development (R&D) activities going on, while Syria and Iran are believed to have weapon programmes in this area. While Israel presumably has relevant R&D chemical projects, Egypt, Syria, and Iran are suspected to have C-weapon programmes (Eriksen and Holøien 2010: 286, Table 1, with the relevant literature), see in this context also IISS (2011).

3 *Military* threats/menaces should be defined when both components – capabilities (plus the options and doctrines associated with them) *and* intentions – do exist at the same time, while a potential threat/menace or a risk is constituted only by one component, capabilities *or* intentions. As developed below, the non-specified term 'menace' is defined in a more comprehensive way: it includes offensive/aggressive foreign policy concepts and behaviour as well as their perceived negative impact on the domestic affairs of other states and on the regional order.

4 Such doubts were corroborated by reports according to which '"five days after the king of Bahrain was quoted in leaked American diplomatic cables as urging Washington to stop Tehran from getting nuclear weapons by whatever means necessary" (*International Herald Tribune*, 4–5 December 2010).

5 According to leaked US diplomatic cables which describe how other Arab partners have secretly lobbied for military action against Iran, the Saudi king was recorded as

> having "frequently exhorted the US to attack Iran to put an end to its nuclear weapons programme," one cable stated. "He told you [Americans] to cut off the head of the snake," said Saudi ambassador to Washington, Adel al-Jubeir, according to a report on Abdullah's meeting with the US general David Petraeus in April 2008.
>
> (*Guardian*, 29 November 2010)

6 Quotations of former Egyptian Foreign Minister Ahmed Maher, in Al-Ahram, 26 March to 1 April 2009.

7 The following figures are based on: Cordesman 2006: 196, Cordesman and Nerguizian 2010: 169, FAS 1999, Feldman and Shapir 2001: 114, Mistry 2003: 101, NTI 2011.

8 The following analysis is above all based on Sariolghalam 2003, 2008, 2010, see in this context also Ehteshami 2010, Hadian and Hormozi 2011, Hadian-Jazy 2004, 2005, Jones 1998, Kubbig 2006, Roshandel 2005.

9 As Russia has decided not to deliver the potent S-300 systems this is unlikely to change soon.

10 Chubin 1994: 48–9, McNaugher 1990, Sick 1987.

11 See also EastWest Institute 2009, Frolov 2007, Senn 2009.

12 See on this the continuous reporting in *Jane's Defence Weekly* and in *Defense News*.

13 This assessment is based on conversations the coordinator of this article had in Cairo from 22 to 25 March 2009.

References

Al Khalifa, Sheikh Khalid Bin Ahmed Bin Mohamed (2009) 'Speech at the 6th IISS Regional Security Summit, the Manama Dialogue', First Plenary Session, Bahrain, 12 December. Online, available at: www.iiss.org/conferences/the-iiss-regional-security-summit/manama-dialogue-2009/plenary-sessions-and-speeches-2009/first-plenary-session/first-plenary-session-sh-khalid-bin-ahmed-bin-mohamed-al-khalifa/ (25 February 2011).

Al Shayji, Abdullah (2010) 'GCC needs New Iran Strategy', *Gulf News*, 1 February.

Al-Ghatam, Mohammed J.K., Al-Sadiq, Abdullah, and Galal, Mohammed Noman (2010) *Bahrain Strategic Report II*, Manama: Bahrain Centre for Studies and Research.

Al-Mani, Saleh (1995) 'Security and Threat Perceptions in Saudi Arabia', in UNIDIR/United Nations Institute for Disarmament Research (ed.) *National Threat Perceptions in the Middle East*, New York: UNIDIR, 83–98.

Alani, Mustafa (2008) 'The Gulf NW and WMD Free Zone: A Track II Initiative', *International Relations*, 22(3): 358–62.

Alshayji, Abdullah K. (2002) 'Mutual Realities, Perceptions, and Impediments between the GCC States and Iran', in Lawrence G. Potter and Gary G. Sick (eds) *Security in the Persian Gulf: Origins, Obstacles, and the Search for Consensus*, New York: Palgrave, 217–37.

Bahgat, Gawdat (2009) 'Egypt and Iran: The 30-Year Estrangement', *Middle East Policy*, 16(4): 47–54.

Brom, Shlomo (2003) 'Military Balances in the Region', in Ephraim Kam and Yiftan S. Shapir (eds) *The Middle East Strategic Balance 2002–2003*, Tel Aviv: Jaffee Center for Strategic Studies, 73–84.

Chubin, Shahram (1994) *Iran's National Security Policy: Intentions, Capabilities & Impact*, Washington, DC: Carnegie Endowment for International Peace.

Cirincione, Joseph, Rajkumar, Miriam, and Wolfsthal, Jon B. (2005) *Deadly Arsenals: Nuclear, Biological, and Chemical Threats*, Washington, DC: Carnegie Endowment for International Peace.

Cobban, Helena (1991) *The Superpowers and the Syrian-Israeli Conflict: Beyond Crisis Management?* New York: Praeger.

Cordesman, Anthony H. (2006) *Arab-Israeli Forces in an Era of Asymmetric Wars*, Westport, CT: Praeger Security International.

Cordesman, Anthony H. and Al-Rodhan, Khalid R. (2007) *Gulf Military Forces in the Era of Asymmetric Wars*, Vol. 1, Westport, CT: Praeger Security International.

Cordesman, Anthony H. and Nerguizian, Aram (2010) *The Arab-Israeli Military Balance: Conventional Realities and Asymmetric Challenges*, Washington, DC: Center for Strategic and International Studies.

Cordesman, Anthony H. and Obaid, Nawaf (2005) *National Security in Saudi Arabia: Threats, Responses, and Challenges*, Westport, CT: Praeger Security International.

Cordesman, Anthony H., Nerguizian, Aram, and Popescu, Ionut C. (2008) *Israel and Syria: The Military Balance and Prospects of War*, Westport, CT: Praeger Security International.

Davis, Jacquelyn K., Perry, Charles M., and Al-Suwaidi, Jamal S. (eds) (1999) *Air/Missile Defense, Counterproliferation and Security Policy Planning: Implications for Collaboration between the United States and the Gulf Co-operation Council Countries*, Abu Dhabi: Emirates Center for Strategic Studies and Research.

DoD/U.S. Department of Defense (2010) *Ballistic Missile Defense Review Report*, Washington, DC: DoD.

Döring, Maurice (2010) *Luftpost aus dem Jemen: Ein fragiler Staat im Kampf gegen den Terror*, Frankfurt am Main: Hessische Stiftung Friedens- und Konfliktforschung/ Peace Research Institute Frankfurt.

EastWest Institute (2009) *Iran's Nuclear and Missile Potential: A Joint Threat Assessment by U.S. and Russian Technical Experts*, New York: EastWest Institute.

Ehteshami, Anoushiravan (2010) 'Iranian Perspectives on the Global Elimination of Nuclear Weapons', *Palestine-Israel Journal of Politics, Economics and Culture*, 16(3/4): 20–9.

Eriksen, Sara Kristina and Holøien, Linda Mari (2010) 'From the Proliferation to Peace: Establishing a WMD-Free Zone in the Middle East', *Nonproliferation Review*, 17(2): 281–99.

FAS/Federation of American Scientists (1999) 'FROG-7A (3R-11, 9K21, 9M21, R-65) FROG-7B (9K52, 9M52, R-70), Luna-M'. Online, available at: www.fas.org/man/dod-101/sys/missile/row/frog-7.htm (25 March 2011).

Feldman, Shai (1997) *Nuclear Weapons and Arms Control in the Middle East*, Cambridge, MA: MIT Press.

Feldman, Shai and Shapir, Yiftah (eds) (2001) *The Middle East Military Balance 2000–2001*, Cambridge, MA: MIT Press.

Forndran, Erhard and Krell, Gert (eds) in cooperation with Hans-Joachim Schmidt (1984) *Kernwaffen im Ost-West-Vergleich: Zur Beurteilung militärischer Potentiale und Fähigkeiten*, Baden-Baden: Nomos.

Frolov, Andrei (2007) 'Iran's Delivery Systems Capabilities', *Security Index*, 13(2): 31–50.

Hadian, Naser and Hormozi, Shani (2011) 'Iran's New Security Environment Imperatives: Counter Containment or Engagement with the US', *Iranian Review of Foreign Affairs*, 1(4): 13–55.

Hadian-Jazy, Nasser (2004) 'Statement', in 108/1 U.S. Congress, Senate, Committee on Foreign Relations, *Iran: Security Threats and U.S. Policy*, Hearing on 28 October 2003, Washington, DC: U.S. Government Printing Office, 40–53.

Hadian-Jazy, Nasser (2005) 'Iran's Probable Attitude to a Gulf Weapons of Mass Destruction Free Zone (GWMDFZ)', in GRC/Gulf Research Center (ed.) *Gulf WMD Free Zone*, Dubai: GRC, 10–11.

Hilal, Khalid, Kuchinsky, Leah, and Tarzi, Amin (2007) 'Israel Sees Growing Missile Threat from Syria', in *WMD Insights*. Online, available at: www.wmdinsights.com/I17/I17_ME2_IsraelSeesGrowing.htm (15 February 2011).

IISS/International Institute for Strategic Studies (2008) *Nuclear Programmes in the Middle East: In the Shadow of Iran*, London: IISS.

IISS/International Institute for Strategic Studies (2010) *Iran's Ballistic Missile Capabilities: A Net Assessment*, London: IISS.

IISS/International Institute for Strategic Studies (2011) *Iran's Nuclear, Chemical and Biological Capabilities: A Net Assessment*, London: IISS.

Jentleson, Bruce W. and Kaye, Dalia Dassa (1998) 'Security Status: Explaining Regional Security Cooperation and Its Limits in the Middle East', *Security Studies*, 8(1): 204–38.

Jones, Peter (1998) 'Iran's Threat Perceptions and Arms Control Policies', *Nonproliferation Review*, 6(1): 39–55.

Knepper, Jennifer (2008) 'Nuclear Weapons in Iranian Strategic Culture', *Comparative Strategy*, 27(5): 451–68.

Kubbig, Bernd W. (1983) *Gleichgewicht oder Überlegenheit: Amerikanische Rüstungskontrollpolitik und das Scheitern von SALT II*, Frankfurt am Main: Campus.

Kubbig, Bernd (1985) '(Re-)Defining and Refining the Criteria for Nuclear Arms Control: Theory and Praxis', *Bulletin of Peace Proposals*, 16(3): 199–223.

Kubbig, Bernd W. (1988) *Amerikanische Rüstungskontrollpolitik: Die innergesellschaftlichen Kräfteverhältnisse in der ersten Amtszeit Reagans (1981–1985)*, Frankfurt am Main: Campus.

Kubbig, Bernd W. (2006) 'Iran and the Nuclear Non-Proliferation Treaty', Briefing Paper, EP-ExPol-B-2005–37, Brussels: European Parliament. Online, available at: www.pedz.uni-mannheim.de/daten/edz-ma/ep/06/pe381.389-en.pdf (25 March 2011).

Landau, Emily B. (2006) *Arms Control in the Middle East: Cooperative Security Dialogue and Regional Constraints*, Eastbourne: Sussex Academic Press.

McNaugher, Thomas L. (1990) 'Ballistic Missiles and Chemical Weapons: The Legacy of the Iran-Iraq War', *International Security*, 15(2): 5–34.

MEMRI/Middle East Media Research Institute (2010) 'Arab Opposition to Amr Moussa's Initiatives for Strengthening Ties with Iran', *Special Dispatch*, No. 2913. Online, available at: www.memritv.org/report/en/4106.htm (25 March 2011).

Mistry, Dinshaw (2003) *Containing Missile Proliferation: Strategic Technology, Security Regimes, and International Cooperation in Arms Control*, Seattle, WA: University of Washington Press.

NTI/Nuclear Threat Initiative (2011) 'Egypt: Missile Overview'. Online, available at: http://nti.org/e_research/profiles/Egypt/Missile/index.html#fn11 (25 March 2011).

President of the United States (2010) 'National Security Strategy', Washington, DC: The White House. Online, available at: www.whitehouse.gov/sites/default/files/rss_viewer/national_security_strategy.pdf (25 March 2011).

Roshandel, Jalil (2005) 'The Nuclear Controversy in the Context of Iran's Evolving Defence Strategy', in Shannon N. Kile (ed.) *Europe and Iran: Perspectives on Non-Proliferaton*, Oxford: Oxford University Press, 47–71.

Rubin, Uzi (2006) *The Global Reach of Iran's Ballistic Missiles*, Tel Aviv: Institute for National Security Studies.

Said, Abdel-Moneim (2009) 'Half a Dozen Charges', *Al-Ahram*, 26 March to 1 April.

Sariolghalam, Mahmood (2003) 'Understanding Iran: Getting Past Stereotypes and Mythology', *Washington Quarterly*, 26(4): 69–82.

Sariolghalam, Mahmood (2008) 'Iran in Search of Itself', *Current History*, 107(713): 425–31.

Sariolghalam, Mahmood (2010) *The Evolution of the State in Iran: A Political Culture Perspective*, Kuwait City: Strategic and Future Studies Center.

Senn, Martin (2009) 'The Arms-Dynamic Pacemaker: Ballistic-Missile Defense in the Middle East', *Middle East Policy*, 16(4): 55–67.

Shapir, Yiftah S. and Brom, Shlomo (2008) 'The Regional Military Balance' in Mark A. Heller (ed.) *The Middle East Strategic Balance 2007–2008*, Tel Aviv: Institute for National Security Studies, 39–48.

Sick, Gary (1987) 'Iran's Quest for Superpower Status', *Foreign Affairs*, 65(4): 697–715.

Solingen, Etel (2007) *Nuclear Logics: Contrasting Paths in East Asia and the Middle East*, Princeton, NJ: Princeton University Press.

Taremi, Kamran (2005) 'Beyond the Axis of Evil: Ballistic Missiles in Iran's Military Thinking', *Security Dialogue*, 36(1): 93–108.

Wendt, Alexander (1992) 'Anarchy Is What States Make of It: The Social Construction of Power Politics', *International Organization*, 46(2): 391–425.

5 Taking Israel's security interests into account

Deterrence policy in a changing strategic environment

Uri Bar-Joseph

Israel's strategy in the light of past, present, and future threats

For more than six decades now, the Jewish state has been involved in an intense armed conflict with state and non-state actors in the Middle East/Gulf. Since its primary interest has been on maintaining the status quo, it has relied heavily on deterrence, both to prevent the emergence of new challenges to its security and to contain existing ones. Political, military, and technological developments during this period have changed the strategic environment, and led to major shifts in Israel's deterrence strategy. This article describes these changes and analyses their implications. The first part briefly discusses the concept of deterrence in Israel's strategic thinking and shows how the changes in its strategic environment led to the declining relevance of one type of deterrence and the growing importance of others. The starting point for the second part of this essay is the fact that the main challenge to Israel's deterrence policy in the future will be non-conventional threats, primarily Iran's nuclear programme, which might compel the Jewish state to rely more heavily on nuclear deterrence. Consequently, I describe the logic behind its nuclear project, explain Israel's policy of opacity, provide a description, on the basis of non-Israeli sources, of the current status of its arsenal, and examine the effect of its image as a nuclear power on the policy of its regional opponents. Finally, the article analyses the main elements of the Jewish state's future deterrence policy vis-à-vis Iran.

Israel's deterrence policy

Although the Jewish state's deterrence policy has been the subject of numerous studies,[1] its role in Israeli strategic thinking remains too often unclear. Until recently its policy comprised four types of deterrence: current, specific, strategic, and cumulative (Bar-Joseph 1998).

- *Current deterrence:* the threat to use force, or the use of exemplary force, to persuade an opponent not to initiate challenges to a given status quo in a low-intensity conflict. This type of deterrence was used primarily against Palestinian guerillas from Jordan and Lebanon, from the 1950s until the mid 1980s. The end of the Palestinian military presence from Lebanon following the 1982 War made this strategy less relevant to the national security policy of the Jewish state, and during the two Palestinian uprisings (the first and the second *Intifada*) it relied more heavily on prevention than deterrence. Following the Israel Defense Forces' (IDF) withdrawal from southern Lebanon in May 2000, current deterrence was used to deter Hezbollah from challenging the security of northern Israel.

- *Specific deterrence:* a strategy aimed at persuading one's opponent to avoid taking specific, well-defined actions short of a general war. Operationally it defines certain 'red lines', or *casus belli*, as boundary lines of vital interests, the crossing of which is highly likely to produce a military response or even lead to the outbreak of war. Past examples of such 'red lines' are Egypt closing the Tiran Straits (near Sharm el-Sheikh, in the Red Sea) to Israeli shipping, or the Iraqi army's advance into Jordan. Peaceful relations with Cairo and Amman and the stable strategic relationship with Damascus have diminished the importance of specific deterrence. Currently, the most important 'red line' involves preventing other regional actors from becoming nuclear powers. Specific deterrence is also known as the Begin doctrine, after Prime Minister Menachem Begin, who ordered the destruction of an Iraqi nuclear reactor in 1981 on the basis of this doctrine. As a deterrence strategy, it was reemployed in 2007 when – allegedly – an Israeli air strike destroyed a Syrian nuclear facility. Currently, the main challenge to this doctrine is Iran's nuclear project.
- *Strategic deterrence:* the threat of military force, primarily in the form of a general war, aimed at persuading the opponent to avoid initiating any type of large-scale threat. Until the end of the 1980s, the main challenge to the security of the Jewish state was an Arab attack using conventional weapons. Consequently, strategic deterrence was the main tool in Israel's deterrence arsenal. This situation started to change towards the end of the 1970s in response to a series of events: the 1979 peace with Egypt; the breakdown of the Eastern Front (the coalition of Syria, Iraq, Jordan, the Palestine Liberation Organization (PLO), and other minor Arab actors) as a result of the Iran–Iraq War; the collapse of the Soviet Union leaving Damascus with no superpower support; and the 1991 Gulf War, which defused much of Baghdad's military strength. The cumulative effect of these developments was to leave the Arab world with no viable conventional option. As a result, strategic deterrence lost its position of seniority in Israel's list of defence priorities.
- *Cumulative deterrence:* the long-term policy which aims at convincing the Arab side that ending the conflict by destroying the Jewish state is either impossible or involves such enormous costs and potential risks

that they exceed any expected benefits. It is backed up by the aggregate threat of Israel's use of force, in the form of its military victories, conventional and nuclear capabilities, and its resolve to use all means at its disposal to ensure its continued existence. Cumulative deterrence aims to thwart the Arab goal of destroying the Jewish state. Israeli strategic thinking is that as a result of this policy, its opponents will ultimately be compelled to make peace with the Jewish state. In recent years there has been an increased Arab readiness to formally accept Israel's right of existence. Egypt and Jordan have accepted it in formal peace treaties, the PLO accepted it in the framework of the Oslo negotiations, and Syria has done so de facto by adopting the formula of peace in exchange for the return of the Golan Heights in all direct and indirect negotiations with the Jewish state. The present consensus within the Arab League is expressed in its 2002 Peace Initiative which calls for normalizing relations with Israel in exchange for a return of the 1967 occupied territories and the establishment of a Palestinian state. This proposal constitutes the ultimate success of cumulative deterrence policy.

While these four traditional types of deterrence have mostly become less relevant as the threat of war from state actors has declined, two new types of challenges have emerged: sub-conventional threats from non-state entities, and nuclear threats, primarily from Iran. Currently, the main challenge to the Jewish state's deterrence involves sub-conventional threats by Hezbollah, Hamas, and potentially other Islamic terrorist organizations. Given their limited resources, such groups cannot pose major challenges to Israeli strategic deterrence,[2] but their improved military capabilities, primarily their extensive use of high-trajectory weapons, creates a new type of threat, which challenges the reputation of the Jewish state for deterrence and may ultimately lead to a massive response. This was the case with the PLO, which had been building up its capability to conduct massive rocket attacks from southern Lebanon since the early 1970s – a threat that led the IDF to launch the First Lebanon War in 1982. Similarly, their inability to put an end to Hezbollah rocket launches in the summer of 2006 led the IDF to use extensive force, in a conflict that more closely resembled a conventional war than a low-intensity conflict. The number of air sorties and artillery rounds, used by the IDF during this confrontation, exceeded that of the 1973 Yom Kippur War, the greatest confrontation in the history of the Israeli–Arab dispute. The IDF even used massive force against a relatively weak Hamas organization in Gaza during "Operation Cast Lead" in December 2008 and January 2009, requiring the mobilization of a number of reserve brigades.

A further challenge to specific deterrence is the increasing military capabilities of non-state actors. In the past, low-intensity conflicts such as those resulting from the sabotage acts by militant Palestinian organizations from

1965–67 or the post Six Day War conflict with the PLO along the borders with Jordan and Lebanon have not involved any specific rules of engagement. But as the ability of such actors to inflict damage upon Israel has increased, so has their vulnerability to punishment. Consequently, both sides have established 'red lines' the violation of which could lead to a massive retaliation. Thus, Palestinians were deterred from even minor acts of violence against the Jewish state between August 1981 and June 1982 because they feared an invasion of southern Lebanon, where the PLO had deployed most of its military power. Similarly, clear rules of engagement limiting both sides' freedom of action were established in the confrontation with Hezbollah during the 1990s and following the Israeli unilateral withdrawal from southern Lebanon in May 2000 (Sobelman 2004). By contrast, the Jewish state failed, for years, to draw 'red lines' in its confrontation with Hamas in Gaza. This was mainly the result of its interest in avoiding a large-scale invasion of the Strip after the 2005 unilateral withdrawal. The consequence was a lingering low-intensity conflict that lasted for years. "Operation Cast Lead", which from the Israeli perspective was a response to cumulative Hamas provocations, demonstrated to both sides the potential costs involved in such an exchange and facilitated the reestablishment of rules of conduct. As a result, the massive use of force in the Second Lebanon War and "Operation Cast Lead" revived specific deterrence policy. However, since deterrence merely buys time in order to allow the settlement of a conflict, an endless continuation of the present status quo is likely to yield a failure again.

The second and the more important challenge to the deterrence policy of the Jewish state is the likelihood that Iran becomes a nuclear power within the next few years. This is the focal point of Israel's strategic thinking, and for good reason. Except for the first weeks of the Arab armies' invasion, immediately after the establishment of the Jewish state in May 1948, Israel has never faced a real existential threat. This situation would change dramatically if Tehran became a nuclear power since the Islamic Republic would have the capability to destroy the Jewish state. Given the magnitude of this new threat, I will now consider its implications by focusing on nuclear deterrence, which would be the main means to deter its counterpart. I shall first discuss central elements of Israel's nuclear buildup and strategy before turning to analyse how it might cope with the threat of an Iranian bomb.

Israel's nuclear deterrence: logic, policy, capabilities, and impact

The logic

The Jewish state, which is regarded, at least since the early 1970s, as the sixth member of the nuclear club, is also its most unique member. Israel is not an officially recognized nuclear power and its programme remains opaque. In addition, its motivation for going nuclear is unique since all other nuclear powers were primarily motivated to do so by the activities of their opponents. In contrast, the driving force behind the project of the Jewish state has been the anxiety of future Arab conventional superiority. David Ben-Gurion feared that in the long run quantitative advantages in manpower, territory, and material resources would give the Arabs the ability to destroy Israel. Ben-Gurion put little trust in US support to neutralize this quantitative advantage. His experience in the Suez/Sinai War in 1956, when joint American–Soviet pressure compelled the Jewish state to give up the occupied territory, seemed to be clear proof that his country could not rely on external support. The fear motivating Ben-Gurion to secretly build the nuclear reactor in Dimona has been vividly described by his former assistants: he used to pace nervously in his office, looking at the big map of the Middle East and saying: "All night I could not sleep. What is Israel? Only a small spot! How can she survive in this Arab world?" (quoted in Bar-Zohar 1978: 1,399).

Such nightmare scenarios received a more organized form in the IDF planning. The worst-case scenario throughout most of the first two decades of the Jewish state's existence was "mikreh hakol" ("scenario of everything"), a coordinated (possibly surprise) all-out conventional Arab attack, against which the relatively small Israeli military would have no viable defence. In the mid 1960s, when building the bomb had become a practical option, strategists drew four 'red lines' whose crossing could trigger its use: an Arab invasion into the *hinterland*, the destruction of the Israeli Air Force (IAF), heavy bombing of its cities, possibly with chemical and biological weapons, and an attack against the Jewish state with nuclear weapons (Cohen 1998: 237). Until today the traumatic memory of the Holocaust is seen as the best evidence of Israel's need for perpetual military preparedness and helps explain the logic of their nuclear ambitions. In order to prevent a second national catastrophe, the Jewish state must be able to deter any threat to its existence. The nuclear project was to become the main component in Israel's strategic deterrence, aimed at convincing the Arabs to avoid challenging its very existence. It was also thought to be the key component in the cumulative deterrence strategy as the *ultima ratio*, a policy ensuring that the Arab annihilation of the Jewish state would effectively lead to the mutual destruction of both parties.

The logic of Israel's nuclear deterrence strategy was tested by three inter related events in mid 1967. The first was the crisis evoked by President Gamal Abdel Nasser of Egypt in mid May, allegedly in response to the intention of the Jewish state to attack Syria. In contrast to his usually prudent policy, Nasser openly challenged Israel's specific deterrence strategy by crossing a number of 'red lines', the most important being the closure of the Tiran Straits to Israeli shipping, which was publicly declared as a *casus belli*. At the height of the crisis, Arab countries unanimously gathered behind Nasser, a unified military command was established, army units reaching from Morocco to Iraq were dispatched to the front, and the whole Arab world was swept up in with a frenzy of popular support for what looked like the imminent destruction of the Jewish state. From Israel's perspective, the rapid escalation of the crisis fit the "scenario of everything". At one stage the IDF general headquarters demanded a preemptive strike, warning that "any delay is an existential gamble" (Oren 2004: 129). The second event was a critical element in the response of the Jewish state to the growing threat. By a number of accounts, about two weeks into the crisis, Israel crossed the nuclear threshold by assembling three nuclear devices. On 3 June, three days before the outbreak of the 1967 War, Shimon Peres, who was in charge of the nuclear project until May 1965, secretly suggested to Prime Minister Levy Eshkol to carry out a demonstrative nuclear test in order to deter the Arabs and prevent a war.[3]

The last event was the outcome of the war. The IDF's swift and decisive military victory was a clear proof, however, that previous fears of conventional inferiority, which constituted the logic for the nuclear programme, were ill founded. In reality, the conventional balance of forces tilted towards the Jewish state while the nuclear project was being launched in the late 1950s and the gap widened further during the 1960s. After 1967, the Arab conventional threat gradually ceased to exist. In other words, the irony of history is that in late May 1967 Israel became nuclear and only a few days later, its justification for doing so had evaporated and never returned. The gradual disappearance of the Arab conventional threat also partially explains the policy of nuclear opacity.

The policy

The policy of nuclear opacity allows the Jewish state to be treated as a nuclear power, enjoying the advantages of nuclear deterrence without officially admitting its status, thus evading the potential costs of adverse publicity associated with crossing the threshold. Nuclear opacity has been a highly successful policy for 50 years now, and some experts consider it to be Israel's most sophisticated strategic innovation (Cohen 2005: 141). But like some other projects, opacity was the product of a series of ad hoc improvisations in response to internal debates and external pressures, all of which took place within a cultural envelope of secrecy in security affairs.

Secrecy has always been a dominant value for the Jewish state. This is a product of a culture with historical roots in a vulnerable, persecuted Eastern European Jewish community and a military culture shaped by underground militia fighting under the Ottoman and British rule in Palestine, where secrecy was necessary to keep military capabilities safe. Even after the establishment of the Israeli state and the transformation of the underground military into a standing regular army, secrecy continued to be central to national security policy. The culture of secrecy reached its zenith when it came to the nuclear project, "the holy of the holies". As Avner Cohen (ibid.: 148) has described it, opacity dominated the discussions even among those who had a legitimate need to know, and certain critical words remained taboo even in the most intimate forums. In addition to the culture of secrecy, two other elements have contributed to the policy of opacity. One is the tacit debate within the leadership of the Jewish state which began shortly after the Dimona project was launched. The pro-nuclear school regarded the bomb as a long-term panacea to Israel's quantitative inferiority to the Arabs, whereas their pro-conventionalist counterparts believed that it was unnecessary, too expensive (on the account of a conventional build-up), and too risky because it could lead to a confrontation with the superpowers. The political compromise between these two approaches called for continuing the project but without publicly crossing the nuclear threshold (Bar-Joseph 1982).

The second and the more decisive factor that kept the bomb in the basement was American pressure. From December 1960, when it became known that the Jewish state was building a nuclear reactor in the Negev, the White House insisted on US inspections to ensure that the reactor was not being used for military objectives. In the course of inspections that lasted for eight years (1961–69), American experts failed to find clear-cut evidence that Israel was pursuing the bomb, primarily since the Jewish state succeeded in hiding the well-concealed separation plant built near the reactor. This game of hide and seek was a source of inconvenience for both sides. Israel was embarrassed by the need to cheat its ally, and the US was frustrated when it learned it was being deceived (Shalom 2005). In 1969, when the Central Intelligence Agency estimated that the Jewish state either possessed nuclear weapons or could build them within a very short period of time, Prime Minister Golda Meir reached a very intimate and secretive understanding with President Nixon who had just entered office: the White House agreed to put an end to the annual visits to Dimona and to stop pressuring the Jewish state to sign the Nuclear Non-Proliferation Treaty (NPT), and in return the Israelis committed themselves not to conduct nuclear tests or publicly disclose their capabilities. These understandings have been respected by all subsequent governments of both countries and they continue to be the main reason for the policy of opacity (Cohen 2010; Cohen and Burr 2006).

The capabilities

Since 1967, when the Jewish state allegedly assembled its first nuclear devices, its stockpile has expanded significantly. Lacking any hard evidence, Israel's supply of nuclear weapons is mainly estimated by calculating the theoretical production capacity of the Dimona reactor. An updated academic study claims that the Jewish state has between 70 and 400 nuclear weapons. Some of these may have a yield of over 100 kt and may also include thermonuclear warheads (Toukan and Cordesman 2009: 9).[4] The upper limit of 400 weapons seems to be based on the Dimona plant producing enough plutonium for about ten warheads a year. The smaller figure is probably an estimation of what Israel might consider a sufficient number of warheads for minimal deterrence.

In the early 1960s Leonard Beaton and John Maddox (1962: 175) suggested that the Jewish state could use Britannia passenger planes of the national carrier El Al to deliver nuclear bombs into enemy territory. Since that time Israel has developed an impressive array of nuclear-capable launching systems, which form a national triad. The first component of the triad consists of three types of surface-to-surface missiles (SSMs): the Jericho-1 (originally the French-made MD-620) with a range of 235–500 km and a payload of 650 kg, which was deployed since the early 1970s (NTI 2010); the solid-propellant Jericho-2 with a range of about 1,500 km and a payload of approximately 1,000 kg,[5] which entered service in the early 1990s and according to American estimates, about 100 of which were deployed until 2004; and a third generation three stage solid propulsion Jericho-3 with a range of 4,800–6,500 km, which is capable of carrying a single 750 kg nuclear warhead or two to three Multiple Independently Targetable Re-entry Vehicle low-yield warheads and is currently under development (Lennox 2007: 82–3).

The second component of the Israeli triad is its aerial component. It includes a fleet of about 80 F-15s (25 of which are F-15I 'Ra'am', the Israeli version of the F-15E) and 300 F-16 fighter-bombers capable, through air refuelling, of reaching any spot in the Middle East or beyond. Finally, the third component of the triad is the naval one, constituting the second strike capability of the Jewish state. Three German-built Dolphin class diesel submarines are the centrepieces of this strategy, each capable of launching cruise missiles from their six single 533 mm torpedo tubes (IISS 2009: 250). Two more submarines of this type have been ordered by the Israeli Navy (*Der Spiegel*, 24 January 2011: 77). According to various sources, the first test of cruise missiles launched from a Dolphin submarine was conducted in the Indian Ocean in 2000. They had a range of 1,500 km and US experts estimated that they were developed based on the air-launched Popeye-turbo cruise missile (FAS 2010). In 2003 it was also reported that the Jewish state had modified the American-made Harpoon cruise missile, outfitting it to carry nuclear warheads (Frantz 2003). Each Dolphin submarine can carry five UGM-84C Harpoon underwater-to-surface guided missiles. Other units – the Eilat type corvette and the Hetz type frigate – are also equipped with this weapon system (IISS 2009: 250). From all indications, since the 1960s Israel has succeeded in building up an impressive arsenal of nuclear warheads and launching systems. With the exception of the US, Russia, China, Britain, and France, no other nation today has similar capabilities.

The impact

The success of the policy of nuclear deterrence is mixed. On the one hand, the build-up of an impressive arsenal, the achievement of de facto nuclear status, and securing American acceptance of it has been a success. On the other hand, nuclear capabilities seem to have played only a minor role in deterring the Arabs from initiating a conventional war against the Jewish state and, ultimately, accepting its right to exist (Maoz 2003). Rather, these goals were achieved by the alternative to nuclear deterrence, i.e. conventional superiority. IDF military superiority turned nuclear deterrence strategies into a secondary and somewhat redundant safety net. Non-conventional deterrence first became relevant during the 1967 War. Although an Egyptian attack on Dimona ranked highest on the list of Israel's strategic fears prior to the war (Grundman 1998: 159–60, 162; Gluska 2004: 227–30), its nuclear capability played only a minor role in how Nasser initiated and handled the conflict. His generals had told him, both before and during the crisis, that Cairo's army could not confront the IDF due to the five years of futile war it had spent in Yemen since 1962 (El-Gamasy 1993: 21–2). A weakened military combined with Soviet pressure and the fear of an American response sufficed to deter Egypt from attacking the Jewish state. The decisive victory of the IDF succeeded in showing that conventional force indeed sufficed to secure the existence of the Jewish state. Its victory, moreover, left Israel in possession of the Sinai, the Gaza Strip, the West Bank, and the Golan Heights, the return of which had, since 1967, become the Arab condition for ending the conflict. Thus, the military victory of the Jewish state set the terms for Arab acceptance of Israel's existence in the region – fulfilling precisely the role that the fathers of the nuclear project envisioned for the bomb.

The 1973 Yom Kippur War, generally considered the most severe challenge to the security of the Jewish state, was the second test of its nuclear deterrence policy. Syria and Egypt had only limited territorial goals in this conflict. While Damascus' interest was in the occupation of the Golan Heights, Cairo attempted to occupy a strip of land up to 10 km east of the Suez Canal in order to jump start a political process. Despite claims that fear of the Israeli bomb led Sadat to limit his territorial goals detailed Egyptian accounts of the war fail to mention this as a factor (Aronson 2007: 200–3; Bundy 1988: 510–11). Instead, the sole reason they give for the limited advance by Egyptian ground forces is their vulnerability to IAF attacks, once they got beyond the air cover provided by the dense air defence system west of the canal.[6] The influence of nuclear deterrence on Damascus is somewhat more evident. The few extant Syrian descriptions of the war show that limited territorial goals were mainly the product of Soviet objections to an attack on the Jewish state and the recognition of the IDF's conventional superiority. Nevertheless, a diplomat of the Asad regime (whose familiarity with this issue is unclear) claimed that the fear of Israel's nuclear capabilities played a role in limiting the war goals (Parker 2001: 102–3, 119). No other Syrian officials have made such claims, rather they have generally tended to ignore the influence of the bomb on the outcome of the Arab–Israeli conflict.

The third test of the nuclear deterrence policy occurred during the 1991 Gulf War, when Iraq launched 39 Scud missiles against the Jewish state. While these attacks did not threaten Israel's existence, they were relevant for its non-conventional deterrence. For some observers, the mere fact that the Jewish state suffered from missile attacks throughout the war without retaliating constituted a deterrence failure (Gordon 1991; Inbar and Sandler 1993), even a nuclear deterrence failure (Maoz 2003: 317). For others nuclear deterrence was credited with effectively preventing Saddam Hussein from using chemical warheads (Baram 1992; Feldman 1991). If Baghdad was indeed deterred from using chemical weapons for fear of an Israeli nuclear retaliation, as an Iraqi defector implicitly claimed after the war (PBS 1996), then this would be the only known case of its kind in the history of the conflict. But it is unclear whether chemical warheads were not launched because they had not been tested or whether they were considered unreliable. The lack of corroborative evidence for the claim that fears of the bomb channelled Baghdad's behaviour means that this explanation remains only one of several possible.

The impact of Israeli non-conventional capabilities on the nuclear policies of regional actors has been limited as well. Iraqi motivation for developing a nuclear programme has changed over time. Until an IAF air strike destroyed the Osiraq reactor near Baghdad in 1981, it was mainly geared towards becoming the dominant power in the Gulf region and gaining the prestige associated with being the first Arab nuclear power. Although Saddam Hussein claimed in 1975 that his intention was to counter Israel's nuclear arsenal (*Al Usbu Al-Arabi*, 8 September 1975), a prudent analysis shows that the capabilities of the Jewish state were secondary in his strategic calculus even if his claim was excellent public relations within the Arab world (Solingen 2007: 146–7). Paradoxically, the destruction of Osiraq, which according to Iraqi scientists was an ineffective means to get to the bomb, allowed Saddam Hussein to secretly initiate an uranium enrichment process, which within ten years brought the country very close to nuclear capability (Hamza 1998; Obeidi and Pitzer 2005). This time, however, the prime target of the project was Iran, its formidable enemy in the war that lasted until 1988. Similarly, Israel's image as a nuclear power played a minor role in the Libyan programme, which was exposed in 2004. The logic behind this project is not entirely clear but it was apparently motivated by the regime's self-survival interest and its pursuit of prestige rather than a reaction to any real or imaginary external threats (Solingen 2007: 226–8).

By contrast, Syria's build-up of weapons of mass destruction (WMD) was a direct response to the threat posed by the Jewish state. Since 1974, when

Damascus received the first shipments of Scud missiles from the Soviet Union, it had come to possess a rather impressive stockpile of strategic means of delivery, including 48 improved Scud-B and Scud-C SSMs, equipped with chemical and possibly biological warheads, and with a range stretching over the whole of Israel (Cordesman 2007; IISS 2009: 266). This arsenal was built in response to the latter's conventional superiority and Syria's growing regional isolation, primarily after the collapse of the Soviet Union. It is quite likely that Damascus believed its WMD provided the strategic capability to deter the Jewish state from the use of nuclear weapons as well. The Syrian project, the centre of which was the reactor destroyed in September 2007, allegedly by the IAF, may have been intended as a minimal response to Israel's nuclear advantage.

Finally, the Iranian programme, which constitutes the most severe challenge to the regional nuclear monopoly of the Jewish state, was driven by the need to counter Iraq's WMD capabilities that had proved so deadly in the 1980–88 War. The fall of Saddam Hussein taught Tehran that a nuclear capability was the ultimate means of deterring the US from taking similar measures against the Islamic Republic, and this became the primary incentive for accelerating the project. In addition, the programme is seen in Iran as proof for scientific prowess, and hence considerations of national pride and prestige were other factors pushing Tehran in this direction. In short, the opposition to Israel is only a secondary consideration motivating the development of Iran's nuclear programme, although publicly the Jewish state, along with the United States, is portrayed as the Islamic Republic's top enemy (Solingen 2007: 165–70, 185–6).

This overview of Israel's deterrence policy over the past 50 years leads to three broad conclusions. First, the Jewish state has succeeded in building formidable nuclear capabilities and an effective non-conventional deterrence policy that allows it to enjoy the benefits of such a strategy without paying the price involved in publicly crossing the threshold. Second, the reality of the Arab–Israeli conflict made nuclear posturing superfluous, since its superiority in conventional weapons has sufficed to deter the regional opponents of the Jewish state from posing any serious military threat to its continued existence. Third, the persisting progress of Iran's nuclear project is likely to pose the most formidable challenge to date to Israeli nuclear deterrence policy. The remainder of this article analyses some of the ways the Jewish state is likely to cope with this threat.

Coping with the Iranian threat

A messianic regime, combining an extreme anti-Israeli rhetoric, a massive support of anti-Israeli militia, primarily Hezbollah and Hamas, denial of the Holocaust, and the rapid development of a large-scale nuclear programme – these factors have turned the Islamic Republic into the Jewish state's strategic nightmare. Israel has responded with a combination of coordinated diplomatic activity to halt Tehran's nuclear project, apparent covert operations aimed at slowing its progress, and preparations, in an almost public manner, for an air strike. This has so far failed to stop Iran from becoming nuclear. Under these circumstances, four different scenarios regarding the future of Tehran's programme can be formulated. First, the Islamic Republic gives up all enrichment activities; second, Iran continues to enrich uranium, but does not build nuclear weapons; third, it becomes an undeclared nuclear power; or fourth, it openly announces that it possesses nuclear weapons.

The question I shall now address is how the Jewish state would respond to each of these cases. From the Israeli perspective, the first of them is the most desirable one, and it is also the best-case scenario for most other regional actors. But the chances that Tehran will unilaterally give up its nuclear project, even in the event of a change of regime are highly unlikely (see the chapter of Gawdat Bahgat *et al.*). The second scenario is far more likely than the first, primarily since the Islamic Republic may consider nuclear threshold status preferable to the public violation of the NPT, to which it is a signatory, and because of the adverse effects of more restrictive sanctions that might be triggered by such a move. From the perspective of the Jewish state, any development in which Iran remains without nuclear capabilities is clearly preferable to one in which it possesses such weapons. The problem, however, is that neither Israel nor the rest of the world for that matter will be in a position to confirm for sure whether Tehran has made the transition from the status of a threshold state to one with a bomb in the basement. Under such circumstances the Jewish state's current policy of demanding international action to prevent the Islamic Republic from becoming nuclear will probably continue.

The final two scenarios both involve Iran crossing the nuclear threshold. From the perspective of the Jewish state, the difference between whether its adversary is a declared or an undeclared nuclear adversary is likely to be minor. For security purposes, the main issue is that a Muslim regime committed to Israel's destruction would have gained the capability to achieve this goal. Although officials of the Jewish state have refused to discuss their response to such a development, fearing it might be interpreted as acceptance, the likelihood of such a scenario requires discussion. The core of any Israeli policy vis-à-vis a nuclear Iran would be nuclear deterrence. If the effectiveness of such a policy is likely to remain sufficiently high independent of whether or not its nuclear arsenal is kept in the basement, the Jewish state may retain its undeclared nuclear status even if Tehran publicly crosses the threshold. But if the Islamic Republic publicly threatened Israel with nuclear destruction, the Jewish state might go public, less for strategic reasons than to calm domestic fears.

The country is usually referred to as a 'one bomb state' since, given its size, a singular nuclear strike could destroy it. Hence, to deter any Iranian temptation to use its nuclear capability in a first strike, the core of Israel's nuclear deterrence strategy would have to be a second strike capability. This is not a novel idea. Already in 1997 a senior official warned:

> Iran should know, and she will, that on the day she will launch against us whatever she has, even if everything will be hit, Israel will have the ability to erase [the Islamic Republic] from the face of the earth, quickly, strongly, and in an inelegant manner.
>
> (quoted in Kaspit 1997)

Since a first strike by Iran would probably be with ballistic missiles, the Jewish state can be expected to focus on an anti-missile strategy, which consists of four layers: attacks on launchers and missiles, boost-phase intercept, interception of incoming missiles, and passive defence (Shapir 2001). At present, the most advanced component in the mix is the interception of incoming ballistic missiles. Israel is a world leader in the development, production, and deployment of missile defence (MD) systems. At the centre of its efforts is the Arrow project, which is capable of intercepting medium- and short-range ballistic missiles. Following tests of the first generation of the Arrow in the early 1990s, two batteries of the second generation, the Arrow-2, became operational on the eve of the Second Gulf War. They cover the Israeli territory from the upper Galilee in the north to the central Negev in the south, including Dimona. Since late 2002, the system has been upgraded several times. In a test in April 2008, its latest version, Arrow-2 block-4, detected and intercepted a target missile that simulated the Iranian Shahab-3, indicating a defence capability against Tehran's longest-range ballistic missile.

Currently the Jewish state and the US are developing the Arrow-3, which will provide an upper-tier component to the MD layout. When deployed, Israel will have a three-layered missile defence architecture: the Arrow-3, capable of intercepting targets at an altitude of about 100 km; the Arrow-2, able to intercept targets at an altitude of 60 km; and the American-made Patriot Pac-3, which is meant to intercept incoming ballistic missiles at lower altitudes. The ability of the Jewish state to defend itself against an Iranian nuclear attack is a stabilizing factor if it contributes to Israel's sense of security and moderates its instincts for preemption. On the other hand, it may also increase the Islamic Republic's fears of a first strike against its nuclear arsenal if Tehran applies the same strategic logic that led the Soviets to oppose Ronald Reagan's Strategic Defense Initiative during the 1980s, i.e. that MD may be effective enough to protect the Jewish state after it crippled Iran's launching capabilities, even if it does not provide an effective means of defence against a first strike.

Since the development of missile defence demanded extensive technological and financial resources, which Israel was able to acquire only with American support, Tehran is unlikely to develop similar capabilities. Although the Islamic Republic invests substantial resources into building such a system,[7] it is questionable whether the programme is effective. Hence, Iran will probably focus on hiding its SSMs within its vast territories, particularly in the eastern part of the country, so that a maximal number of them would survive an Israeli first strike. In addition, Tehran may resort to other means of delivery such as cruise missiles or having terrorist groups smuggle nuclear devices into the Jewish state that by their nature cannot be deterred. Evidence that the Islamic Republic cooperates with terrorists in the nuclear field would therefore heighten Israeli fears that Iran intends to use the bomb in an offensive manner and might lead to a preemptive strike by the Jewish state.

To summarize, a future balance of threat would match up two unequal opponents: Israel has (according to foreign sources) a sophisticated nuclear arsenal and means of delivery, a highly developed missile defence layout, and a well-established command and control system. Tehran would have a far smaller and less sophisticated nuclear arsenal, less developed means of delivery, hardly any active defence capabilities, and an inexperienced command and control system. But despite these advantages, Israel's status as a 'one bomb state' might still leave it in a weak position. As far as an Israeli–Iranian missile disarmament agreement is concerned two problems arise. First, missiles are the main means by which Tehran can strike the Jewish state, while the latter has other options, such as the use of advanced fighter-bombers by which to strike the Islamic Republic. Consequently, Tehran is unlikely to give up its stockpile of SSMs. Second, since both states have other opponents against which they might need their missile arsenal, a bilateral agreement is a practical option for neither of them. In Israel's case, a regional accord might be acceptable. For Iran, which considers its strategic arsenal as a means by which to deter the US, this constitutes an obstacle.

Given the current state of bilateral relations, a solution of the conflict about Tehran's nuclear programme that the Jewish state can accept is unlikely in the near future. Therefore the most effective means to reduce the burden on Israel's deterrence seems to be a political one: significant progress in the Arab–Israeli peace process. A settlement with Syria would weaken the strategic cooperation between Damascus and Tehran and hinder the Islamic Republic's ability to provide military support to Hezbollah in Lebanon. A settlement with the Palestinians acceptable to the majority of the people in the West Bank and Gaza would neutralize much of Iran's ideological justification for the destruction of the Jewish state since even the fundamentalist regime in Tehran would find it difficult to be more committed to Palestine than the Palestinians themselves. Indeed, at least once President Ahmadinejad was quoted as saying, in response to a question about the Islamic Republic's stand with regard to a two-state solution, that: "If they [the Palestinians] want to keep the Zionists, they can stay. ... Whatever the people decide, we will respect it. I mean, it's very much in correspondence with our proposal to allow Palestinian people to decide through free referendums" (quoted in Tatchell 2008).

A withdrawal from the territories, which Israel perceives as necessary for its security, is also likely to drive the United States and European countries to compensate the Jewish state for the growing sense of insecurity. This might be done in two ways: first, by the sale of advanced weapon systems that would widen the quality gap between the IDF and the Arab armies as well as Iran; and, second, by upgrading and by formalizing security guarantees. The latter could be accomplished in a number of ways including the signing of a formal American–Israeli defence pact or accepting the Jewish state into the North Atlantic Treaty Organization (NATO). Already a formal agreement with the United States or NATO would make nuclear deterrence vis-à-vis a nuclear Iran less important.[8] In short, a settlement with Syria and the Palestinians, which might be based on lines similar to the Arab Peace Initiative of 2002, would transform the Arab–Israeli conflict into a secondary issue in Middle Eastern politics, thus turning the conflicts between the Muslim Sunni and Shia and those between conservative Arabs and radical Islamists into the main axis of regional rivalry. Strengthening American and NATO commitments to the security of the Jewish state, would result in a major reduction of the burden on Israeli nuclear and conventional deterrence, enabling the country to live in a far less threatening environment.

Notes

1 See for example Almog (2004), Jervis *et al.* (1985), Lieberman (1994), Mearsheimer (1983), Shimshoni (1988), Stein (1996).
2 The launching of 4,200 rockets during the Second Lebanon War in 2006 caused 54 Israeli casualties and over 12,000 rockets that were launched from the Gaza Strip since 2001 caused only about 20 casualties (Harel 2009).
3 See Bar-Zohar (2006: 401), Cohen (1998: 275), Peres (1995: 166).
4 According to IISS (2009: 249), Israel has up to 200 nuclear warheads.
5 Feickert (2004: 36); see also the Table compiled by Michael Haas and Bernd W. Kubbig in the Appendix. According to the IISS (2009: 249), Israel deployed in 2009 about 100 missiles of both the Jericho-1 and Jericho-2 type.
6 See, for example, the memoirs of Egypt's then-Chief of Staff (El Shazly 1980: 24–9).
7 Recently, for example, Iranian officials claimed that Iran built "hi-tech missile defense systems capable of tracing and intercepting Cruise Missiles" (quoted in *Jerusalem Post*, 7 September 2009).
8 In light of American apprehension of an Israeli strike against Iran's nuclear facilities American experts suggested to extend a formal defence treaty with Israel in order to dissuade it from such a move (Simon 2009: 7).

References

Almog, Doron (2004) 'Cumulative Deterrence and the War on Terrorism', *Parameters*, 34(4): 4–19.
Aronson, Shlomo (2007) 'Defending the Holy Land: A Critical Analysis of Israel's Security and Foreign Policy (Review), by Zeev Maoz' *Shofar: An Interdisciplinary Journal of Jewish Studies*, 25(4): 199–201.

Baram, Amatzia (1992) 'Israeli Deterrence, Iraqi Response', *Orbis*, 36(3): 397–410.

Bar-Joseph, Uri (1982) 'The Hidden Debate: The Formation of Nuclear Doctrines in the Middle East', *Journal of Strategic Studies*, 5(2): 205–35.

Bar-Joseph, Uri (1998) 'Variations on a Theme: The Conceptualization of Deterrence in Israeli Strategic Thinking', *Security Studies*, 7(3): 145–81.

Bar-Zohar, Michael (1978) *Ben Gurion: A Political Biography*, Hebrew, Tel Aviv: Am Oved.

Bar-Zohar, Michael (2006) *Phoenix: Shimon Peres: The Biography*, Hebrew, Tel Aviv: Yedioth Ahronoth.

Beaton, Leonard and Maddox, John (1962) *The Spread of Nuclear Weapons*, London: Chatto & Windus.

Bundy, McGeorge (1988) *Danger and Survival: Choices about the Bomb in the First Fifty Years*, New York: Random House.

Cohen, Avner (1998) *Israel and the Bomb*, New York: Columbia University Press.

Cohen, Avner (2005) *The Last Taboo*, Hebrew, Or Yehuda: Kineret.

Cohen, Avner (2010) *The Worst-Kept Secret: Israel's Bargain with the Bomb*, New York: Columbia University Press.

Cohen, Avner and Burr, William (2006) 'Israel Crosses the Threshold', *Bulletin of the Atomic Scientists*, 62(3): 22–30.

Cordesman, Anthony H. (2007) *The Israeli 'Nuclear Reactor Strike' and Syrian Weapons of Mass Destruction: A Background Analysis*, Washington, DC: Center for Strategic and International Studies. Online, available at: http://csis.org/files/media/csis/pubs/071024_syriannucl_weapcontext.pdf (25 March 2011).

El Shazly, Saad (1980) *The Crossing of the Suez*, San Francisco, CA: American Mideast Research.

El-Gamasy, Mohamed Abdel Ghani (1993) *The October War: Memoirs of Field Marshal El-Gamasy of Egypt*, Cairo: American University in Cairo Press.

FAS/Federation of American Scientists (2010) *Popeye Turbo*, Washington, DC: FAS. Online, available at: www.fas.org/nuke/guide/israel/missile/popeye-t.htm (25 March 2011).

Feickert, Andrew (2004) *Missile Survey: Ballistic and Cruise Missiles of Foreign Countries*, Washington, DC: Congressional Research Service.

Feldman, Shai (1991) 'Israeli Deterrence: The Test of the Gulf War', in Joseph Alpher (ed.) *War in the Gulf: Implications for Israel*, Hebrew, Tel Aviv: Jaffee Center for Strategic Studies.

Frantz, Douglas (2003) 'Israel's Arsenal is Point of Contention', *Los Angeles Times*, 12 October.

Gluska, Ami (2004) *Eshkol, Give the Order!* Hebrew, Tel Aviv: Maarachot.

Gordon, Shmuel (1991) 'First Lessons from the War in the Gulf', in Nathan Shaham (ed.) *War in the Gulf*, Hebrew, Tel Aviv: Sifriat Poalim, 93–113.

Grundman, Moshe (1998) *Cautious Assessment: Writings by Aharon Yariv*, Hebrew, Tel Aviv: Maarachot.

Hamza, Khidhir (1998) 'Inside Saddam's Secret Nuclear Program', *Bulletin of the Atomic Scientists*, 54(5): 26–33.

Harel, Amos (2009) 'A Slight Hit in the Rear', Hebrew, *Haaretz*, 20 November.

Inbar, Efraim and Sandler, Shmuel (1993) 'Israel's Deterrence Ability: A New Assessment', Hebrew, *Maarachot*, 328: 9–19.

IISS/International Institute for Strategic Studies (2009) *The Military Balance 2009*, London: Routledge.

Jervis, Robert, Lebow, Richard N., and Stein, Janice G. (1985) *Psychology and Deterrence*, Baltimore, MD: Johns Hopkins University Press.

Kaspit, Ben (1997) 'Open Questions: The Iranian Threat and the Iraqi Danger', *Maariv*, 7 November.

Lennox, Duncan (ed.) (2007) *Jane's Strategic Weapons Systems*, No. 46, Bracknell: IHS Global Limited.

Lieberman, Elli (1994) 'The Rational Deterrence Theory Debate: Is the Dependant Variable Elusive?' *Security Studies*, 3(3): 384–427.

Maoz, Zeev (2003) 'The Mixed Blessing of Israel's Nuclear Policy', *International Security*, 28(2): 44–77.

Mearsheimer, John J. (1983) *Conventional Deterrence*, Ithaca, NY: Cornell University Press.

NTI/Nuclear Threat Initiative (2010) 'Israel Profile: Missile Overview'. Online, available at: www.nti.org/e_research/profiles/Israel/Missile/index.html (25 March 2011).

Obeidi, Mahdi and Pitzer, Kurt (2005) *The Bomb in My Garden: The Secrets of Saddam's Nuclear Mastermind*, Hoboken, NJ: Wiley.

Oren, Michael (2004) *Six Days of War: June 1967 and the Making of the Modern Middle East*, Hebrew, Or Yehuda: Kineret.

Parker, Richard B. (ed.) (2001) *The October War: A Retrospective*, Gainesville, FL: University of Florida Press.

PBS/Public Broadcasting Service (1996) 'Interview with General Wafic Al Samarrai, Head of Iraqi Military Intelligence'. Online, available at: www.pbs.org/wgbh/pages/frontline/gulf/oral/samarrai/1.html (25 March 2011).

Peres, Shimon (1995) *Battling for Peace*, London: Weidenfeld and Nicholson.

Shalom, Zaki (2005) *Israel's Nuclear Option: Behind the Scenes Diplomacy between Dimona and Washington*, Eastbourne: Sussex Academic Press.

Shapir, Yiftah S. (2001) 'Non-Conventional Solutions for Non-Conventional Dilemmas?' *Journal of Strategic Studies*, 24(2): 147–75.

Shimshoni, Jonathan (1988) *Israel and Conventional Deterrence: Border Warfare from 1953 to 1970*, Ithaca, NY: Cornell University Press.

Simon, Steven (2009) *An Israeli Strike on Iran*, New York: Council on Foreign Relations.

Sobelman, Daniel (2004) *New Rules of the Game: Israel and the Hizbollah after the Withdrawal from Lebanon*, Tel Aviv: Jaffee Center for Strategic Studies.

Solingen, Etel (2007) *Nuclear Logics: Contrasting Paths in East Asia and the Middle East*, Princeton, NJ: Princeton University Press.

Stein, Janice G. (1996) 'Deterrence and Learning in an Enduring Rivalry: Egypt and Israel, 1948–73', *Security Studies*, 6(1): 104–52.

Tatchell, Peter (2008) 'Ahmadinejad Accepts Israel's Right to Exist', *Guardian*, 29 September.

Toukan, Abdullah and Cordesman, Anthony H. (2009) *Study on a Possible Israeli Strike on Iran's Nuclear Development Facilities*, Washington, DC: Center for Strategic and International Studies.

6 Advancing the control of weapons of mass destruction

An incremental approach

Gawdat Bahgat, Avner Cohen, Sven-Eric Fikenscher, Giorgio Franceschini, Bernd W. Kubbig, and Patricia Lewis

The WMD problématique

As emphasized in the Introduction to this volume, conventionally tipped missiles deserve to be studied as a research and policy area in their own right. By the same token many Arab states and Iran consider missile systems as a viable countermeasure against Israel's nuclear and conventional supremacy. Therefore the reduction and elimination of missiles cannot be achieved unless parallel steps are taken to reduce and eventually eliminate weapons of mass destruction (WMD), i.e. nuclear (N-), biological (B-), and chemical (C-)weapons. They cannot only be delivered by missiles, but their inclusion in a future arms control agenda would allow for a trade-off between the proponents of missile and nuclear disarmament. This explains why this article focuses primarily but not exclusively on the nuclear issue. However, including biological and chemical weapons in the discussion has the advantage of avoiding a too narrow concentration on the nuclear question, which proved to be unproductive in the past, the Arms Control and Regional Security talks in the early 1990s being a case in point (see the chapter of Emily B. Landau and Dalia Dassa Kaye). The ultimate goal of a Weapons of Mass Destruction Free Zone (WMDFZ) notwithstanding, our emphasis is on a more moderate step-by-step process which parallels our suggested incremental path in the missile area. This gradual design recognizes the gap between the lacklustre performance of states on the nuclear question and the perennial endorsement of a WMDFZ that so far has remained confined to the level of concepts and rhetoric. However, if initiated and conducted successfully, the Middle East Conference on the establishment of a WMDFZ, including its delivery vehicles, which the Nuclear Non-Profileration Treaty (NPT) Review gathering envisaged in May 2010, could make a difference in this respect.

The building blocks of the incremental process include: first, advancing the ratification process of the *Comprehensive Test Ban Treaty* (CTBT) which on the international level has become a symbol for taking the commitment to disarmament seriously; second, broadening the support of the

Additional Protocol which expands the scope of inspections and therefore strengthens the Nuclear Non-Proliferation Treaty; third, negotiating a regional *Fissile Material Cut-off Treaty* (FMCT) that could aim at largely prohibiting the production of fissile material for nuclear explosive purposes, leaving existing stockpiles untouched; fourth, reaching *(regional) fuel cycle arrangements* which include different schemes (WMDC 2006: 73–7), among them setting up an international centre for producing nuclear fuel for countries that accept non-proliferation and verification norms and might even forego the option of uranium enrichment. At the time of this writing such a plan is – on the global level – realized by the International Atomic Energy Agency (IAEA); Kuwait and the United Arab Emirates have financially supported the envisaged centre. Fifth, the negotiation and implementation of a zone free from B- and C-weapons.

In our multi-phase approach the first step turns on the postures and positions of the three major players in the region – Egypt, Israel, and Iran – especially with respect to nuclear weapons.[1] Then we will explore the opportunities for decision-makers in the three countries to move towards implementing the individual building blocks. In this chapter, the Introduction, the Iran profile, and the conclusions were co-authored by Sven-Eric Fikenscher, Giorgio Franceschini, and Bernd W. Kubbig. Gawdat Bahgat contributed the section on Egypt, while the profile of Israel was written by Avner Cohen and Patricia Lewis.

Egypt's weapons of mass destruction posture

Cairo had, and probably still has, multiple reasons to produce and develop WMD. In 1948, 1956, 1967, and 1973 Egypt went to war with Israel, which had successfully manufactured nuclear weapons in the late 1960s/early 1970s. The most populous country in the Middle East, with approximately 80 million people, and the cultural centre of the Arab world, Egypt claims, with strong justification, a prominent and leading role in the region. Israel's nuclear weapons and Iran's aspirations pose serious challenges to its prestige. Finally, compared with most other regional powers, it has a reasonably well-developed human and technological infrastructure. Thus, Cairo has arguably both the motives and the means to develop WMD. Indeed, in the late 1950s and early 1960s (leading up to the 1967 War with Israel) Egypt, under President Gamal Abdel Nasser, pursued an active strategy aimed at acquiring nuclear weapons to counter Israeli ambitions[2] and to expand and strengthen its status as the leader of the Arab world.

However, the country's nuclear programme was frozen after the defeat in the 1967 War, since the conflict had deprived it of vital financial resources. Cairo's signing of the NPT in 1968 signalled a significant shift in its nuclear strategy. Thirteen years later (February 1981), Egypt ratified the treaty. This decision was driven by the realization that joining the accord was the only way of obtaining the needed technology to launch Cairo's civilian nuclear power programme. Hence, instead of seeking to match Israel in this area, Egypt has sought to build a strong conventional military power and to develop (undeclared) chemical weapons capabilities. Additionally, Cairo has employed diplomatic means to pressure Israel to join the NPT, to give up its nuclear weapons, and to place its facilities under the full scope of IAEA safeguards. This provides the backdrop for understanding Egypt's proposals for the establishment of a Nuclear Weapon Free Zone (NWFZ) or a WMDFZ. Cairo's commitment to the first concept dates back to the early 1970s when the NPT was opened for signature and it has remained consistent since then. In 1974 Egypt – together with Iran – introduced the first proposal to denuclearize the Middle East to the United Nations General Assembly.

Israel has been largely unresponsive to these proposals and produced its own version in 1980. After intense diplomatic bargaining led by Cairo, Israel decided to drop its resolution and to join the consensus on the Egyptian draft (Karem 1988: 106–7) which since then has been adopted unanimously each year without a vote. Frustrated with the lack of progress on the creation of an NWFZ and the continuing military imbalance in the region, President Hosni Mubarak introduced a new initiative in 1990 to broaden the concept of an NWFZ to include B- and C-weapons. Mubarak's proposal to deal with *all* WMD was largely a response to the potential regional implications of Saddam Hussein's threat to burn half of Israel with chemical weapons if attacked. In addition to calling for the prohibition of B-, C-, and N-weapons, the initiative urged the implementation of verification measures to ensure full compliance by all parties.

The intense confrontation over Iran's nuclear programme since the early 2000s has further fuelled Egyptian efforts to address the proliferation issue and prompted Cairo to press for a WMDFZ within the framework of the NPT Review Conference in 2010. Tehran's nuclear project is important in an indirect way. The Islamic Republic does not pose a national security threat to Egypt and, unlike Israel, Cairo and Tehran have never gone to war against each other nor has Iran ever occupied any part of Egyptian territory. Rather, Cairo has a twofold concern about Tehran's nuclear programme. First, a nuclear Iran might become more assertive in pursuing its regional agenda and supporting Hezbollah in Lebanon and Hamas in Gaza. Along with the kings of Saudi Arabia and Jordan, then-President Mubarak has echoed his concern about the rise of Shiism in the Middle East/Gulf and was worried about the close relationship of the Islamic Republic and Hamas to the Muslim Brotherhood, the largest and strongest Egyptian political group. Second, Tehran's nuclear project underscores Cairo's inferior status with respect to nuclear arms and technology in general. Egypt has never sat comfortably with the fact that Israel has a nuclear monopoly in the region. A nuclear Iran would further highlight its non-conventional inferiority.

Tehran's nuclear ambitions, however, are not likely to prompt Cairo – even after the 'Arab Spring' – to develop its own nuclear weapon programme. This is due to the fact that Iran does not pose a sufficient security threat to trigger such a reaction. Having managed to live with a nuclear Israel for four decades, Egypt will probably – although uncomfortably – be able to adjust to a nuclear Islamic Republic (if it materializes). Cairo's leaders are likely to keep their diplomatic efforts to establish a WMDFZ in the entire Middle East. However, Egyptian officials have struggled to articulate a coherent stance on Iran's nuclear programme. Former President Mubarak once rejected a potential expansion of the US nuclear umbrella to include the Gulf states and his own country. He stressed that such a concept "would imply accepting foreign troops and experts on our land and an implicit acceptance of regional nuclear powers. We do not accept that" (quoted in Mahdy 2010). As to whether Cairo would go nuclear if Iran does, Nabil Fahmy, the former Ambassador to the United States (1999–2008), has argued that such a quid pro quo is not the only option. Egypt, Fahmy stressed, can pursue its national security interests by "taking political measures, balancing with other weapons systems, by limiting commitments to agreements and by trying to have a … response" (quoted in Crail and Pomper 2008). Since the Islamic Republic does not have nuclear weapon capabilities, these options are only speculative.

Nevertheless, three concrete elements can be identified in Cairo's position. First, Egyptian officials have repeatedly confirmed that like other signatories of the NPT, Tehran has the right to benefit from the peaceful use of nuclear energy provided it can prove its project is exclusively civilian. Second, Iran's nuclear dossier needs to be dealt with politically, and not militarily. Cairo believes that military action against the Islamic Republic would further destabilize the region. Third, in order to avoid double standards, a comprehensive approach to nuclear disarmament should be adopted that includes both Iran and Israel and leads to the establishment of a WMDFZ in the entire Middle East/Gulf. Regarding the stability of Egypt's position on nuclear arms control, there is some uncertainty surrounding Mubarak's political successors. Will future governments adopt a different stance on nuclear weapons? Historical cases suggest that new leaders seek to gain popularity and legitimacy by addressing issues of national prestige. Such a possibility cannot be ruled out. However, by vigorously pursuing the goal of a WMDFZ, Cairo has gained an international reputation which it would lose by abandoning this agenda.

To sum up, Egypt does not face an external security threat significant enough to trigger a nuclear weapon programme since making peace with Israel in the late 1970s. Cairo's relations with Israel and Iran can be characterized as rivalries, but a military confrontation seems remote and the financial price for going nuclear would be very high. Against this backdrop it is highly unlikely that the leaders in post-Mubarak Egypt will seek to develop nuclear weapons. Cairo has so far refused to ratify the

Additional Protocol as well as the CTBT and opposes a FMCT as well as fuel bank agreements, since all the proposals fail to address Israel's nuclear superiority. In addition, a regional FMCT would, from Egypt's point of view, encourage official recognition of Israel as a nuclear power. Its overall goal continues to be to pressure its neighbour into giving up its nuclear arsenal and join the NPT. Cairo's officials claim that Israel's refusal to take such action has further fuelled regional tensions and instability. In 2008, the Arab League even vowed that its members would withdraw from the NPT if Israel officially announced it had nuclear weapons (*Haaretz*, 5 March 2008).

Meanwhile, like other Middle Eastern states, Egypt has embarked on an ambitious project to expand its civilian nuclear power infrastructure. In 2006 Mubarak announced that several civilian nuclear plants for electricity generation would be built. Cairo's pursuit of nuclear power raises a strategic dilemma regarding enrichment and reprocessing. With the exception of Iran, which is currently developing such capabilities, and Israel, which already has these technologies, no other state in the Middle East has the capacity for enrichment and reprocessing. Egypt's officials have argued that the NPT does not prohibit nuclear activities, including enrichment, as long as these activities remain peaceful and under the supervision of the IAEA. Against this backdrop, the international community has increasingly promoted a multilateral fuel cycle arrangement that would oblige participating states to forego the option of enrichment. This is one of the reasons why Egypt rejects it. Israeli nuclear capabilities also explain why Cairo, the most vocal promoter of a zone free of WMD, remains a holdout to the Biological and Toxin Weapons Convention (BTWC) and the Chemical Weapons Convention (CWC) and maintains a stance of deliberate ambiguity on biological and chemical arms. Although Egypt acceded to the Geneva Protocol (NTI 2011a) and signed the BTWC, it has refused to ratify the BTWC.

In addition, Israel's stance on the control of B- and C-weapons is also cited as a motive. As Ambassador Sameh Shoukry (2006), then the Permanent Representative to the United Nations explained, "Israel not only sits on an arsenal of nuclear weapons, but has consistently refused to sign the B[T]WC." There is no corroborated open source evidence of any organized Egyptian biological weapons related research activity (NTI 2011b). On the other hand, there is serious concern about Cairo's fulfilment of its obligations under the Geneva Protocol. It is widely believed that the Egyptian armed forces employed bombs and artillery shells filled with phosgene and mustard agents in support of South Yemen against royalist troops in North Yemen during the civil war (1963–67). Cairo also uses its reported but officially unconfirmed chemical weapons capability to pressure Israel for a comprehensive agreement banning all WMD in the region (Shoham 1998: 56). In pursuing this objective, it has refused to sign the CWC.

Israel and the NWFZ in the Middle East: tiptoeing down a "long corridor"[3]

Soon after the signing of the 1979 Peace Treaty with Egypt, Israel joined Cairo and Tehran at the United Nations in supporting a consensus on a non-binding General Assembly resolution calling for the establishment of an NWFZ in the Middle East. Ever since Foreign Minister Yitzhak Shamir first presented Israel's vision of an NWFZ, Israeli representatives have on various occasions and in various ways repeatedly voiced their support for such a long-term goal. However, Israel has always insisted that the negotiation of an NWFZ can only be initiated *after* a fundamental transformation of the Arab–Israeli conflict and therefore in essence only after peace has been established; hence the mantra is: first peace, then the zone. Behind the rhetoric of the zone, Israel thus far seems firmly committed to maintaining its undeclared regional nuclear monopoly, viewing it as an indispensable and fundamental feature of its national security, and relying on its policy of nuclear opacity as the means of keeping discussion of these matters off limits. The nuclear issue, which is viewed as an existential one within Israel, is a matter about which there is such a consensus across the elite political spectrum that no government has yet dared to question it. Far from dismantling its nuclear programme, Israel is not even considering the option.

To understand how the Israeli government thinks about the NWFZ proposal one needs to look at the domestic and external origins of its position. In the mid and late 1970s, in the wake of the 1973 War, some Israeli leaders (primarily, and publicly, former Minister of Defense Moshe Dayan and, less publicly, his successor Shimon Peres) were proposing that Israel should move into an open nuclear deterrence posture (Cohen 2010: 59–87). In response, the opposition camp, (primarily Prime Minister Yitzhak Rabin and Foreign Minister Yigal Allon) wanted to block this option. An Israeli commitment to the concept of an NWFZ seemed to them a way to strengthen the more cautious policy of nuclear opacity. There was also the realistic concern that Israel's declaration that it will 'not be the first to introduce nuclear weapons to the region' (Cohen 1998: 207–15) – which is part and parcel of its doctrine of nuclear opacity – would be insufficient to divert external criticism away from it failing to join the NPT. In the 1970s Israel recognized it was not in a position to simply reject the NPT as irrelevant and inapplicable to its situation, while at the same time relying on the accord to keep the rest of the region free from nuclear weapons. It realized the need to offer some positive stance on nuclear disarmament. Both Rabin and Allon thought that the idea of the NWFZ as a long-term goal would be an appropriate response to both domestic and external pressures on Israel's opacity.

It was under the leadership of Shalheveth Freier, Director General of the Israeli Atomic Energy Commission from 1971–76, that opacity was codified as an official national doctrine. Freier formulated the Israeli vision of a Middle East free of nuclear weapons as a "long corridor" (quoted in Cohen 2010: 42) in time punctuated by stages along which regional states would have to traverse to make it a reality. The ultimate objective of an NWFZ from the start has been linked to Israel's long-standing demand for full and mutual political recognition among all the states of the region. The genius of the NWFZ proposal is that it is a diplomatic instrument which eases the pressure on Israel to sign the NPT, without the requirement to make any concrete commitments. It is indeed a *vision* of a Middle East without nuclear weapons, but one that imposes no requirements whatsoever on Israel unless and until some substantial political requirements regarding Arab belligerency and recognition are met. Until that "long corridor" is travelled, Israel sees no need to accept limits on its freedom of action in the nuclear field.

From the Israeli perspective, it is of utmost importance to treat the nuclear issue in a regional framework. The diplomatic advantages over the universal setting of the NPT seem obvious.

1 It allows Israel to build and present a positive image of itself as a proponent of non-proliferation in contrast to the picture presented by its rejection of the NPT. Not only does the NWFZ present an alternative to the NPT, but it allows Israel to explain why it was compelled to reject the accord (due to the existential threats from its neighbours), without openly acknowledging anything about its nuclear activities.
2 The emphasis on a broad regional vision gives Israel a great deal of political flexibility; its details need to be directly negotiated among all relevant parties over time in order to become a real blueprint rather than a fixed treaty.
3 The vision of the NWFZ, unlike the NPT, means that Israel is able to directly link the nuclear issue with the Arab–Israeli conflict, and in doing so ties the nuclear problem to the central issues of war and peace, recognition, and legitimacy.
4 Perhaps most importantly, the NWFZ vision does not require Israel to respond, either in deed or in words, until there are substantial changes in the political arena. Until they take place, Israel continues with its policy of opacity.
5 Endorsing the vision of an NWFZ allowed Prime Minister Rabin (and subsequently Prime Minister Menachem Begin) to signal that Israel remained committed to opacity. This has meant that the country does not believe in open nuclear deterrence as the key to security.

There is a certain schizophrenia that characterizes Israel's attitude towards nuclear disarmament, in particular its commitment to an NWFZ. While Israel claims to support this vision as one of the conditions for a peaceful Middle East, under its commitment to opacity the country has thus far effectively refused to become engaged in efforts to move incrementally towards the establishment of such a zone. Effectively, as a consequence of opacity, Israel is unable to synchronize the political and the nuclear clocks. So long as the country is firmly committed to the policy of ambiguity, it is resigned to a self-imposed silence which precludes constructive discussions of the elimination of nuclear weapons. However, its stance on incremental steps towards WMD control indicates at least some, albeit very limited, leeway for compromise. Israel, which has consistently remained outside the NPT and therefore is not obliged to adopt full scope safeguards including the Additional Protocol, signed the CTBT in October 1996 and has expressed no substantive reservations about it – joining the treaty does not conflict with Israel's overall nuclear policy. On the contrary, signing the CTBT has often been presented by Israeli officials as one of the central pillars of Israel's current policy on arms control, a policy that highlights increasing cooperation with international bodies and adherence to multilateral agreements, while being consistent with the opacity policy.

With regard to the FMCT, Israel has maintained an ambivalent attitude. It keeps as low a profile as possible within the framework of the Geneva Conference on Disarmament (CD), where negotiations on an FMCT have remained stalled since the late 1990s, assuming that it would be wiser to let others impede the negotiating process. This strategy has generally paid off because other countries have consistently blocked progress on the fissile material cut-off issue in Geneva. In 1998 then-Prime Minister Benjamin Netanyahu, under heavy American pressure, agreed to join the consensus in the CD to begin negotiations; however, there was no commitment from any state to conclude them and indeed the talks fell apart a few months later. In 2006, in an extraordinary departure from decades of understanding and practice, the US tabled a draft mandate and treaty that attempted to define the meaning of fissile material for weapons purposes and without any provision for effective verification. Israel took a strong vocal stand against the American draft. It has never been clear as to why Israel took such a strong stance – and wanted it well understood by other countries both inside and outside the Western Group in the CD that it was opposed to the lax verification proposals. Since then, however, in the face of Iran's nuclear ambitions, Israel has become more open and explicit in its opposition to the FMCT.

The Israeli concerns about the FMCT seem to be both political and technical. On the political front, the main aspect is that such an agreement would constitute a 'slippery slope' towards (premature) nuclear disarmament. Technically, it would be very difficult for Israel to maintain its policy of opacity under an FMCT, especially as the treaty would have to contain provisions for transparency and verification. While Israel could continue to produce tritium (the Dimona reactor is believed to produce tritium via neutron irradiation of lithium-6 targets) as a party to the FMCT – fusion materials would not be banned under the proposed accord – it would have to agree to intrusive verification measures to ensure that the reactor was not being used to produce plutonium. These measures could expose past production, making it impossible to maintain opacity.

The proposals for multinational fuel assurances or international and regional fuel cycle arrangements have not been officially endorsed by Israel, although Gideon Frank, then the Director General of the Israeli Atomic Energy Commission, was a member of the independent international Expert Group on Multilateral Approaches to the Nuclear Fuel Cycle that was commissioned by the Director General of the IAEA. Currently, however, the issue of multinational fuel assurances appears to have fallen victim to the policy of opacity. So, rather than the full support for such measures, although they are assumed to favour Israel, there are only tepid public expressions of interest and no clear official position. Israel has signed but not ratified the CWC and opinion varies widely as to whether it still maintains an operational C-weapons capability. There is controversy over the alleged use of white phosphorous in Gaza in 2009, which is banned from use as a weapon against civilians but is not banned if used solely for the purposes of a smokescreen or illumination. Although Israel has yet to sign and ratify the BTWC, most experts believe that they do not have an offensive B-weapons programme, although they have most likely developed – as have many other countries – an allowable defensive capability (Cohen 2001: 40–1).

Iran's WMD policy between domestic infighting and external pressure

Iran's nuclear programme started in 1957 under the auspices of a global "Atoms for Peace" euphoria.[4] As a staunch ally of the US, Reza Shah Pahlevi benefited from generous transfers of American technology and know-how. Tehran's cooperation with Western states gained momentum in the early 1970s when the Shah announced his plan to install some 20 nuclear power reactors for civil energy production by the end of the century. Although Iran had signed the NPT in 1968 on the very day it was opened for signature, there was permanent suspicion about the Shah's military aspirations. Such worries receded for a while after the Islamic Revolution in 1979, since Ayatollah Khomeini initially seemed to show little interest in nuclear technology. He terminated most cooperation programmes with other states and a substantial number of nuclear experts fled the country in the years after the revolution. But Khomeini gave up his reluctance and restarted nuclear activities in the mid 1980s in view of the devastating experience during the First Gulf War (1980–88) of being attacked with chemical weapons by Iraq and enjoying no solidarity from the international community. Another factor was the American-led invasion of neighbouring Iraq in 1991 which led the regime in Tehran to conclude that (only) nuclear weapons are an adequate means of deterrence (Hashim 1994: 32–4; Taremi 2005: 102).

International doubts regarding a clandestine nuclear programme for achieving a nuclear weapons capability have always persisted and turned into strong suspicions since the disclosure in 2002 of an undeclared underground enrichment facility at Natanz and of a heavy water reactor at Arak; the revelation in 2009 of a secret uranium enrichment site under construction near Qom fuelled those suspicions. Since 2002 Iran's activities have become a matter of great controversy reflecting the strong involvement of foreign actors, and increasingly deep divisions among the heterogeneous elites in Tehran, highlighting the crucial importance of the vacillating domestic power constellations (see below). In its talks with the EU-3 (France, Great Britain, and Germany) from 2003–05, the reformist government of Mohammad Khatami made certain concessions. It suspended the enrichment activities, and it agreed to apply the – signed (December 2003) but not ratified – Additional Protocol. Khatami had won the prior agreement of the Supreme National Security Council (SNSC) which is chaired by the President and composed of the representatives of all institutions in the security apparatus, and – even more importantly – of Supreme Leader Ali Khamenei, the "single most powerful figure" (Ganji 2008: 45) in the entire foreign policy decision-making process,[5] to whom the SNSC reports (Hadian and Hormozi 2011: 30–1).

Yet the suspension of the enrichment activities did not last for long. The newly elected (June 2005) ultra-conservative President Mahmoud Ahmadinejad began enriching uranium again – and enjoyed the support of the Supreme Leader. When the IAEA Board of Governors decided in early 2006 to "report" the Islamic Republic to the UN Security Council, Tehran informed the Vienna control body that it would immediately stop the voluntarily expanded inspection under the Additional Protocol. To justify its nuclear activities, Iran has argued increasingly legalistically pointing to Art. IV of the NPT which expresses the "inalienable right" of all NPT members to use nuclear energy for peaceful purposes and has stressed that it has no intention whatsoever of acquiring a nuclear weapons capability (Kubbig 2006: 6–7). But in practice the Islamic Republic has failed to report a number of sensitive nuclear activities to the IAEA, and even after its failures were exposed in 2002, it has not been able to provide sufficient guarantees that its nuclear programme is for exclusively peaceful purposes. The IAEA reports variously highlight "the possible existence in Iran of past or current undisclosed nuclear related activities involving military related organizations, including activities related to the development of a nuclear payload for a missile" (IAEA 2011: 8).

The resolutions of the UN Security Council with which the international community has responded to Tehran's behaviour prohibit enrichment, reprocessing, and heavy water related activities. These 'smart sanctions' attempt to target the regime as well as individuals and organizations that are behind the nuclear and missile project, mainly by an almost all-encompassing arms trade ban. But punitive measures against Iran by the United States unilaterally and bilaterally by pressuring its allies to curtail trading and banking activities, have turned out to be a real challenge for the Islamic Republic. American decision-makers, aware of the loopholes of their sanctions, decided to sharpen existing measures by targeting the Iranian financial sector and the export of gasoline to the Islamic Republic. The intention is to focus on its dependence on foreign investments and its inability to refine oil – about 40 per cent of Iranian gasoline is imported. Since 2006 financial institutes have been specifically targeted by the U.S. Department of the Treasury, which has built up an impressive financial intelligence division that is secretly putting pressure on foreign banks by contacting and threatening to limit their access to the American market (Kubbig in cooperation with Fikenscher 2007).

In early 2011, U.S. Secretary of State Hillary Clinton stressed for the first time that Tehran's ability to produce a nuclear weapon had been delayed by sanctions (*International Herald Tribune*, 11 January 2011). The week before the departing director of Israel's intelligence service (Mossad) had stated that he believed that the Islamic Republic would not be able to have nuclear capability before 2015 "following a series of failures that had set back its program by several years ... largely because of the Mossad's efforts" (*Haaretz*, 7 January 2011, see also *Haaretz*, 3 June 2011), supposedly computer sabotage ('stuxnet'). Although the Iranian government did not admit that cyber worms had attacked the Natanz centrifuge facility, it has acknowledged that its nuclear plants have been subject to cyber attacks. This may explain the removal of some 1,000 malfunctioning centrifuges (Albright *et al.* 2010) which meanwhile might have been replaced.[6]

A reliable assessment of the effects both of sanctions and obvious sabotage activities as well as targeted killings is at this time of writing not possible, but there can be no doubt that sanctions have negatively affected the overall economy and especially the energy sector. To date what can be said is that the strategy to influence the calculus of the fractionated elite, by splitting it up and thus making it more willing to compromise and cooperate in the nuclear field, has not been successful. On the contrary, at a meeting with six major powers (the five permanent members of the UN Security Council plus Germany) in January 2011, the position of Tehran was less flexible, more demanding, and more assertive than in the past. The Iranian chief representative Saeed Jalili reportedly insisted on preconditions including at least a partial suspension of sanctions before discussing the details of further talks.

More specifically, the Islamic Republic was no longer interested in a fuel swap initially offered in October 2009 as a confidence-building measure to reduce Iran's stock of low-enriched uranium of some 1,600 kg by 75 per cent (1,200 kg, less than required to build a bomb) for a Tehran reactor. Yet meanwhile the Islamic Republic has nearly three times the amount of low-enriched uranium plus about 40 kg of uranium enriched to

19.75 per cent. Jalili reportedly stated that his country was no longer interested in an external fuel source for the reactor because it had found its own source of uranium and could produce the fuel itself (*International Herald Tribune*, 22/23 and 25 January 2011). The hardened Iranian position indicates that even if the United States continues in isolating the Islamic Republic in the economic and energy related sector, this does not necessarily translate into Tehran's decision to comply with international demands concerning its nuclear programme.

What exactly accounts for Iran's nuclear aspirations and its reluctance to cooperate fully with the IAEA? Basically, the interests that have been identified and led to the acceleration of the missile programmes after the First Gulf War (see the article of Bernd W. Kubbig *et al.*) apply even more so to the nuclear field. The same holds true for the multitude of domestic and regional factors explaining Tehran's nuclear ambitions in the 1990s and after the toppling of its arch enemy Saddam Hussein in 2003: the wish to demonstrate the capacity to master a so-called cutting edge technology, and associated with it, "national pride" (Mousavian 2010). The Islamic Republic's nuclear ambitions, too, have been dominated by both its security concerns and regional aspirations:

> For most analysts, Iran faces a real security dilemma in its tense relations with the U.S.... The strong U.S. military presence on Iran's doorstep is certainly a concern.... Iran is said to have regional ambitions and, from a predominant Arab perspective, it is also seen to be entertaining unreasonable claims to the region's political-security agendas hitherto regarded as Arab concerns.
>
> (Ehteshami 2010: 21–3)[7]

In Tehran's nuclear policies there are 'red lines', areas of common ground, and of deep disagreements, long-term strategic objectives as well as short-term tactics which display its erratic policy of cooperation and non-cooperation, threatening overtures, and subsequent talks. It is not always possible to differentiate reliably between all these dimensions. Both continuity and fluctuation are due to the structure of the decision-making processes with its inherent deep divisions, ongoing struggles with a clear tendency in the last years: those ultra-conservatives, led by President Ahmadinejad, whose constituency are the country's poor and uneducated, sidelined the reformists, but – at this point of writing – have lost some influence at the expense of the likewise ultra-conservative clerics, led by the Supreme Leader (*Guardian*, 6 March 2011; *International Herald Tribune*, 23 June 2011). Interestingly Ahmadinejad – unlike the clerics – showed some pragmatism in October 2009 when he was domestically weakened by the green movement which was the major driving force behind the unprecedented uprising in the summer of 2009 and has been crushed now: his acceptance of the above-mentioned nuclear swap with the six states aimed at consolidating his power, whereas the reformers Mir Hossein Moussavi and Mehdi Karroubi, who intended to thwart the President's plans, accused Ahmadinejad of selling out the national interest.

Both with respect to its declaratory policy and its formal membership to the relevant WMD treaties, the position of Iran looks unproblematic: Tehran has solemnly declared that it "considers acquiring, developing, and using WMD as inhumane, immoral, illegal, and [to be] against its very basic principles" (Islamic Republic of Iran 2005: 2). While Iran had ratified both the NPT (1970) and the BTWC (1973) under the rule of Reza Shah Pahlevi, the CWC was signed (1993) and ratified (1997) later. Recent claims on chemical and biological weapon activities within the Islamic Republic cannot be determined and may have been exaggerated (IISS 2011: 95–118).[8]

Regarding the building blocks in the nuclear realm, it is a signatory of the CTBT, but has refused to ratify it for reasons which are largely similar to Egypt's. In addition, Tehran points to the insufficient commitment of nuclear weapon possessors to the goal of nuclear disarmament under Art. VI of the NPT. Even more importantly, since for the Islamic Republic the threat of a military confrontation with nuclear powers, i.e. the United States and Israel, is much more realistic, Iran links its refusal to ratify the CTBT to the current security doctrines of the nuclear weapon states; they do not provide sufficient guarantees for non-nuclear weapon states to be exempt from nuclear attacks in case of an armed conflict (Islamic Republic of Iran 2001) – here the 'no first use' dimension enters the debate. Indeed, neither Israel's nor the United States' nuclear policies provide any guarantee to Tehran in this respect, as the latest U.S. Nuclear Posture Review of 2010 indirectly underscores. It pledges to refrain from threatening or using nuclear weapons only "against non-nuclear weapon states that are party to the NPT and in compliance with their nuclear non-proliferation obligations" (DoD 2010: 46).

The Islamic Republic's change in its attitude towards the actual application and/or ratification of the Additional Protocol will depend on the development of a mutually satisfactory policy with the major powers, especially the United States. The Iranian government has demonstrated in the past that it is not interested in exclusively enriching its fuel abroad; Tehran rejected the Russian offers in this regard. Yet it may be in principle open for regional fuel cycle centres if they do not impinge on its national enrichment capabilities. The position on a regional FMCT remains unclear and so does its explicit attitude towards a WMDFZ. If the rejection of the idea of a sub-regional WMDFZ in the Gulf as proposed by all six members of the GCC is any guide (see the article of Bernd W. Kubbig *et al.*), then in the final analysis regional security arrangements, especially a modus vivendi between Washington and Tehran, become vital.

The future of nuclear negotiations, both with the six powers (on the nuclear programme) and with regional states (on the WMDFZ) will depend to a large degree on the power constellations within Iran. If the ultra-conservative clerics further increase their influence and if the divisions continue, prospects for the ratification of important treaties in the nuclear realm – the CTBT and the Additional Protocol – are dim. The same holds true for the acceptance of a to-be-negotiated FMCT. Yet even for a more moderate Iranian government there is little political room to manoeuvre in the nuclear field. It could exclude the weaponization option, i.e. the designing and building of nuclear devices suited for its missiles, and probably head-off full nuclear capability (actual breakout capability without building the bomb). But any government is likely to retain a partial nuclear capability, i.e. especially some enrichment capability (maybe reduced to 3,300 centrifuges at the Natanz facility) for face-saving reasons. The zero enrichment option is not on the table for Iran.

Conclusions

Eliminating WMD from the Middle East remains an arduous task and requires the concerted will of all countries in the region. We take the view that the current impasse, caused mainly by Israel's de facto nuclear status and Iran's nuclear challenge, is detrimental for all parties in the Middle East/Gulf and represents a non-sustainable – and potentially explosive – situation. Without providing a blueprint for negotiators and recognizing that a WMDFZ is the ultimate and maximum objective, we suggest several pragmatic and practical steps in that direction. First, it is imperative that Egypt, Israel, and Iran ratify the CTBT. This would not only give momentum to the accord which has been in limbo for a number of years, but would also increase the security of all countries in the region. Israel could gain from a CTBT ratification in two ways: on the one hand, it could show its commitment to confidence-building by demonstrating that its well-known reservations towards the NPT do not extend to all multilateral arms control treaties. On the other hand, such a move might reduce Iran's and Egypt's reservations and bring them closer to ratification. Thus, Israel could facilitate the acceptance of a further non-proliferation and disarmament commitment after the NPT by two key countries in the region.

In order to finally overcome the objections of Cairo and Tehran, additional measures of confidence-building would be instrumental: if the nuclear weapon states were in principle open to provide unambiguous negative security guarantees to all non-nuclear weapon states in the Middle East, a request which was put forward by Iran in the context of the CTBT, this would be a big step forward. Once a confidence-building process, finally including such negative security guarantees, has been initiated, ratifying the CTBT should not constitute an insurmountable obstacle for either Egypt or the Islamic Republic, since the accord does not include new restrictions for non-nuclear weapon countries of the NPT. Regarding

Iran the six powers would probably also have to rethink the punitive elements (sanctions) of their overall approach and offer economic incentives in exchange for Tehran's decision to ratify the accord. By doing so Iran could demonstrate good faith in the peaceful purpose of its contested nuclear programme and contribute to realizing its declared policy goal of an NWFZ in the Middle East/Gulf. The strategic advantage of a Nuclear Test Free Zone (Goldschmidt 2010; see also Lumpe 1996) must be balanced against an objection raised by Israel (and also by Tehran): the treaty requires Israel to be part of the Middle East and South Asia working group within the CTBT Organization, which comprises a number of countries that do not have diplomatic relations with Israel and sometimes openly question its right to exist, among them the Islamic Republic of Iran. As both states have also commonly joined other multilateral institutions pragmatic ways can certainly be found to solve the problem, and as with most issues raised within the working group, these are rather technical than political.

Second, significantly advancing the control of WMD in the region requires Israel to come clear about its nuclear weapons at some point, i.e. to declare openly its possession of nuclear weapons. Transparency is a prerequisite for a future FMCT, for a responsible Israeli nuclear declaratory policy, and for a number of confidence-building measures on the way to a zone free of nuclear weapons. "Bringing Israel's Bomb Out of the Basement" (Cohen and Miller 2010) requires a courageous rethinking by the elites both in its own country and in the Arab League, as the country profiles of Israel and Egypt have shown. The former must abandon its traditional nuclear policy while the latter must recall its threat to collectively withdraw from the NPT should Israel admit to having a nuclear weapons arsenal (*Haaretz*, 5 March 2008). Israel could express its commitment to pursue negotiations in good faith on serious steps towards nuclear disarmament and reassure countries in the region that the sole purpose of its arsenal is to deter a nuclear attack by issuing a 'no first use' pledge. These highly sensitive issues can only be carried out after intense consultations between Israel and the Arab League. Needless to say that all this will have to be done in a way that is compatible with their security.

Third, these steps might pave the way for improving the overall level of transparency regarding the regional nuclear programmes and help to build confidence on this issue. This entails different obligations for Israel and the other states in the Middle East. Whereas for Israel it is not clear how the transparency issue might play out in detail, for all other countries the NPT provides a straightforward option: all nuclear activities should be monitored by the IAEA through truly twenty-first century standards of verification as codified in the Additional Protocol. If Tehran were to take such a step it would enhance the credibility of the Iranians and therefore mitigate the nuclear crisis. Accepting the Additional Protocol would probably be even more important than the suspension of enrichment activities, since it would serve a double purpose. On the one hand, it would show the good faith of the Islamic Republic and send a clear signal to the international community that it has nothing to hide. On the other, it could allow the IAEA, the Security Council, and the international community to come to trust the Iranian nuclear programme in the longer term and eventually lay the foundation for some technological cooperation. Ratifying the Additional Protocol is not an unreasonable demand, as the Islamic Republic was already willing to accept it on a voluntary basis from 2003 to 2006. It is therefore mainly a matter of political will and ultimately an issue of Iranian domestic policy. Similar considerations also hold for Egypt, a country embarking on an ambitious nuclear energy programme and going through a political transformation. In order to prevent the next nuclear crisis, Cairo would be well advised to ratify the Additional Protocol soon. If the domestic situation does not allow a formal ratification, the governments of both states could always opt for a voluntary implementation of the Additional Protocol as an interim step.

Fourth, confidence in the peaceful nature of nuclear programmes would be enhanced, if all countries renounced the pursuit of sensitive nuclear activities on their territory. The current discussion on multilateral fuel banks shows a promising way forward on how to pursue peaceful nuclear energy in a region characterized by mistrust and instability. A Middle Eastern fuel bank could serve multiple purposes: it would provide an incentive to abandon (or not even to begin) domestic attempts to produce nuclear fuel by states in the region and allow commercial nuclear energy programmes to be launched quicker. In the Iranian case, it would diffuse most of the tensions around its nuclear projects, if Tehran were to stop – or at least curtail – its enrichment activities and instead purchase fuel from a regional fuel bank. Unrealistic as it may seem presently, such a strategy (assured fuel supply in exchange for cuts in the enrichment programme) cannot be ruled out in principle, since most domestic players in the Islamic Republic share a common interest in this energy source. If offered attractive and serious technological cooperation from states outside the region, the Iranian reformists and pragmatists may be willing to compromise on sensitive nuclear issues. Hence, a regional fuel bank would help both the Iranian nuclear energy programme and those countries in the region that are concerned about Tehran's nuclear aspirations.

Fifth, concerning the BTWC and the CWC, which bans the development, production, stockpiling, and other activities of certain agents or toxins, Egypt and Israel could consider the ratification of at least one convention. If their reservations are rather political than strategic they could be induced to ratify the regimes once sufficient progress has been made in confidence-building (Israel) or towards the goal of a WMDFZ (Egypt). In case the strategic reservations of both countries and not the political ones are dominating, they could explore whether they provide a credible 'no first use' pledge for all types of WMD (Friedman *et al.* 2011). In sum, the above-mentioned building blocks, incorporating confidence-building and arms controls measures would be formidable steps on the long way to the elimination of all WMD in the Middle East/Gulf.

Notes

1 We have not included Syria in our country related analysis because the investigations regarding its possible nuclear activities did not lead to clear results at this writing. The context is the following: in early September 2007 the Israeli Air Force demolished a nuclear reactor suitable for producing plutonium, built at al-Kibar near the Euphrates river with North Korean technology and staff. The IAEA in Vienna demanded that Damascus provides information about the site, but Syria stalled, and in the following months it replaced the soil at the location. Remains of materials and equipment were transported to undisclosed venues. Only in June 2008 did Damascus allow IAEA inspectors to visit the site, where they were able to obtain soil samples and found traces of uranium. Since then the regime has interrupted its cooperation with the Vienna control body (*Süddeutsche Zeitung*, 2 December 2010).
2 In December 1960 Israel's Prime Minister David Ben Gurion announced that his country was constructing a nuclear research reactor at Dimona.
3 For more details, see Cohen (1998, 2010).
4 For concise overviews of Iran's nuclear activities, see Albright and Stricker (2010), Chubin (2010), Solingen (2007: 164–86), IISS (2011: 7–46). A comprehensive history of the Islamic Republic's nuclear activities which adequately takes the Iranian perspective into account has yet to be written.
5 See also Hadian and Hormozi (2011).
6 IAEA Director General Yukiya Amano responded to the question of how badly Iran's centrifuge programme was affected by the 'stuxnet' cyber worm from 2009 as follows: "Iran is somehow producing uranium enriched to 3.5 percent and 20 percent. They are producing it steadily, constantly" (quoted in *Washington Post*, 14 February 2011).
7 See also the literature mentioned in note 8 in the chapter of Bernd W. Kubbig *et al.*
8 See also Eriksen and Holøien (2010: 286, Table 1, with the relevant literature).

References

Albright, David and Stricker, Andrea (2010) 'Iran's Nuclear Program', in Robin Wright (ed.) *The Iran Primer: Power, Politics, and U.S. Policy*, Washington, DC: United States Institute of Peace, 77–81.

Albright, David, Brannan, Paul, and Walrond, Christina (2010) *Did Stuxnet Take Out 1,000 Centrifuges at the Natanz Enrichment Plant? Preliminary Assessment*, Washington, DC: Institute for Science and International Security.

Chubin, Shahram (2010) 'The Politics of Iran's Nuclear Program', in Robin Wright (ed.) *The Iran Primer: Power, Politics, and U.S. Policy*, Washington, DC: United States Institute of Peace, 82–5.

Cohen, Avner (1998) *Israel and the Bomb*, New York: Columbia University Press.

Cohen, Avner (2001) 'Israel and Chemical/Biological Weapons: History, Deterrence, and Arms Control', *Nonproliferation Review*, 8(3): 27–53.

Cohen, Avner (2010) *The Worst-Kept Secret: Israel's Bargain with the Bomb*, New York: Columbia University Press.

Cohen, Avner and Miller, Marvin (2010) 'Bringing Israel's Bomb out of the Basement', *New York Times*, 25 August.

Crail, Peter and Pomper, Miles A. (2008) 'Interview with Nabil Fahmy, Egyptian Ambassador to the United States', *Arms Control Today*, 38(9). Online, available at: www.armscontrol.org/node/3303/print (25 March 2011).

DoD/U.S. Department of Defense (2010) *Nuclear Posture Review Report*, Washington, DC: DoD.

Ehteshami, Anoushiravan (2010) 'Iranian Perspectives on the Global Elimination of Nuclear Weapons', *Palestine-Israel Journal of Politics, Economics and Culture*, 16(3/4): 20–9.

Eriksen, Sara Kristina and Holøien, Linda Mari (2010) 'From Proliferation to Peace: Establishing a WMD-Free Zone in the Middle East', *Nonproliferation Review*, 17(2): 281–99.

Friedman, David, Landau, Emily B., Asculai, Ephraim, Malz-Ginzburg, Tamar, and Evron, Yair (2011) 'WMD No-First-Use in the Middle East: A Way to Move Forward in 2012?' *Bulletin of the Atomic Scientists*, 7 February. Online, available at: www.thebulletin.org/web-edition/op-eds/wmd-no-first-use-the-middle-east-way-to-move-forward-2012 (25 March 2011).

Ganji, Akbar (2008) 'The Latter-Day Sultan: Power and Politics in Iran', *Foreign Affairs*, 87(6): 45–66.

Goldschmidt, Pierre (2010) *Let's Start with a Nuclear-Test-Free Zone in the Middle East*, Washington, DC: Carnegie Endowment for International Peace. Online, available at: www.carnegieendowment.org/publications/?fa=view&id=40711 (25 March 2011).

Hadian, Naser and Hormozi, Shani (2011) 'Iran's New Security Environment Imperatives: Counter Containment or Engagement with the US', *Iranian Review of Foreign Affairs*, 1(4): 13–55.

Hashim, Ahmed Salah (1994) *Iranian National Security Policies under the Islamic Republic: New Defense Thinking and Growing Military Capabilities*, Washington, DC: Henry L. Stimson Center.

IAEA/International Atomic Energy Agency Board of Governors (2011) 'Implementation of the NPT Safeguards Agreement and Relevant Provisions of the Security Council Resolutions in the Islamic Republic of Iran', Report by the Director-General, GOV/2011/7, Vienna: IAEA.

IISS/International Institute for Strategic Studies (2011) *IISS Strategic Dossier: Iran's Nuclear, Biological and Chemical Capabilities: A Net Assessment*, London: IISS.

Islamic Republic of Iran (2001) 'Statement of H.E. Dr. M. Javad Zarif at the 2001 Conference to Facilitate the Entry into Force of the CTBT', 11 November. Online, available at: www.un.org/webcast/ctbt/statements/iranE.htm (25 March 2011).

Islamic Republic of Iran (2005) 'Report of the Islamic Republic of Iran to the Security Council Committee Established Pursuant to Resolution 1540 (2004) on the Implementation of Security Council Resolution 1540 (2004)', S/AC.44/2004/(02)/105. Online, available at: www.nti.org/db/1540/pdfs/1540_iran_20050228.pdf (15 March 2011).

Karem, Mahmoud (1988) *A Nuclear-Weapon-Free Zone in the Middle East: Problems and Prospects*, Westport, CT: Greenwood Press.

Kubbig, Bernd W. (2006) 'Iran and the Nuclear Non-Proliferation Treaty', Briefing Paper, EP-ExPol-B-2005–37, Brussels: European Parliament. Online, available at: www.pedz.uni-mannheim.de/daten/edz-ma/ep/06/pe381.389-en.pdf (25 March 2011).

Kubbig, Bernd W. in cooperation with Sven-Eric Fikenscher (2007) *Internationale Sanktionen gegen den Iran: Erfolgsbedingungen und Effektivität*, Frankfurt am Main: Hessische Stiftung Friedens- und Konfliktforschung/Peace Research Institute Frankfurt.

Lumpe, Lora (1996) 'A Flight Test Ban as a Tool for Curbing Ballistic Missile Proliferation', in Peter Hayes (ed.) *Space Power Interests*, Boulder, CO: Westview Press, 146–82.

Mahdy, Fareed (2010) 'Egypt Rejects U.S. Nuclear Umbrella', International Press Service. Online, available at: http://ipsnews.net/print.asp?idnews=48156 (15 February 2011).

Mousavian, Hossein (2010) 'Ein Plädoyer für Verhandlungen', *Frankfurter Allgemeine Zeitung*, 4 December.

NTI/Nuclear Threat Initiative (2011a) 'Chemical Overview'. Online, available at: www.nti.org/e_research/profiles/Egypt/Chemical/index.html (25 March 2011).

NTI/Nuclear Threat Initiative (2011b) 'Egypt Profile'. Online, available at: www.nti.org/e_research/profiles/Egypt/index.html (25 March 2011).

Shoham, Dany (1998) 'Chemical and Biological Weapons in Egypt', *Nonproliferation Review*, 5(3): 48–58.

Shoukry, Sameh (2006) 'Statement of Ambassador Sameh Shoukry, Permanent Representative of Egypt to the United Nations', Sixth Review Conference of the Biological Weapons Convention, Geneva: United Nations.

Solingen, Etel (2007) *Nuclear Logics: Contrasting Paths in East Asia and the Middle East*, Princeton, NJ: Princeton University Press.

Taremi, Kamran (2005) 'Beyond the Axis of Evil: Ballistic Missiles in Iran's Military Thinking', *Security Dialogue*, 36(1): 93–108.

WMDC/Weapons of Mass Destruction Commission (2006) *Weapons of Terror: Freeing the World of Nuclear, Biological and Chemical Arms*, Stockholm: WMDC.

7 From (potential) spoilers to guarantors?

The role of external powers

Sven-Eric Fikenscher, Sameh Aboul-Enein,[1]
Gülden Ayman, Anton Khlopkov, and Hui Zhang

Extra-regional actors and missile disarmament in the Middle East

In 1987 mounting concerns – mainly of Western states – about ongoing proliferation activities led to the establishment of the Missile Technology Control Regime (MTCR). The MTCR has established some supply-side focused rules. It prohibits the export of missiles able to carry a 500 kg payload over 300 km, while largely ignoring the demand-side, i.e. the interests and motivations behind missile projects. The second international agreement dealing with the dangers posed by missiles, the Hague Code of Conduct Against Ballistic Missile Proliferation, has not rectified this shortcoming. Nevertheless, real progress in non-proliferation has been made especially in the years following the establishment of the MTCR. In the MTCR's first decade, Argentina, Brazil, Iraq, Libya, South Africa, South Korea, Syria, and Taiwan were thwarted from advancing their missile agendas. Since then, however, India, Iran, North Korea, and Pakistan have launched medium-range missiles and several other states have expanded their research activities (Mistry 2003). So obviously outlawing the export of delivery systems can, though in many cases not sufficiently, curb their spread.

In the Middle East/Gulf, a region that has generally remained outside the MTCR, proliferation has largely depended on extra-regional supply. Egypt, Iraq, Yemen, Syria, and Iran mainly imported (Soviet) Scuds (Frolov 2007: 32). North Korea as well as China have also been highly involved in the transfer of missile technology to the Middle East, Pyongyang being the major supplier of Iran and Beijing being one of the major sources for the Saudi programme. The United States provided numerous anti-air and anti-missile capabilities to its allies. With regard to the aims of the present volume, namely to explore the idea of (complete) missile disarmament, this history testifies to the importance of examining the role of proliferators and analysing how they might be convinced to stop their deliveries. But the overall question regarding the role of external powers is an even more complicated one. The regional activities of the major proliferating countries that are feeding the arms race do not stop at the transfer of weapon systems. The US military presence and its willingness to use force or the threat of force against its regional enemies, thereby creating demand for such weapons, is a case in point. As a result, we analyse all relevant policies of external powers that (might) undermine regional disarmament in the missile sector and ask about the conditions under which they would support missile control and ultimately a Missile Free Zone (MFZ) so as to change their role from (potential) spoilers to that of guarantors of an MFZ. In this regard we follow an all-encompassing missile definition (see the Introduction by Bernd W. Kubbig) and will finally – based on the analysis of the external powers' positions – examine whether it needs to be narrowed.

First, we describe the boundaries of the zone and evaluate the implications of its membership for the role of extra-regional powers operating military bases in close proximity of zonal states. Then, in order to examine the role of external actors as (potential) spoilers, we consider the lessons which can be learned from the establishment of the Nuclear Weapon Free Zones (NWFZs) in Latin America and the South Pacific. We will use these results to evaluate those policies of extra-regional countries that (might) undermine missile disarmament in the Middle East, before turning to the issue of exports. In the following step we examine the motives behind such policies in order to conclude if and how they can be changed. We will focus on the states which are the most active suppliers of missile technology *and* have the power and willingness to influence regional arms dynamics, the United States, Russia, and China.[2]

These countries are not only major providers of missile technology, but also key players in the region. Washington has an enormous military presence in the Middle East and remains the major peace broker. Moscow is also involved in various diplomatic initiatives such as the Middle East Quartet and has occasionally played a special role in negotiations on the Iranian nuclear programme, while Beijing has stepped up its 'energy diplomacy' within the envisaged zone. We will neither consider North Korea, which, although it is an influential supplier, lacks political clout in the Middle East, nor will we do an in-depth analysis of European countries which are no longer major missile exporters.[3]

The scope of the zone: where to draw the line?

Since our research interest is based on a distinction between regional and extra-regional players, the first question is how to define the boundaries of a Missile Free Zone in the Middle East. Unfortunately, similar attempts of the United Nations with regard to the proposed establishment of an NWFZ or a Weapons of Mass Destruction Free Zone (WMDFZ) are of little help. Each year since 1974 the General Assembly has duly passed resolutions calling for an NWFZ for "the region of the Middle East"[4] without further specification. The Final Documents which were passed by the 1995 and 2010 Review Conferences on the Nuclear Non-Proliferation Treaty (NPT) both include a section on the conflict region, without addressing the central geographical issue. The two agreements simply support the General Assembly resolutions by calling for a "zone free of nuclear and all other weapons of mass destruction and their delivery systems" but remain non-specific about its location, only exhorting all "regional parties of [the] Middle East" to follow their call (Review Conference 2010: 30).

The landmark UN study on the establishment of an NWFZ, worked out by James Leonard, Jan Prawitz, and Benjamin Sanders (1991), passed by the General Assembly in 1990, is the notable exception to other resolutions, since it provides some guidance on a definition of the region. The three authors reject fixed boundaries and instead suggest that only "core countries" whose participation is indispensable for facilitating arms control should be charter members of the accord. With regard to its final form Leonard *et al.* (ibid.: 20) offer two possibilities. The first, based on a study of the International Atomic Energy Agency, includes "the area extending from the Libyan Arab Jamahiriya in the West, to the Islamic Republic of Iran in the East, and from Syria in the North to the People's Republic of Yemen in the South". The second includes "all State members of the League of Arab States…, the Islamic Republic of Iran and Israel".

Both proposals are compelling as far as the ultimate boundaries of a zone are concerned. In the foreseeable future, however, they are too far-reaching, because they include actors that cannot be regarded as "core countries" since they are not at the centre of the conflict formations such as the Arab states west of Egypt, where, in addition, important disarmament successes have been reached since 1990. The establishment of an NWFZ on the entire continent in 1996 and the discontinuation of Libya's WMD programmes in 2003 have undoubtedly been the most important milestones. In addition, the missile related capabilities of the North African states are rather limited. Even the range of Tripoli's missiles is relatively short – they are only able to reach some parts of the neighbouring countries – and its attempts to acquire the North Korean Nodong with a range of more than 1,000 km have failed (GlobalSecurity.org 2010). Therefore our conception of a (preliminary) MFZ covers the territory from Egypt in the west to Iran in the east and from Syria in the north to Yemen in the south (see Map 1), including those states that have been identified in this volume as being of central importance for regional conflict dynamics (see the articles of Bernd W. Kubbig *et al.* and of Uri Bar-Joseph).

Nevertheless, this definition raises further questions regarding US involvement in the conflict formations and its military presence in Afghanistan and Turkey. Putting the two countries under zonal restrictions might convince American foes to join an arms control accord, as long as the disarmament obligations apply to Washington's local military presence as well. However, this could also complicate the process, since Kabul and

Ankara would have to accept the same restrictions as those which apply to the zonal states. This would not be acceptable for Turkey which opposes participating in regional disarmament accords since it does not consider itself to be a part of the Middle East. Instead a promising solution could focus solely on the role of Washington. A separate agreement between the members of the MFZ and the United States in which the US foregoes (the option of) deploying missiles and nuclear weapons in Afghanistan and Turkey – where Washington still stores nuclear arms at Incirlik air base (Kristensen 2010) – would address Iranian fears of being encircled by American troops. The agreed-upon deployment of the missile defence (MD) radar system in Turkey might be affected by such an accord. But Ankara would still be allowed to import MD systems from the US as long as they are not operated by American personnel.

However, such a deal would still allow Washington to use Turkish and Afghan military bases, a source of concern for Tehran, since an air attack against its nuclear facilities might be initiated from either country. Iran, which has mainly relied on ballistic missiles as a deterrent against the United States (Taremi 2005), can hardly be expected to commit to the terms of an MFZ if American troops are not withdrawn in substantial numbers from both states and if there are not significant limitations on the presence of American fighter jets in the area. The same holds true for the deployment of naval forces carrying missiles, especially cruise missiles, in close proximity to the proposed zone. However, extending an arms control agreement to international waters is problematic, since the Law of the Sea grants each country the right of free navigation (Prawitz 2008: 334). So obviously an MFZ would require negative security assurances to the signatories that entail guarantees not to use or threaten to use the weapons banned in the zonal states. Such declarations have been a major component of the NWFZ agreements in other regions to which we will turn in the following section.

The past as prologue: external powers and the establishment of NWFZs

Zonal disarmament concepts have so far focused on nuclear weapons. In this context we touch upon two central agreements: the Treaty of Tlatelolco (Treaty for the Prohibition of Nuclear Weapons in Latin America and the Caribbean) and the Treaty of Rarotonga (South Pacific Nuclear Free Zone). Both of these accords highlight the importance of external powers. In addition to requiring negative security assurances to the member states, the Additional Protocols oblige extra-regional actors in possession of nuclear arsenals to refrain from any action that would violate the agreements or contribute to their violation, i.e. proliferation or testing activities – restrictions which some of the external powers did not accept. As a result, they were reluctant to support and sign the treaties, hindering their enactment. So the establishment of both accords can give us some insights about how extra-regional actors may undermine the implementation of a zone free of a specific weapon category.

The idea of creating an NWFZ in Latin America was born in the context of the Cuban Missile Crisis which awakened regional governments to the looming threat of nuclear war. The Treaty of Tlatelolco was negotiated in the late 1960s and was quickly signed and ratified by many Latin American countries (Redick 1981), but its unanimous acceptance in the region was only possible more than 30 years later, when Cuba officially joined the accord. Some of the conflicts preventing the treaty from being accepted by all Latin American and declared nuclear weapon states were related to the role of the United States. Washington indicated its general support to the Preparatory Commission of the Tlatelolco Treaty, but strongly opposed the inclusion of the Virgin Islands and the Commonwealth of Puerto Rico in the envisaged zone until a comprehensive disarmament agreement with the other nuclear powers had been reached. The United States considered the Virgin Islands part of its territory and insisted on exercising control over Puerto Rico, one of its so-called insular areas. The inclusion of Guantanamo, Cuba, where a major American military base is located, was conditioned on Havana's participation which in turn, supported by the Soviet Union, demanded an end to the "illegal detention" of its territory and the denuclearization of all US bases in Latin America. Washington only offered to make the Panama Canal part of the zone on the condition that its transit rights would not be affected, i.e. guaranteeing that nuclear weapons can be shipped or flown to allies, while remaining under own control, a point on which it finally prevailed (Robinson 1970).

In the case of the Rarotonga Treaty the positions of some external actors were even more problematic. The treaty was the collective product of the South Pacific Forum, a regional international organization crafted in the wake of Paris' atmospheric nuclear testing in French Polynesia in the 1980s. Not surprisingly Paris, which was determined to sustain and improve its nuclear forces and to conduct future tests, was one of the major spoilers (Power 1986: 465–7). The United States also opposed the agreement with the transit question again looming large. From Washington's point of view, the treaty was not compatible with the ANZUS[5] defence accord between Australia, New Zealand, and itself, which, according to the American reading, guaranteed transit rights to the US. Since Washington's nuclear presence in the region was considered vital to its overall deterrence concept, the United States declared that it would not accept any restrictions in this regard (ibid.). However, in 1996 Washington finally signed the Additional Protocol, but has yet to ratify it.

In both cases, the NWFZ agreements were based upon regional initiatives, motivated either by the fear of being caught up in a nuclear war between the two superpowers (Tlatelolco) or an interest in protecting the environment from nuclear tests (Rarotonga). External powers were in many cases obstacles to progress, although they ultimately agreed to the accords. Additionally, except for France's testing activities beneath the Mururoa Atoll and the transit of American nuclear armed ships in Latin America and the Southern Pacific, both regions were free of nuclear weapons, Argentina and Brazil pursuing of programmes in this area notwithstanding. But the situation with regard to an MFZ in the Middle East is strikingly different since all regional actors possess the weapons at stake and therefore would have to relinquish an essential part of their military capabilities. The realization of an MFZ would therefore first and foremost depend on the commitment of the main regional players.

Nevertheless, a number of lessons about the role of external powers can be learned from the Tlatelolco and Rarotonga experiences. Extra-regional actors in possession of the weapons banned from the zone can hinder the disarmament efforts by, first, refusing to accept that territories under their control become part of the zone, second, maintaining military bases within the region (American positions with regard to Tlatelolco), third, insisting on their right to use parts of the zone for transit purposes (American position with regard to Tlatelolco and Rarotonga), and, fourth, refusing to give up testing activities (French position with regard to Rarotonga). In the following section these four challenges for the realization of zonal disarmament will be applied to the establishment of an MFZ in the Middle East/Gulf.

The lessons from Tlatelolco and Rarotonga: challenges for a Middle Eastern MFZ

Territorial disputes and testing are obviously the least problematic issues for an MFZ in the conflict region. Since the United States transferred authority over Iraq back to Baghdad, there are no territories under the authority of extra-regional powers in the Middle East/Gulf. Hence this provides no reason for external actors to refuse support of an MFZ. The same holds true for missile testing. Nuclear tests harm the environment and are potentially dangerous for the people living near the testing area, so there is a tendency to use far-removed and largely uninhabitated locations. Missile tests, on the other hand, can equally be conducted on one's own territory. In fact choosing a distant location may cause more harm than good, since neighbouring states may perceive the test as an aggressive act. So it is hardly surprising that external powers have never tested their own missile systems in the Middle East.[6]

The issue of transit rights is different. American warships and aircraft carriers that are equipped with cruise missiles routinely cross the Suez Canal. Since these weapon systems might be used in a future military confrontation, for instance with Iran, Washington's adversaries can be expected to reject significantly scaling back their missile arsenals. However, the overall American military presence in the Middle East/Gulf would be even more problematic for its rivals. Unlike Russia and China, the recent debate among Moscow's desicion-makers about setting up a naval base in Syria notwithstanding, the United States deploys missiles and troops in the region. Shortly before the outbreak of the Second Gulf War General Colin Powell, then-Chairman of the Joint Chiefs of Staff, had decided that Patriot Pac-2s were needed to protect American troops and Washington's allies. Production was accelerated – the system was still under development – and 424 Pac-2s were deployed in Israel and Saudi Arabia by the start of "Operation Desert Storm" (Sherman 2003). While Israel relied on Pac-2s to protect its population centres, in Saudi Arabia the Patriot systems were used to defend American soldiers. After the war, these weapons were still used for troop protection in the region. In 2007 the existing systems in Kuwait and Qatar were replaced by the more advanced Pac-3s (Capaccio 2007).

These weapons could become an obstacle for an MFZ, since Iran, fearing that its deterrence concept is weakened, might be tempted to step up its missile programmes, so that the Patriots could not intercept its delivery systems, especially since the Pac-3 is considered to be much more accurate than the older Pac-2 (see the article of Dennis M. Gormley *et al.*). To be sure, the deployment of US missile defence systems is hardly the aspect of American military activities in the Middle East/Gulf that its foes consider the most troublesome. Yet the overall troop presence in Arab countries, mainly in the Gulf, is a greater concern, above all of Iran (Taremi 2005: 96–103). Washington has systematically stepped up its military activities in the region after the Iraqi invasion of Kuwait in 1990 (Yetiv 2008: 28–9). According to the U.S. Department of Defense, the United States – at the time of this writing – uses military facilities in Iraq, Saudi Arabia, Kuwait, the United Arab Emirates (UAE), Bahrain, Qatar, Oman, and Egypt (see Map 2).[7] Since the military bases in the Gulf could support another major military operation – in 1991 and 2003 such campaigns were successfully initiated against Iraq – Iran would probably insist on a large-scale winding down of US military presence before agreeing to reduce its missile arsenal.

The alliance factor: US–Israeli missile defence cooperation

Another aspect of Washington's security policy that, although not related to the Tlatelolco and Rarotonga experiences, could easily contribute to regional missile ambitions and therefore be detrimental for the zonal concept, is the support for Israeli missile defence. The military cooperation between Israel and the United States has included such efforts since the 1980s. In a Memorandum of Understanding signed in 1986, Washington pledged to support the development of an anti-missile shield for the Middle Eastern state. One year later, Israel's ambitious MD programme Homa (wall) was initiated (Piper 2002), and in 1995 the Arrow-2, jointly

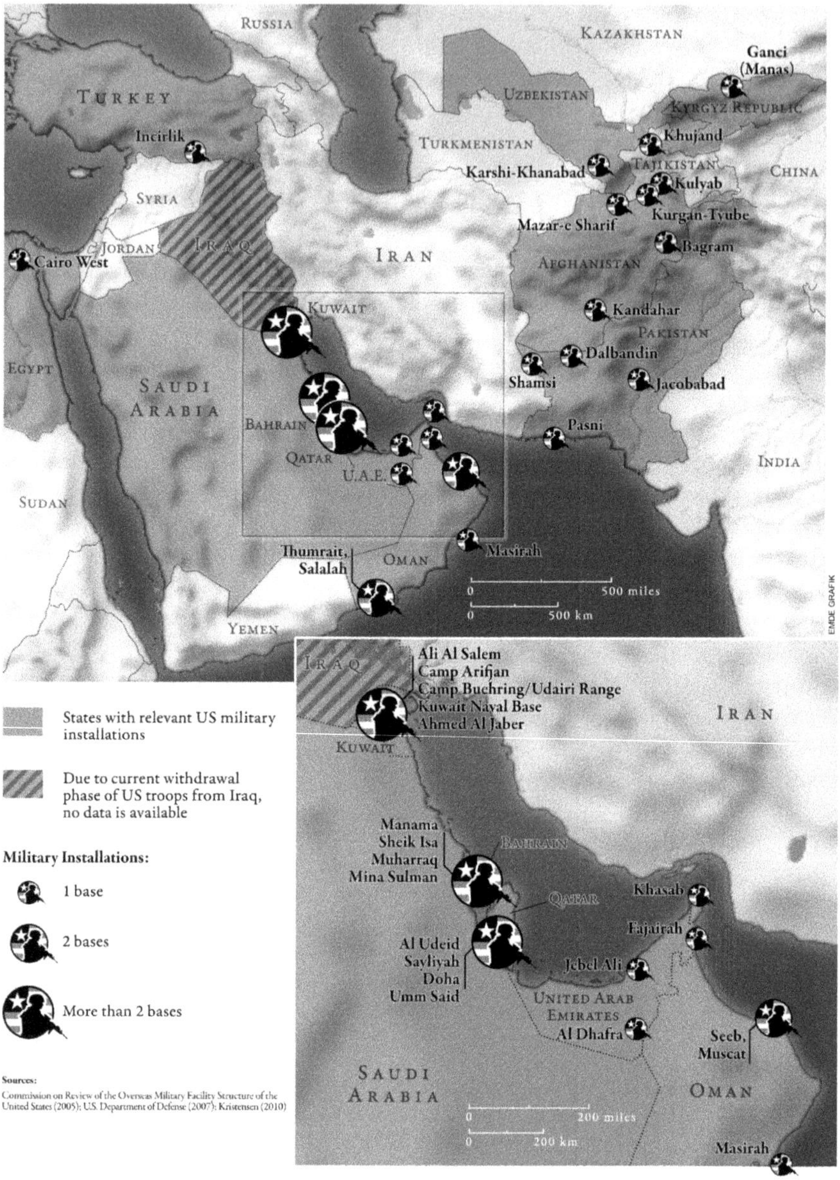

Map 2 US military presence in the broader Middle East.

produced by Boeing and Israel Aerospace Industries (IAI) (Sharp 2010: 10), was tested for the first time and later declared operational. Israel manages the system independently, but receives financial support from the United States for its maintenance (ibid.: 12).

Currently IAI is working on a new version of this weapon, the Arrow-3, whose development will be fully funded by Washington (Katz 2010). Another Israeli project, David's Sling/Magic Wand, which is currently in the production cycle, is co-developed, co-financed, and envisaged to be co-operated by the United States (Sharp 2010: 10–12).[8] In view of the fact that the financial support does only account for a comparatively small part of the overall American military aid which Israel could easily spend on its own, deepening the technological cooperation would cement Washington's role in regional missile (defence) policies. The deployment of the US X-Band radar system has also contributed to Israel's MD capabilities (ibid.: 11). The danger is again that rival states which rely upon missile deterrence consider MD programmes a challenge to their security strategy, which may result in an even more intense (offence–defence) arms race in the Middle East/Gulf.

The Arrow project – the Arrow-2 has a detection range of 500 km (Army-technology.com 2010) – was driven by the Israeli idea of a theatre MD system capable of intercepting the Arab R-17/SS-1c Scud-B as well as the 9M79/SS-21 Scarab, the longer-range Iraqi Al-Hussein, and the Saudi DF-3/CSS-2 (NTI 2005). So the Arab countries, which consider ballistic missiles a powerful deterrent, may be tempted, depending on the Arrow's feasibility and their relations with Israel, to speed up their missile programmes. The same holds true for Iran, while David's Sling/Magic Wand which is envisaged to intercept long-range rockets and slower-flying cruise missiles with ranges of 40–300 km, such as large parts of Hezbollah's arsenal, could convince the Lebanese militia to pursue its rocket projects with renewed vigour.

External powers as game changers? The historic role of the United States, Russia, and China as missile suppliers

Washington's arms exports to the Middle East/Gulf were on the rise due to the Nixon Doctrine of 1969 which was aimed at increasing its allies' responsibility for their own security (Yetiv 2008: 30). In the following decades anti-air missiles accounted for most of the American missile deliveries to the region, providing Bahrain, Egypt, Iran, Israel, Jordan, Kuwait, Saudi Arabia, and the United Arab Emirates with state-of-the-art technology (SIPRI 2010a). However, such capabilities are not as destabilizing as missiles fired at ground targets (see the chapter of Dennis M. Gormley *et al.*). The main recipient of US missile technology was Israel. After the 1973 War, Washington supplied the Middle Eastern state with 150–200 Lance surface-to-surface missiles (SSMs) with a range of 130 km

(Navias 1993: 70); a deal that was, in the regional context, unparalleled until 2000 when the United States agreed to sell MGM-140A ATACMS (165 km), the replacement system for the Lance, to Bahrain. Israel (along with various Arab states) also belonged to the importers of the sophisticated RGM-84 Harpoon anti-ship missile (120 km), which it may have modified to carry nuclear warheads (NTI 2008). In general, the US has not changed its missile export behaviour over the years. The new elements were the above-mentioned transfer of the MGM-140A ATACMS and the delivery of Pac-2 missile defence systems to Kuwait and Saudi Arabia (see the Table compiled by Michael Haas and Bernd W. Kubbig in the Appendix).

The Soviet Union's active penetration of the Middle East/Gulf arms market began in the mid 1950s with the so-called 'Czech deal', in which Egypt signed a contract for the purchase of weapons and military hardware via Czechoslovakia. Starting in the 1960s, the Soviet Union supplied missile systems to its Cold War allies in the Middle East, including Egypt, Iraq, Syria, and Yemen (South and North). As in the American case the exports mainly included anti-air capabilities, especially surface-to-air missiles of which Moscow was the leading manufacturer. In addition, the more problematic SSMs were also part of the Soviet deliveries to the region. The sale of the Luna/Frog-7 (70 km) in the 1970s was a case in point. So was the transfer of the R-17/SS-1c Scud-B (300 km) in the 1970s and 1980s. Syria and North Yemen (the latter did not receive the aforementioned systems) were also supplied with 9M79/SS-21 Scarabs (70 km) which were to replace the older Luna/Frog-7s (Barabanov 2007; SIPRI 2010a; see also the Table compiled by Michael Haas and Bernd W. Kubbig in the Appendix). Due to the Soviet Union's collapse, Moscow's ties with its traditional clients deteriorated and the amount of military deliveries dropped dramatically. But in the early 2000s Russia managed to gradually restore its position as an arms exporter in the region, mainly affecting weapon deals with Egypt, Iran, Syria, and the UAE. More recent air defence systems and anti-tank missiles were at the heart of these agreements.

China's major missile related transfers to the Middle East were conducted in the 1980s and early 1990s (Kan 2011; NTI 2007). The People's Republic reportedly transferred missiles and related technology to a number of Middle Eastern/Gulf countries, including Iraq, Iran, Egypt, Saudi Arabia, Syria, and the UAE (NTI 2007). The main beneficiary of these exports was Tehran. Since the mid 1980s, Beijing has sold various anti-air capabilities, anti-ship cruise missiles – the HY-2/SY-1A/CSS-N-2s (95 km) being the most advanced system (also sold to Egypt and Iraq) – as well as components, technology, and materials thereof to Iran (NTI 2007; SIPRI 2010a). As far as SSMs are concerned, Beijing agreed in 1989 to allow Tehran to produce M-7/CSS-8s (150 km) in licence. In addition, the purchase of M-11/CSS-7/DF-11s (300 km) and M-9/CSS-6/DF-15s (600 km) was negotiated in the late 1980s and early 1990s, but the talks never led to concrete results. News media reports suggested that China has also delivered a range of guidance systems as well as telemetry equipment used in flight tests to Iran in the mid and late 1990s and the Clinton administration concluded that missile components were shipped to the American foe at that time too (Kan 2011: 21–2). Among the weapon deals with other regional states in the surface-to-surface realm the agreement with Damascus about the transfer of M-9/CSS-6/DF-15s in 1991 which, however, was never realized and especially the supply of medium-range ballistic missiles to Saudi Arabia in 1987/88 stood out. Riyadh was provided with 50 DF-3/CSS-2s (2,800km) (SIPRI 2010a; see also the Table compiled by Michael Haas and Bernd W. Kubbig in the Appendix).

A new era? Recent trends in missile related exports to the region

In the last years the main focus of Washington's arms export policy – continuing transfers of anti-air capabilities notwithstanding – was on missile defence. The Bush administration's mid 2007 decision to increase American military aid to the member states of the Gulf Cooperation Council to about $20 billion over ten years paved the way for far-reaching MD related arms transfers that were negotiated shortly thereafter. The United States was asked by Kuwait and the UAE to provide them with massive stockpiles of missile defence systems and related technology. Both countries set their sights on modern Pac-3 technology. Kuwait ordered 80 missiles and technology to upgrade 60 Pac-2 missiles and radars worth more than $1.3 billion. The UAE requested entire Pac-3 systems. Besides that the UAE also asked for the import of the Terminal High-Altitude Area Defense (THAAD) system, including three fire units with 147 missiles as well as nine launchers plus radar sets. The entire MD deals between the United States and the UAE amount to about $16 billion (Blanchard and Grimmett 2008: 12, 15–17). The supply of MD systems or components such as equipment to update existing capabilities could also be part of a major American–Saudi arms deal under which Riyadh plans to import US weapons for an estimated cost of $60 billion over ten years.

The Saudi Kingdom will obviously mainly procure fighter jets and helicopters, but it remains to be seen what the final results of the negotiations will be. Washington is said to encourage Riyadh to purchase missile defence systems such as THAAD (Entous 2010). However, the recent American focus on MD does not mean that the transfer of other missile systems, besides anti-air capabilities, can be ruled out. Israel and Egypt have both requested the sale of the updated Harpoon Block II anti-ship missile (280km), while the UAE and Jordan approached Washington in order to be provided with the High Mobility Artillery Rocket System, called HIMARS. The UAE has also asked for the purchase of the modernized

MGM-140A ATACM with a range of up to 300 km, so has Bahrain. Congress has already been notified of all possible deals (DSCA 2011), a necessary step in order to finalize them, but obviously no supplies have taken place so far.

Russia, on the other hand, has abandoned its previous role as a major supplier of missile systems. In Moscow's overall weapon exports, the share of missiles significantly decreased. The Middle East was no exception. Cooperation with Iran came to an end and despite the fact that the contract with Syria about the sale of the Iskander-E/SS-26 Stone (280 km) had already been signed, then-President of Russia, Vladimir Putin, refused to supply them (Vorobyev and Gelman 2005). Although there were rumours about the conclusion of the deal in the following years, Moscow stood by its decision not to deliver the weapon system at the time of writing. The UAE's interest to obtain Iskander and Club-M multi-purpose coastal missiles (275 km) also did not materialize so far. Nevertheless, Moscow renewed its commitment to supply Damascus with P-800 Yakhont anti-ship cruise missiles (300 km).

Under the terms of the treaty that was concluded between both countries in 2007, Russia will transfer the weapons and could be paid as much as $300 million (IndiaDefence.com 2010). The envisaged transfer of this system, despite not being carved in stone at the time of writing and although the weapons are not intended for use against land targets, signals the readiness of the Medvedev administration still to export entire missile systems of major military significance. In addition, Moscow – still one of the world's major designer, manufacturer, and exporter of all classes of air defence missile systems – mainly focused on large-scale projects in this sector in recent years. It held negotiations with a variety of regional states about their supply. Additionally, S-75s (SA-2 Guidelines), S-125s (5V27/SA-3B Goas), S-200VEs (SA-5 Gammons), and other air defence systems that were formerly supplied are now being modernized in Egypt and Syria (Rybas 2008: 36–62, 338).

China's export policy to the region changed almost similarly. In the case of Beijing there is not even compelling evidence for the sale of any missile system in the last years. In 1998 the People's Republic confirmed that it had halted all transfers of anti-ship cruise missiles to Tehran and would not assist it in upgrading its current inventory (NTI 2003). However, 40 C-701/FL-8 were still delivered between 2001 and 2004, although these systems have a very limited range (20 km). In addition, Iran has been able to produce different anti-ship cruise missiles among them the C-802 (120 km) in licence under the terms of various agreements which have been negotiated prior to this pledge. Saudi Arabia has approached China and tried to initiate the purchase of newer, more advanced missiles as possible replacements of the DF-3/CSS-2, but Beijing, despite rumours about an upgrade of the older delivery system, has obviously not agreed to such a deal.

Nevertheless, the supply of missile *technology* that continued is another story. However, in recent years the involvement of the government in Beijing has remained particularly unclear. Statements from US officials – traditionally the most critical of the role of the People's Republic – suggest that Chinese entities have increasingly operated on their own (Kan 2011: 56–7). Between 2001 and 2010 the United States sanctioned companies and individuals from the People's Republic between 49 and 62 times[9] because of their alleged involvement in the transfer of technology to Iran that might be used for the development of missiles (in some cases also for weapons of mass destruction) (ibid.: 71–5). Until the overthrow of Saddam Hussein, Iraq was considered to be another recipient of missile technology and in February 2011 a report to the U.S. Congress which is worked out on a regular basis mentioned for the first time that entities from the People's Republic were involved in transferring missile related technology to Syria (ibid.: 23, 55).

Between commerce and commitment: the motives of external powers

Since the end of the Second World War the United States has pursued three main goals in the region: first, to ensure access to the rich oil and gas resources, second, to guarantee the security of Israel, and, third, to prevent any rival hegemony from rising (Covarrubias and Lansford: xvii–xviii; Zunes 1994). In recent decades, another objective of American policy has developed: securing peace between Israel and its neighbours, although this has sometimes been pursued with less vigour (Covarrubias and Lansford: xviii). Its three major goals all contributed to Washington's policies that might undermine missile disarmament. The interest in assuring its access to the Middle East/Gulf's oil and gas resources has certainly been one of the major reasons for the constant presence of American troops – protected by Patriot missile defence systems – in the region. It was not a matter of coincidence that the first US forces were deployed to the Kingdom of Saudi Arabia (Yetiv 2008: 28). The relations of the two countries – the world's largest exporter (Riyadh) and the world's largest importer of oil (Washington) – were characterized by a high degree of interdependence. Explorations and developments in Saudi Arabia have even been mainly carried out by American companies (Bahgat 2008: 66–9). Concerns about the security of oil-exporting Gulf states in view of the coming Second Gulf War surely contributed to strengthening American military presence in the region in 1990. The same holds true for the tensions with Iran that facilitated the construction of the new United States Central Command Forward Headquarters in Qatar. Additionally, the alliances with countries that harbour large energy resources have been strengthened by the transfer of American missile technology.

The willingness to protect Israel was demonstrated by the latter's role as the main recipient of American missile technology as well as US financial and technical support for its missile defence programmes. Security sector cooperation between the two countries is largely based on two Memoranda of Understanding, signed in 1981 and 1983, in which both parties pledged to coordinate their strategic planning and war-fighting capabilities. The intention was to allow Israel to maintain a "qualitative military edge" in the region. Unlike other recipients of military aid, the Middle Eastern state may use about one-quarter of its funds to purchase equipment from home-based manufacturers. Its status as a "major non-NATO ally" assures it preferential treatment in bidding for American defence contracts and enables the country to acquire equipment at reduced cost or for nothing at all (Thomas 2007: 4–5). Since the 1950s, however, Washington's most important goal has been to make sure that no rival power would become too influential in the Middle East/Gulf. Led by advocates of the policy of containment, mainly in the State Department, the US largely followed realist thinking and used its leverage in the region in order to forge alliances against its adversaries.

During the East–West conflict its main goal was to contain the Soviet Union. That is why Iran belonged to the first countries that imported American missile technology. In 1979 the Islamic Revolution prompted the Carter administration to apply the same approach to Tehran (Gold 1988: 119), paving the way for further exports to US allies in the Arab world. After the Second Gulf War, President Clinton pursued a policy of "dual containment" towards Iran and Iraq (Tanter 1998: 77), which is now applied solely to Tehran. The containment of Iran was also the rationale behind the decision to strengthen the military ties with the Arab Gulf countries, including the MD agreement with the UAE and Kuwait, although then-Under Secretary of State for Political Affairs Nicholas Burns was somewhat hesitant to admit that. According to him Iran was "a factor in this, but it wasn't the overriding factor" (quoted in Porth 2007). Burns' boss, Condoleezza Rice, was more outspoken about the reasons for the arms transfers and pointed to "a broader strategy to counter the negative influences of Al-Qaeda, Hezbollah, Syria, and Iran" (quoted in ibid.). One of the American negotiators was even more candid when he told CNN, speaking on the condition of anonymity, that this was "all about Iran" (quoted in CNN 2007).

The Soviet's arms export strategy was strongly driven by the desire to counterbalance the US by making deliveries to 'countries of socialist orientation'. The Middle East was no exception. Following the collapse of the Soviet Union and the economic chaos of the early 1990s, Moscow's export policy was now no longer strategically, but commercially motivated. At the time of the Soviet Union's demise, Russia's defence companies had lost most of their governmental contracts or were not being paid on time. Expanding exports was the only chance for most of them to avoid bankruptcy; this paved the way for (intensifying) military cooperation with countries such as Iran and the UAE. Despite the economic pressure to initiate far-reaching sales, missiles were not at the centre of the deals. In the wake of the Soviet–American talks on eliminating intermediate-range missiles, Russia has restricted its deliveries and joined the Missile Technology Control Regime in 1995. Additionally, the Putin and Medvedev administrations consider stability in the Middle East/Gulf vital to Moscow's interest. The fact that Russia and the regional countries have a common interest in coordinating their activities in the global energy market – together Tehran and Moscow control about 20 per cent of the global oil and almost half the world's gas reserves (Kaim 2008: 267) – is frequently cited as a motive for weapon sales.

In fact energy-based interests have mainly contributed to deepening trade relations with a variety of Middle East states and stepping up diplomatic initiatives in order to ensure influence. Vladimir Putin's visit to Saudi Arabia, Qatar, and Jordan in 2007 – the first ever of a Russian/Soviet leader to the Gulf – was a case in point (ibid.). The new foreign policy concept – approved by President Dmitriy Medvedev on 12 July 2008 – clearly states that "Russia intends to further develop relations with ... leading regional powers in the bilateral and multilateral formats" (quoted in *Rossiyskaya Gazeta*, 12 July 2008). Expanding economic and political ties with all the countries that harbour the largest energy resources such as Saudi Arabia, Iraq, Iran, and the UAE is hardly compatible with exporting highly destabilizing weapon systems, especially missiles, to them, since the above-mentioned Arab states and Tehran have confrontational relations.

In addition, Moscow has become increasingly concerned about the security of its southern borders: Islamist fundamentalism is on the rise in South Caucasus as well as in the Muslim parts of Russia, and many of these organizations rely on support from the Middle East/Gulf. Moreover, the acquisition of short- and medium-range missiles by Iran and Saudi Arabia is a major cause of concern because of the territorial proximity of those countries to the Russian border. Although Moscow considers American claims about Iran's nuclear programme exaggerated it agreed to the UN arms embargo against Tehran and lived up to it. United Nations Security Council Resolution 1727, passed in December 2006, banned the export of items and technology related to delivery systems capable of carrying at least a 500 kg payload to a range of at least 300 km. In June 2010, Resolution 1929 prohibited the supply of all weapon categories listed in the UN Register of Conventional Arms, including missiles with a range of at least 25 km, and related spare parts (SIPRI 2010b). In view of the growing concern of Russia's Arab partners and American pressure, S-300 anti-air systems were not supplied to Iran (*Reuters*, 23 September 2011). And if, despite all these points, the export of missiles to the region is taken into account for economic reasons, it is still doubtful that Moscow is willing to risk a major confrontation with the United States and Israel – the Putin administration was not and prevented the sale of the Iskander. The Medvedev government's renewal of the commitment to stick to an earlier agreement with Syria about the transfer of the Yakhont is the notable exception to Russia's overall arms export strategy. Air defence systems, on the other hand, are still supplied to Arab states, since they are not considered a threat to regional stability and belong to the most advanced products of Moscow's military companies that still depend on their export.

The Chinese government largely ignored the proliferation activities of its ailing state-owned defence sector, which had to find ways to generate revenues due to the economic reforms of the late 1970s. During this period the government promoted rather than controlled weapon exports (Fan 2008), facilitating missile transfers to the Middle East/Gulf. Most of them were realized after Beijing had initiated the first market-oriented reforms. The People's Republic was a latecomer to many international arms control and disarmament treaties and regimes. It was not involved in the negotiations of the MTCR in the 1980s and initially refused to adhere to the agreements' guidelines. This changed in the early 1990s, when China, opting for expanding its economic relations, was actively involved in arms control and non-proliferation negotiations and generally decided to adhere to global norms and agreements, including those that focus on the missile sector. In particular, Beijing pledged to abide by the MTCR's guidelines in 1991, despite its status as a non-member state, and even applied to join the agreement in 2004. In 1994 the People's Republic and the United States negotiated a joint statement on missile proliferation in which China agreed to ban the supply of all MTCR class missiles. In 2000 Beijing issued another declaration, this time promising to strengthen missile export control regulations and to work out a control list, which, based on the MTCR's guidelines, was completed in 2001.

The reluctance to export missile technology to the Middle East/Gulf was also rooted in the People's Republic's increasing energy needs: presently the country receives half of its oil imports from the region. Since 1993, when China became a net importer of oil, its policies towards the Middle East have been mainly dominated by the intention to secure the supply of oil and gas (Lu and Chen 2007: 82).[10] The increasing dependence on the region's energy market could be seen as an incentive for the export of missile related equipment to energy-rich countries, but in fact China is primarily interested in not endangering its business relations with all the different suppliers, by and large preferring stability. The proliferation of missiles or missile technology is generally considered to undermine this goal. Nevertheless, China – almost like Russia – is reluctant to lobby for non-proliferation, if this might endanger its relations with the oil- and gas-exporting actors. This is one reason why US requests for tougher sanctions against Iran have frequently been rejected or watered down by Beijing, although it finally agreed to establish and (finally) strengthen the successive UN arms embargoes. Given the fact that Iran can hardly afford to interrupt its oil sales, since it depends on generating revenues this way, and in view of the low degree of recent arms exports, China did not have much to lose in this regard.

In addition, the People's Republic considers missile exports to American foes in the region still as a potential response to US arms sales, especially missiles and missile defence systems, to Taiwan. China sees these deals as an illegitimate interference in internal affairs, since it insists that the island is a province of the People's Republic. Beijing is also seriously concerned about the American deployment of missile defence systems and the growing MD cooperation between Washington and Tokyo, fearing that such systems could neutralize its deterrence capabilities. The same holds true for the Ground-based Midcourse Defense system of the United States, since it might intercept China's very small number of intercontinental-range ballistic missiles capable of reaching the American homeland (Podvig and Zhang 2008). Despite its newly created rules for arms transfers, Beijing could not and cannot guarantee that its defence industry lives up to these laws, because it lacks sufficient enforcement infrastructure and resources.[11] The decentralization could further decrease the ability of the government to exercise control over military companies. This is particularly troublesome, since some of them, driven by economic interest, are reluctant to abide by the new laws. Missile technology, although probably not entire systems, may therefore still make its way to the conflict region.

A forward strategy for missile disarmament: under which conditions would the external powers support an MFZ?

The United States, Russia, and China would be destined to play an influential role in a Middle East/Gulf MFZ. The success of an MFZ, ranging from Egypt in the west and Iran in the east as well as from Syria in the north to Yemen in the south, would depend on the willingness of the US to refrain from deploying missiles in close proximity to the region, to substantially limit its military presence in close proximity of the region, and to provide negative security assurances to the members of the MFZ. In addition, as the Treaties of Tlatelolco and Rarotonga have shown, territories under external control, military bases, transit rights, as well as testing activities can be important issues by which extra-regional powers have an influence on the success or failure of zonal disarmament. In the Middle East, Washington could spoil an MFZ in part because of controversies on these aspects. While territorial disputes and testing issues would probably not arise, the United States does have military facilities in many influential Arab countries, especially in the Gulf, and has deployed missile defence systems to protect its most important bases and its allies. This might fuel the threat perceptions of Iran whose deterrence strategy mainly relies on missile capabilities. Washington's decision to co-finance the most important Israeli MD systems could be perceived by Tehran as an extension of this threat. Israel's Arab neighbours could also be worried about the MD capabilities. The success of an MFZ would also depend on the prevention of missile transfers, a challenge that involves the roles of Russia and China, too.

Based on our analysis of the motives of external powers, the question of how they might be convinced to abandon counterproductive policies and to support an MFZ in the Middle East/Gulf can be answered as follows: as for American military presence, the challenge would not only be to convince the United States to give up its regional MD systems, but to negotiate an agreement that leads to a substantial withdrawal of American armed forces from the region, mainly from the Gulf. A security accord between Washington and Tehran including conditions for the withdrawal of American troops that is mutually satisfactory would have to be worked out. Both the United States and Iran could in principle agree to limit the Iranian meddling, while significantly reducing American forces. In addition, the US would probably insist on severely limiting Tehran's nuclear and missile programmes in parallel, if not in advance of such steps. A bilateral accord would therefore have to include arms control measures. As long as Iran has not abandoned its missile ambitions, the basic rationale for multilateral MD cooperation between Kuwait, the UAE, and Washington continues to exist and the United States can hardly be expected to change its approach. As far as the American support for Israeli missile defence is concerned, limiting Washington's financial input would – given the small portion of the overall US military aid which is spent on MD and Israel's capacity to finance the systems alone – probably be largely symbolic. However, the capabilities themselves can be expected to be an issue in disarmament negotiations. American security gurantees might lead to a compromise in this regard.

Regarding the sale of missile technology one can conclude that the three powers were instrumental in providing many of the regional actors with their current capabilities. Nevertheless, their export activities remained largely focused on anti-air missiles that are not considered to be as destabilizing as SSMs. Especially the United States and the Soviet Union/Russia have vigorously pursued the export of their related systems to a variety of Middle Eastern/Gulf countries and remain committed to this cause. Therefore, at least Washington's and Moscow's support for an MFZ would probably depend on the precise definition of the term 'missile'. An all-encompassing definition that includes anti-air capabilities would be problematic for both actors, since they would be concerned to lose the market. The transfers of SSMs mainly included the sale of short-range missiles or related technology. The notable exception was China's sale of DF-3/CSS-2s to Saudi Arabia; Beijing had not pledged to comply with the MTCR at that time. This demonstrates the willingness of external powers to comply with international commitments concerning missile control and their general interest – however motivated – in regional stability. The American decision-making process in the context of entering the Middle East missile game with the sale of Hawk air defence systems to Israel is a telling example. The proponents of the deal mainly argued that the sale "would fill an important gap" in the capabilities of an ally without significantly contributing to a "shift in the balance of military power" between regional adversaries (quoted in Bass 2003: 164). However, Washington's continued export of combat aircraft would probably be a challenging issue in MFZ negotiations, since missiles are considered to be countermeasures against such weapons (see the article of Bernd W. Kubbig *et al.*).

Unlike the American missile (defence) transfers, Beijing's and Moscow's exports to the Middle East/Gulf have already decreased significantly since the early and mid 1990s. At that time Russia signed the MTCR and Beijing had begun to modify its arms transfer policies. Both actors started to redefine their strategy towards the region on the basis of a new interest in stabilizing arms dynamics, motivated either by the interest in energy related cooperation with Middle Eastern actors and the danger of being affected by the ongoing conflicts (Russia) or the need for the steady import of oil and gas (China). Both countries have pledged to abide by MTCR regulations, fostering norm construction in the missile sector. As a result, one can expect Moscow and Beijing to support the overall goal of missile disarmament, if not an MFZ, since as long as it is feasible and facilitates trust among regional actors, it advances their common interest in preventing an escalation of the existing tensions. From a Russian point of view far-reaching missile reductions would also contribute to implementing its initiative to globalize the Intermediate-Range Nuclear Forces Treaty since a number of the countries in the region possess missiles that would fall under this accord.

With regard to China, currently a possible motive for future missile sales to the Middle East is the conflict with the United States about the MD related arms deals with Taiwan and Japan. Hence Beijing's support for an MFZ would be facilitated if the US were to limit its weapon sales, especially of missile defence systems, to these states. In addition, mutual approaches to missile export control must be found. Washington has imposed a variety of sanctions against Chinese companies that export sensitive technology to Iran, while Beijing rejects such measures as an extra-territorial use of US law. Cooperative steps which might pave the way for overcoming this stalemate could include confidence-building measures, including assurances that the American missile defence programme is not targeted at the People's Republic and, most importantly, mutual efforts to strengthen the Chinese implementation of its missile transfer regulations. Washington already has valuable experience in this regard, and both countries could coordinate their export controls, while effectively enhancing non-proliferation measures. The United States could implement measures such as providing special training for customs, border controls, and enforcement officers in order to assist China in detecting and interdicting unauthorized missile technology transfers (Zaborsky 2005).

In summary, the external powers, US, Russia, and China, are instrumental for achieving missile disarmament in the Middle East. Realizing such a goal requires a future regional security accord that takes their interests into account. Hence negotiating an agreement that limits missiles in the Middle East/Gulf would pose an even more serious diplomatic challenge than the previous articles in this volume on regional dynamics suggest. Nevertheless, there is a history to build upon. The MTCR is now accepted by the three major external actors as a legally binding agreement to which they adhere. As a result, effective export controls of WMD-capable missiles could therefore be established. The resolution of the 2010 NPT Review Conference presents a new opportunity to build on this achievement. The Final Document mentions that Russia, the United Kingdom, and the United States (the depositories of the NPT), along with the UN Secretary General, will convene a conference of Middle Eastern states in 2012 on the establishment of a WMDFZ, including the means for their delivery. Together with China they might support the Finnish facilitator who is tasked with preparing the conference by advocating – as an interim step towards an MFZ/WMDFZ – the ban of WMD-capable missiles.

Notes

1 Sameh Aboul-Enein is a diplomat and scholar and contributes his views solely in his personal capacity.
2 The People's Republic has virtually ended all its missile exports, but has become the main licenser of missile systems.
3 The overall export activities of major extra-regional suppliers will be analysed in the article of Dinshaw Mistry and Mark Smith that focuses on the success of the existing missile control measures.
4 See the first resolution (3263 XXIX) of this kind (UNGA 1974).
5 ANZUS is an abbreviation of the participants' names: Australia, New Zealand, and the United States.
6 On the importance of a missile test ban as an important step towards disarmament, see the article of Dinshaw Mistry and Mark Smith; see also Aboul-Enein and Gopalaswamy (2009).
7 See Commission on Review of the Overseas Military Facility Structure of the United States (2005), DoD (2007), Kristensen (2010).
8 American defence giant Raytheon manufactures the system together with the Israeli company Rafael Advanced Defense Systems. In Fiscal Year 2010 the U.S. Congress allocated $80 million for the project (Sharp 2010: 10–12).
9 In some cases the Iran, North Korea and Syria Nonproliferation Act was applied, leaving it open which of the three countries was the business partner of the Chinese entities.
10 The ties with Saudi Arabia and Iran that provide approximately 30 per cent of Beijing's overall oil imports were intensified – in 2004 a $100 billion deal with Tehran was signed, assuring the import of 270 million tons of liquefied natural gas over a 25 year period and a stake of 50 per cent in the development of the Yadavaran oilfield. The People's Republic has also expressed its interest in direct pipeline access to Iran via Kazakhstan (Zhao 2004).

11 This assessment is corroborated by the unclassified "Comprehensive Report of the Special Advisor to the DCI on Iraq's WMD", issued on 30 September 2004. The document states that "[i]t is likely that newly privatized state-owned companies were willing to circumvent export controls and official U.N. monitoring to supply prohibited goods". However, "there is no evidence to suggest Chinese Government complicity in supplying prohibited goods to Iraq". The report even concludes that the government in Beijing tried to interrupt the transfers (Kan 2011: 55).

References

Aboul-Enein, Sameh and Gopalaswamy, Bharath (2009) 'The Missile Regime: Verification, Test Bans and Free Zones', *Disarmament Forum*, 4: 43–52.

Army-technology.com (2010) 'Arrow 2 Theatre Ballistic Missile Defence System, Israel'. Online, available at: www.army-technology.com/projects/arrow2/ (25 March 2011).

Bahgat, Gawdat (2008) 'United States Oil Diplomacy in the Persian Gulf', in Markus Kaim (ed.) *Great Powers and Regional Order: The United States and the Persian Gulf*, Aldershot: Ashgate Publishing Limited, 53–69.

Barabanov, Mikhail (2007) From Nasser to Nasrallah, *Kommersant Dengi*, Russian, 15 January.

Bass, Warren (2003) *Support Any Friend: Kennedy's Middle East and the Making of the U.S.–Israeli Alliance*, Oxford: Oxford University Press.

Blanchard, Christopher M. and Grimmett, Richard F. (2008) *The Gulf Security Dialogue and Related Arms Sale Proposals*, Washington, DC: Congressional Research Service.

Capaccio, Tony (2007) 'U.S. Upgrades Patriot Missiles in Qatar, Kuwait: Changes to Boost Security, Reassure Allies, Bush Says', *San Diego Union-Tribune*, 12 January. Online, available at: www.signonsandiego.com/uniontrib/20070112/news_1n12patriots.html (25 March 2011).

CNN (2007) 'Rice: U.S. Military Deals with Mideast Promote Stability'. Online, available at: http://edition.cnn.com/2007/POLITICS/07/30/mideast.USarms/index.html (15 February 2011).

Commission on Review of Overseas Military Facility Structure of the United States (2005) 'Executive Summary'. Online, available at: www.fas.org/irp/agency/dod/obc.pdf (25 March 2011).

Covarrubias, Jack and Lansford, Tom (2007) 'Preface', in Jack Covarrubias and Tom Lansford (eds) *Strategic Interests in the Middle East: Opposition and Support for US Foreign Policy*, Aldershot: Ashgate Publishing Limited, xv–xix.

DoD/U.S. Department of Defense (2007) *Base Structure Report Fiscal Year 2007 Baseline: A Summary of the DoD's Real Property Inventory*, Washington DC: DoD. Online, available at: www.defense.gov/pubs/bsr_2007_baseline.pdf (25 March 2011).

DSCA/U.S. Defense Security Cooperation Agency (2011) '36(b) Arms Sales Notifications'. Online, available at: www.dsca.osd.mil/pressreleases/36-b/36b_index.htm (25 March 2011).

Entous, Adam (2010) 'Saudi Arms Deal Advances: White House to Notify Congress Soon of $60 Billion Package, Largest Ever for U.S.', *Wall Street Journal*, 12 September. Online, available at: http://online.wsj.com/article/SB10001424052748704621204575488361149625050.html (25 March 2011)

Fan, Jishe (2008) 'China Nonproliferation Export Control System: Evolution and Revolution'. Online, available at: http://belfercenter.ksg.harvard.edu/files/Jishe_Fan_Export_Control_MTA_Talk_1.pdf (25 March 2011).

Foreign and Defense Policy Council (no year indicated) 'Russia and the "Wider Middle East"', Russian. Online, available at: www.svop.ru/upload/images/Russia-ME.doc (25 March 2011).

Frolov, Andrei (2007) 'Iran's Delivery Systems Capabilities', *Security Index*, Vol. 13, 82(2): 31–50.

GlobalSecurity.org (2010) 'Weapons of Mass Destruction (WMD): Libyan Missiles'. Online, available at: www.globalsecurity.org/wmd/world/libya/missile.htm (25 March 2011).

Gold, Dore (1988) 'Toward the Carter Doctrine: The Evolution of American Power Projection Policies in the Middle East, 1947–1980', in Steven L. Spiegel, Mark A. Heller, and Jacob Goldberg (eds) *The Soviet-American Competition in the Middle East*, Lexington, MA: Lexington Books, 113–23.

Grinevsky, Oleg (2002) *Scenario for the World War III*, Russian, Moscow: OLMA-Press.

IndiaDefence.com (2010) 'Russia to Deliver 72 Supersonic Anti-Ship Cruise Missiles to Syria'. Online, available at: www.india-defence.com/reports-4537 (25 March 2011).

Kaim, Markus (2008) 'U.S. Gulf Policy: Challenges for the Next U.S. Administration', in Markus Kaim (ed.) *Great Powers and Regional Order: The United States and the Persian Gulf*, Aldershot: Ashgate Publishing Limited, 263–9.

Kan, Shirley A. (2011) *China and Proliferation of Weapons of Mass Destruction and Missiles: Policy Issues*, Washington, DC: Congressional Research Service.

Katz, Yaakov (2010) 'U.S. to Fully Fund Arrow 3 System: Cost of Ballistic Missile Defense System Likely to Reach $100 Million', *Jerusalem Post*, 25 July. Online, available at: www.jpost.com/Home/Article.aspx?id=182562 (25 March 2011).

Kristensen, Hans M. (2010) *Status of U.S. Nuclear Weapons in Europe 2010*, Washington, DC: Federation of American Scientists. Online, available at: www.fas.org/programs/ssp/nukes/images/euronukes2010.pdf (25 March 2011).

Leonard, James, Prawitz, Jan, and Sanders, Benjamin (1991) 'Effective and Verifiable Measures Which Would Facilitate the Establishment of a Nuclear-Weapon-Free Zone in the Middle East', Report of the Secretary-General, A/45/435, New York: United Nations Department for Disarmament Affairs.

Lu, Chunlong and Chen, Jie (2007) 'China's Middle East Policy since the Post-Mao Reform', in Jack Covarrubias and Tom Lansford (eds) *Strategic Interests in the Middle East: Opposition and Support for US Foreign Policy*, Aldershot: Ashgate Publishing Limited, 81–95.

Mistry, Dinshaw (2003) 'Beyond the MTCR: Building a Comprehensive Regime to Contain Ballistic Missile Proliferation', *International Security*, 27(4): 119–49.

Navias, Martin S. (1993) *Going Ballistic: The Build-up of Missiles in the Middle East*, London: Brassey's Ltd.

NTI/Nuclear Threat Initiative (2003) 'China's Missile Exports and Assistance to Iran'. Online, available at: www.nti.org/db/china/miranpos.htm (25 March 2011).

NTI/Nuclear Threat Initiative (2005) 'Israel Profile: Missile Chronology 1990–1999'. Online, available at: www.nti.org/e_research/profiles/Israel/Missile/3571_5219.html (15 February 2011).

NTI/Nuclear Threat Initiative (2007) 'China's Missile Exports and Assistance to the Middle East'. Online, available at: www.nti.org/db/China/mmepos.htm (15 February 2011).

NTI/Nuclear Threat Initiative (2008) 'Israel Profile: Missile Overview'. Online, available at: www.nti.org/e_research/profiles/Israel/Missile/index.html#fn4 (15 February 2011).

Piper, Gerhard (2002) 'Israels Raketenabwehrprogramme', Berlin Information-center for Transatlantic Security. Online, available at: www.bits.de/public/articles/ami/ami1002–2.htm (25 March 2011).

Podvig, Pavel and Zhang, Hui (2008) *Russian and Chinese Responses to U.S. Military Plans in Space*, Cambridge, MA: American Academy of Arts and Sciences.

Porth, Jacquelyn S. (2007) 'Gulf Region Arms Sales Support Stability: U.S. Congress Still Must Approve Large Weapon Purchases'. Online, available at: www.america.gov/st/washfile-english/2007/July/20070730145144sjhtrop0.5523493.html (25 March 2011).

Power, Paul F. (1986) 'The South Pacific Nuclear-Weapon-Free Zone', *Pacific Affairs*, 59(3): 455–75.

Prawitz, Jan (2008) 'A Brief Sketch of the Proposed Zone', *International Relations*, 22(3): 331–7.

Redick, John R. (1981) 'The Tlatelolco Regime and Nonproliferation in Latin America', *International Organization*, 35(1): 103–34.

Review Conference of the Parties to the Treaty on the Non-Proliferation of Nuclear Weapons (2010) 'Final Document', Vol. 1. Online, available at: www.un.org/ga/search/view_doc.asp?symbol=NPT/CONF.2010/50%20%28VOL.I%29 (25 March 2011).

Robinson, Davis R. (1970) 'The Treaty of Tlatelolco and the United States: A Latin American Nuclear Free Zone', *American Journal of International Law*, 64(2): 282–309.

Rybas, Alexander (2008) *Russia's Military-Technical Cooperation with Foreign States*, Russian, Moscow: Nauka.

Sharp, Jeremy M. (2010) *U.S. Foreign Aid to Israel*, Washington, DC: Congressional Research Service.

Sherman, J. Daniel (2003) 'Patriot PAC-2 Development and Deployment in the Gulf War – Lessons Learned', *Acquisition Review Quarterly*, 10(1). Online, available at: www.findarticles.com/p/articles/mi_m0JZX/is_1_10/ai_99753429/pg_13?tag=artBody;col1 (25 March 2011).

SIPRI (2010a) 'SIPRI Arms Transfers Database'. Online, available at: http://armstrade.sipri.org/armstrade/page/trade_register.php (25 March 2011).

SIPRI (2010b) 'UN Embargo on Iran'. Online, available at: www.sipri.org/databases/embargoes/un_arms_embargoes/iran (25 March 2011).

Tanter, Raymond (1998) *Rogue Regimes: Terrorism and Proliferation*, Basingstoke: Macmillan.

Taremi, Kamran (2005) 'Beyond the Axis of Evil: Ballistic Missiles in Iran's Military Thinking', *Security Dialogue*, 36(1): 93–108.

Thomas, Michael (2007) *American Policy toward Israel: The Power and Limits of Beliefs*, London: Routledge.

UNGA/United Nations General Assembly (1974) 'Establishment of a Nuclear-Weapon-Free Zone in the Region of the Middle East', Resolution 3263 XXIX, New York: United Nations.

Vorobyev, Vladislav and Gelman, Zakhar (2005) 'Putin Has Managed to Overcome a Wall', Russian, *Rossiyskaya Gazeta*, 25 April.

Yetiv, Steve A. (2008) *The Absence of Grand Strategy: The United States in the Persian Gulf, 1972–2005*, Baltimore, MD: Johns Hopkins University Press.

Zaborski, Victor (2005) *Bring China Closer to MTCR*, Washington, DC: Center for International Trade and Security.

Zhao, Renfeng (2004) 'Iran Prefers China for Oil Exploration Projects', *China Business Weekly*, 9 November. Online, available at: www.chinadaily.com.cn/english/doc/2004–11/09/content_390435.htm (15 February 2011).

Zunes, Stephen (1994) 'The Roots of U.S. Middle East Policy and the Need for Alternatives', in Elise Boulding (ed.) *Building Peace in the Middle East: Challenges for States and Civil Society*, Boulder, CO: Lynne Rienner, 173–85.

8 The verification challenge

Concepts, requirements, and technologies

Jürgen Scheffran, Bharath Gopalaswamy, Dennis M. Gormley, Bernd W. Kubbig, Uzi Rubin, and Hartwig Spitzer

Defining the task

In the Middle East missiles of all kinds are more prevalent, more frequently used, and more destabilizing than anywhere else in the world. In addition, some of the region's actors have acquired sufficient technology and infrastructure from abroad and have the national human resources to develop and manufacture their own missiles – whether free-flying unguided rockets, ballistic and cruise missiles, or space launch vehicles (SLVs). At the same time, massive imports of those rockets and missiles as well as their materials, components, and complete systems continue apace. Consequently, the gradual transformation of the Middle East/Gulf into a Missile Free Zone (MFZ) is an ambitious and formidable task. The final goal of an MFZ would involve numerous (non-)governmental entities, dozens of operational sites, test ranges and manufacturing plants, hundreds of industrial enterprises, thousands of kilometres of porous borders, and tens of thousands of rockets and missiles. In a region riven with historical animosities, ethnic prejudice, and religious strife, this becomes an enormous challenge. The ultimate objective of an MFZ as well as the step-by-step approach towards its realization will hinge upon a system of effective verification measures. They are indispensable for providing the chronically suspicious protagonists with the necessary quantum of confidence/trust[1] to assume that the other parties are willing to comply with the agreed-upon reduction steps and the MFZ itself.

Yet the challenge is also a conceptual one, since the literature on Nuclear Weapon/Weapons of Mass Destruction Free Zones (NWFZ/WMDFZ)[2] has largely ignored the verification issue as it relates to missiles.[3] Nevertheless, this article does not need to start from scratch, since the verification problem was part of regional arms control agreements in the 1970s. In addition, where helpful we will refer to those definitions, standards, and experiences that formed part of the concept of arms control during the East–West conflict. This contribution thus outlines the role of verification and its requirements, describes the potential and limits of existing technologies, and presents (mainly cooperative) verification measures as necessary (yet limited) confidence- and security-building measures (CSBMs) on the incremental way to an MFZ. We conclude that the enormous challenges can adequately be met, but that satisfactory results which are politically acceptable to the relevant parties presuppose a readiness by all sides to embark on this long and bumpy journey.

Arms control concepts and the role of verification

Verification can be defined as a process designed to convince the parties involved that the stipulations of a (zonal) treaty/regime are being observed, and that all parties are complying with their obligations. The verification process consists usually of two major elements: the gathering or fact-finding of data (monitoring) and the interpretation (evaluation) of the information collected. The first component implies the capability to reliably and accurately determine whether a treaty has been violated and to provide timely warning of this fact. It requires having several reliable and sensitive means of gathering information and of knowing where and what to look for. The element of interpretation is mainly a matter of judgement, subject to political consideration, as the collected data is often ambiguous and its assessment inconclusive.[4] According to an authoritative source such as the former U.S. Joint Chief of Staff General David Jones, "much of intelligence is science, but a lot of it is art. ... In a lot of cases we are dealing with percentages substantially below 100 percent ... It is not a black and white case, in our judgment" (quoted in Wieß 1989: 35).

Since a perfect level of confidence cannot be achieved in practice, standards of adequacy or effectiveness are necessary, as well as cooperative measures such as established counting rules regarding weapons and delivery vehicles, agreed-upon prohibited activities that would deliberately conceal, camouflage, or encrypt data needed to verify an accord, and joint consultative mechanisms as well as forums in order to address controversial compliance related judgements; on-site inspections are of special importance. Such cooperative verification measures in a bilateral or multilateral setting complement unilateral ones such as human sources (spies, emigrants, and defectors) as well as National Technical Means (NTMs – see below).

Since the central task of verification is to reduce the risk of cheating, defining the term 'risk' becomes crucial. In the American–Soviet context, risk was seen as the product of two elements: the probability that a given violation would escape detection, and the magnitude of the threat to US (or Soviet) security should such a violation occur. As a result, the official American thinking on verification recognized the importance of assessing the military significance of a treaty violation and the ability of the United States to detect such cheating "in sufficient time to take whatever action the situation required".[5] A major task for a verification system in the specific regional context of the Middle East would be to agree on the term 'military significance', above all by concretizing the two paramount arms control objectives of strategic/arms race and crisis stability. In this context verification should contribute to slowing down the arms race or even reducing arsenals, preventing surprise attacks, and denying the other side the ability to launch a disarming first strike to gain political advantage. In concurrence with the insights of other articles in this volume, we think it would make sense to focus attention on those delivery vehicles which are considered by one or more parties as especially destabilizing.

By the same token, the history of arms control during the entire Cold War is not only ample proof of the central role of verification, but also of the perennial fight both within the United States (Krass 1989: 17; see also Wieß 1989: 66–9) and between East and West about how much verification is enough. This is true for nuclear arms control approaches and the Treaty on Conventional Armed Forces in Europe (CFE). They illustrate how difficult it is to verify limitations on weapons and force levels (Dunay 1991). At the same time, they demonstrate that successful solutions can be developed which lead to cooperation and build trust among adversaries. Broadly speaking, these positive results were achieved primarily through a combination of NTMs (mainly satellites, but also surveillance aircraft) and on-site inspections.

Insofar as qualitative provisions are included in the various approaches (e.g. how to verify a modernization permitting up to a 5 per cent change in certain parameters) the verification task gets even more complicated (Wieß 1989: 158–66). There is no consensus in the literature on whether a comprehensive 'zero' agreement such as the Intermediate-Range Nuclear Forces Treaty (INF) is easier to verify than a limited and fixed number.[6] Despite the premise that trust is antithetical to the verification endeavour – "Let's not trust them, let's verify"[7] – "verification cannot act as surrogate for trust. … [S]ome initial trust must be present if a verification system is to preserve arms control agreements" (Krass 1985: 750). In the US–Soviet context, Allan Krass (ibid.) spelled out four necessary conditions for starting and maintaining such a process:

> First, we must assume that the Soviets have an objective interest in the success and preservation of agreements. Second, we must be willing to be reassured by evidence of Soviet compliance. Third, we must maintain confidence in our own intelligence capabilities and recognize that the more militarily significant a violation is, the less likely the Soviets can keep it hidden. Finally, we must treat ambiguities, technical violations, and misunderstandings in a calm, businesslike, and confidential manner on the assumption that the other side has an interest in clarifying and correcting any incidents of noncompliance.

Krass' conditions cannot be applied without qualification to the Middle East context, however: no agreement will be signed, ratified, and implemented if it is perceived as damaging to national security and as a result meets political resistance at home. The complex and difficult process of 'selling' a negotiated agreement domestically by demonstrating its robustness and credibility is needed – this task may require being able to point to inspections as an important element of verification (Ledermann 1994: 148). In the entire process of gradual arms reductions leading to an MFZ, the role of verification will be to generate trust in the effectiveness of any negotiated agreements that are reached. At the same time, one needs to distinguish between legitimate control and inspection measures on the one hand and steps that are regarded as illegal espionage, on the other. As Jeffrey Richelson (quoted in Wieß 1989: 21) has pointedly noted: "Today's espionage is tomorrow's verification." This touches upon the delicate aspect of intrusiveness which has to be tackled in a satisfactory way as a condition for the political acceptance of any arms control agreement on the domestic level.

The Cold War experience with verification is still useful as a frame of reference when it comes to those phases of a gradual arms control process towards an MFZ involving military arsenals, especially the most destabilizing ones. However, for the (probably initial) phase in which verification is a crucial element of arms control agreements designed as CSBMs – and not as reductions affecting military arsenals – one can turn to the history of the Middle East itself. A "complex verification regime" (Lederman 1994: 141) was part of four arms control agreements: the Separation of Forces Agreement between Egypt and Israel of January 1974 (Sinai I), the Interim Agreement of September 1975 (Sinai II), the Peace Treaty of March 1979 between both countries, and the Syrian-Israeli Separation of Forces Agreement (Golan Heights Agreement) of May 1974. Each of them included different zonal aspects in terms of delineated demilitarized buffer areas plus thinned out territories of limited military deployment (forces/weapons). A comprehensive verification regime was constructed to monitor their execution (ibid.: 139).

As far as the necessary cooperation between regional actors, external parties, and their technologies and verification methods are concerned the situation has always been complex. A good example is the Sinai II Agreement. The accord was monitored by Israel, Egypt, the United Nations Emergency Force (UNEF) and the U.S. Sinai Support Mission. Washington conducted reconnaissance flights over the zones to check deployments. "Israel, Egypt, and UNEF could also request an earlier reconnaissance mission as a challenge inspection provision" (ibid.: 140). In addition, up to 200 US civilian technicians provided early warning detection, identification, and reporting tasks with respect to any unexpected Egyptian and Israeli force movement by monitoring three watch stations at the Mitla and Gidi Passes (Mandell 1987: 6). Both Egypt and Israel were allowed to use their available NTMs (flight missions, reconnaissance aircraft) according to an agreed-upon schedule (Lederman 1994: 140).

The UNEF "also played a significant role" (ibid.: 141). Its 4,000 men controlled access to the buffer zone and verified compliance with troop and armament levels in the limited force area as well as the neutrality of the buffer zone by permanent observation and on-site inspections. As to the verification principles and rules, the provisions in the Interim Agreement were implemented in an equal, symmetric, and mutual/reciprocal way. Three examples highlight the confidence- and security-building character of Sinai II: first, both Egypt and Israel were entitled to operate reconnaissance aircraft up to the middle line of the buffer zone according to the agreed-upon schedule (two aircraft, up to seven times a week, at altitudes greater than 4,750 metres). Second, when Israel balked at foregoing its electronic surveillance station at the western end of the Gidi Pass, this problem was resolved by agreeing that Egypt would be permitted to build and maintain a similar station at the eastern end of the pass (Mandell 1987: 6). Third, American civilian observers verified the proper monitoring of the Israeli and Egyptian early warning stations in the buffer zone, the results of which were reported to both states and to the UN (Lederman 1994: 140).

The reached level of confidence was seen by both parties as obviously satisfactory. A Joint Commission (later the so-called Liaison System) was created which served well to resolve ambiguities or disputed situations. The overall record of compliance with the Sinai Agreements was good. According to Mandell (1987: 19), for "the most part, all violations were deemed to be small infractions, easily detected, identified and corrected". The positive compliance record implies that there was an understanding of what would have constituted a militarily significant violation. To sum up, any mainly cooperative verification system with its components of monitoring and interpreting the collected data will need workable, agreed-upon definitions, standards of adequacy/efficacy, technologies, procedures, and forums as necessary conditions for an incremental reduction process towards politically acceptable MFZ related agreements. Both the East–West conflict and regional examples (including the important role of third parties) provide resources for developing a successful model for any future missile control and disarmament effort in the Middle East.

Means of verification

An effective verification system requires both technical and human means for monitoring and inspection. Technological instruments help in systematically collecting, analysing, storing, and rapidly disseminating information. They can operate consistently as compared to inspectors and can be designed – if ground-based – to detect treaty-relevant information. For example, in the INF Treaty, missile canisters were x-rayed to determine missile types; however, the limited resolution of the resulting images did not enable other relevant sensitive information to be revealed (Steinberg

1997). The question is: how far and to what extent can technology be relied on to provide credible, trust-building verification measures? In the following we will discuss in a more general way the utility and limitations of various technologies as well as the role of on-site inspections in establishing a mandatory system of verification for reaching consensus on the way towards an MFZ across the entire Middle East/Gulf and for the MFZ itself. Our main conclusion will be: while aerial, satellite, and ground-based surveillance can play a supportive role, human inspections will have to be the backbone of a sufficiently credible and effective monitoring system.

Space-based surveillance

Satellites carry ground-observing, primarily optical sensors which operate mainly at wavelengths in the visible band but also in infrared bands. Recently, synthetic aperture radar (SAR) sensors have been added to this repertoire. Optical resolution can range from about 0.1 m, in the case of the large optical sensors of US military satellites, to about 0.5–1 m in the case of commercial ones. Resolutions of infrared and SAR sensors are poorer than those of visible light ones. But thermal infrared and synthetic aperture radar devices are not encumbered by lighting and weather conditions, and they can recover terrain images at night. When choosing an appropriate wavelength, SAR sensors are even able to penetrate cloud cover. Multi-spectral optical sensors are now available to furnish an extra measure of sophistication to space-based surveillance systems, for instance, for detecting camouflage.

Space-based NTMs – a synonym of intelligence surveillance satellites – have been used for treaty verification since the 1970s. These instruments, which have relatively low operating costs, helped to preserve stability and prevent conflicts under the conditions of limited transparency during the Cold War (Potter 1980). Their major advantage is non-intrusiveness, since they do not violate national sovereign airspaces. Yet, reconnaissance satellites have some serious constraints. First, they are limited to exterior views and cannot observe the interiors of buildings and facilities. While multi-spectral optics and radar technology can provide fine details and important information on material compositions, they cannot penetrate even thin overhead screens that are now routinely used to hide ground activities from space surveillance. Second, as resolution – concerning optical and infrared sensors – is inversely proportional to the field of view, high-resolution observation from space is akin to looking at the target through a long thin straw and thus provides a veritable 'tunnel vision'. A known site can be observed in great detail, while undeclared activities can escape observation for a considerable amount of time. Third, since satellites travel in Keplerian orbits, their passage is predictable over a given segment of territory so that an ambitious country can take precautions against disclosure of illegal activity in time, making them amenable to spoofing.

Airborne surveillance

Observation from (un)manned aircraft carrying optical and radar devices has significant advantages over space-based surveillance but suffers from the basic problem of intrusiveness. Since aircraft operate at altitudes that are a fraction of those of satellites (i.e. they pass much closer to their targets), the resolution of their imagery is potentially much higher. Like satellites, surveillance aircraft carry visible light, infrared, and SAR sensors. Unlike satellites, however, airborne optical devices are capable of panoramic photography and can obtain high-resolution optical and thermal imagery of wide swaths of terrain on both sides of the flight path. The future may see an even more varied repertoire of airborne surveillance sensors with unprecedented capabilities. Gravity anomaly sensors could observe covert underground activities, magnetic anomaly devices might be able to sense underground reinforced concrete construction, radio-frequency sensors could map underground power cables, and high sensitivity chemical 'sniffers' will perhaps reveal the presence of chemicals such as rocket fuels as well as chemical weapons or explosives. These types of exotic sensors are presently under laboratory development, and some of them could emerge operationally within a decade.

On the downside, airborne surveillance suffers from the same drawback as space surveillance by being limited to exterior views without the ability to penetrate thin overhead screens set up to hide covert activity. Moreover, this technology is expensive to operate and – this is its greatest disadvantage – is palpably intrusive. Unlike satellites, surveillance aircraft must fly inside the sovereign airspaces of the target countries, consent for which has to be obtained with a flight plan filed in advance. The Treaty on Open Skies provides a good example of the complications and potential of airborne monitoring (Spitzer 2009).

Ground-based surveillance

Compared to space-based and airborne means of observation, ground-based surveillance offers more able sensing capabilities while increasing also the level of intrusiveness. Its role in treaty verification is to assist human inspections by providing the continuous monitoring of facilities against treaty violations as well as a less intrusive discrimination between allowed and non-permitted goods moving into and out of treaty related facilities. Gate controls of vehicles, persons, packages, or other equipment are increasingly common in security areas such as airports or nuclear facilities using, for instance, metal detectors or x-rays. As demonstrated by the INF verification procedures, non-intrusive measurement devices can be used to gain knowledge of the general internal structure of the missile and its payload without revealing sensitive information on its design or causing irreversible damage. These methods have a high probability of discovery, limited costs, and good transportability, and they should have minimal intrusion and easy handling. Ground-based sensors for remotely controlled monitoring of distant sites are usually optical, but can also include thermal and acoustic as well as seismic variants (Altmann 2005).

They are usually derived from the home security, anti-burglary industry, cheap, and easy to obtain. Although inspection techniques are dependent on the goodwill and permission of the inspected entity, it is possible to employ safety measures – such as periodic 'alive' signals using secret codes every minute – to make sure that violations would become known fast. Using infrasound sensors is similar to devices employed by the International Monitoring System of the Preparatory Commission of the Comprehensive Test Ban Treaty Organization. These instruments detect sound waves travelling through the atmosphere, which can be identified over long observational ranges, a method adopted to observe nuclear and large chemical explosions during the 1950s and 1960s. Using infrasound for monitoring rocket launches has the benefit of building on an already established verification system which might help in synergizing and reducing costs (Gopalaswamy 2009).

Human operated ground-based sensors to assist in the inspection of the flow of goods in and out of plants and ports of entry are plentiful; they are rapidly becoming more sophisticated due to the urgent requirements of homeland defence. They can potentially detect a wide range of signals, including sound and (ultra-)sound ones, light (visible, ultra-violet, infrared), electrical and magnetical fields, penetrating radiation (e.g. x-rays, gamma-radiation, and neutrons), gravitation, and released chemical substances. Future inspectors could peer into the contents of cars and trucks, and persons can be examined at short distance for bodily carried goods. Drawbacks of ground-based surveillance are high operating costs due to its labour intensiveness and its high degree of intrusiveness which makes it dependent on the goodwill and cooperation of the target entity. The first deficit can be addressed by more automation, the second one by limiting the intrusiveness by design. Nevertheless, the impressive range of available and emerging technologies promise that comprehensive and reliable ground-based surveillance is quite feasible for missile disarmament.

Human inspections

As in other fields of human endeavour, technology per se cannot substitute completely for physical human involvement. Likewise, monitoring technologies, advanced as they might be, cannot be expected to replace human inspections and mutually agreed-upon data exchanges. At best, technology could focus, assist, and enhance them. During the Cold War, the added value of on-site inspection was doubted perhaps for political reasons but also because of a desire not to overload the entire arms control process with additional intrusive examination requirements. The opposite view is taken now, and the benefits of human inspection have been demonstrated on many occasions, most noteworthy in monitoring the missile plants of the Saddam Hussein regime in Iraq after their shutdown was mandated by the UN Security Council. The verification of the INF Treaty allowed continued manufacturing of intercontinental missiles and thereby necessitated the examination of and discrimination between missile components leaving the gates of their production plants. This kind of control was used to determine the existence, type, and number of missiles being produced. In a multilateral context, the International Atomic Energy Agency has considerable experience with such inspections. Altogether it can be safely assumed that on-site inspections could provide the most trustworthy tool of verifying the step-by-step process towards an MFZ and of verifying the compliance with a zone related agreement itself.

Verification challenges: space launchers and unmanned aerial vehicles

During a gradual missile reduction process and especially in the context of an MFZ any verification system needs to tackle two fundamental problems: first, how to effectively verify that a long-range missile programme is aimed at legitimate commercial space activities – and not used as a disguise for military purposes, and, second, can one adequately accomplish this in view of the specific problems posed by the unique characteristics of cruise missiles and unmanned aerial vehicles (UAVs)? Complications will ensue if some actors seek to import complete space launch vehicles or components for their own indigenous SLV development. One of the possible routes for clandestine missile activities would be to exploit the dual use potential of space launch technology (Scheffran 1995: 156–64, 1997: 310–26) involving large multistage rockets launched from the ground with high-thrust engines, stage separation techniques, protective covers around the payload, as well as guidance and telemetry systems. Despite their similarities, space launchers and ballistic missiles differ in their basing modes, testing procedures, payloads, flight trajectories, as well as the guidance and re-entry systems that could be used to distinguish them. While early generations of space launchers and ballistic missiles were similar enough to be used interchangeably, their development trajectories diverged with the increased sophistication and the differing requirements of rockets for civilian and military purposes.

The manufacturing of SLVs often requires the same infrastructure as ballistic missiles – in fact, some SLVs contain complete subsystems of operational ballistic missiles, such as Iran's Safir, which uses the module and tankage of the Shahab-3 missile for its first stage propulsion (IISS 2010: 11, 133). Prohibiting SLVs along with ballistic missiles would shut down existing production lines, and verification would be required only to make sure that the lines remain closed. In case of permitting SLVs the verification task is more complicated, since one will have to deal with active production lines and make sure that the quantities produced and deliveries made are strictly for SLVs, which are not stockpiled in warehouses, but actually launched. In addition, cruise missiles introduce another level of complication for any verification system. If they are recallable they are outside the missile definition according to the UN Panel's 2002 report (see the Introduction by Bernd W. Kubbig).

A cruise missile denotes a UAV, usually jet propelled, that can traverse long distances at low altitudes and deliver a conventional, nuclear, biological, or chemical warhead to a predesignated point on the ground with high accuracy. Most types of cruise missiles are specially designed for this purpose, but manned aircraft can be converted to a cruise missile by replacing the human crew with automated navigation and flight control systems. Originally, non-weapon UAVs were largely relegated to reconnaissance and target-drone roles. More recently, reconnaissance UAVs, such as the American Predator, have been adapted for conventional weapons delivery. Indeed, perhaps the simplest such adaptation, would be the addition of a small sprayer that could enable the effective dissemination of a biological payload such as dry anthrax. Assuming that kind of anthrax was available, such a transformation could be accomplished fairly quickly, making detection highly problematic. The distinction between UAVs that can be used for all kinds of attacks and those used only for surveillance is a quite difficult but not impossible one to draw because even here there are observable differences (wing mounts, sprayers) that could be detected by inspections.

Another verification challenge relates to the unique characteristics of UAVs. The engines needed to permit sustained aerodynamic flight do not produce anything comparable to the infrared signature of ballistic missiles, making detection from geosynchronous early warning satellites orbiting the earth at a distance of 36,000 km practically impossible. UAVs are much easier to conceal than ballistic missiles during the development phase because they could be intermingled with civilian or military aircraft. Possibly, manned and unmanned airborne surveillance could conceivably compensate for some of the limitations inherent in space-based detection, yet, in the end, UAV verification will have to depend on a system of challenge and perhaps even surprise inspections of both aircraft and UAV research, development, and production facilities as well as sites deemed to be involved in suspicious activities.

Cooperatively verifying gradual arms reductions on the way to an MFZ

Although cooperative measures have become an integral part of the entire verification system developed during and after the Cold War, the Middle East related literature on the topic is largely sceptical about them. From an Israeli perspective there seems to be a high degree of trust in human sources as well as in mutual and joint Arab–Israeli verification regimes, whereas "the Arabs generally prefer international or third party mechanisms of verification" (Lederman 1994: 157). This raises a central issue: would countries such as Syria and Iran as vital members of the gradual process towards the envisaged MFZ accept such involvement by the United States with its superior technology? In the following we outline two areas of cooperative verification in accordance with the phases suggested in the article on Caps and Bans; the phases are not necessarily 'sequential', since we assume that CBM agreements and arms control agreements run "at least part of the time in parallel" (ibid.: 155).

First area: Open Skies as a useful model for confidence- and security-building in the Middle East context

Open skies type agreements were successfully negotiated and implemented at the end of the Cold War both in a bilateral context (Hungary and Romania) and in a multilateral setting involving what today are 34 states parties. The treaty implementation contributes to the monitoring of compliance with major arms control accords like the CFE, the 1991 Strategic Arms Reduction Treaty (START), and the New START Treaty.[8] The architecture of the multilateral and cooperative Treaty on Open Skies is based especially on the following principles: unlimited territorial access for observation flights with film and video cameras at 30 cm resolution,[9] thermal infrared sensors at 50 cm resolution, and sharing of the image data between the observed and the observing party; a system of agreed flight quotas; and costly cooperative execution of the flights with joint teams (with a flight timeline allowing for an element of surprise).

It can be assumed that Open Skies flights are being used regularly to monitor missile sites of the states parties. The resolution of 0.3 m allowed by the treaty is fully adequate to identify such locations precisely. The future use of digital aerial cameras with high (14 or 16 bit) sensitivity will enhance the capability to take useful images under conditions of reduced lighting, e.g. while flying under the clouds. Despite the historic successes of using cooperative third party aerial monitoring in the Middle East/Gulf it is hard to imagine under the present political conditions that an Open Skies regime or external party observation would be welcomed by all states of the region as part of a verification system of incremental missile reductions towards an MFZ and of an MFZ itself. There are four major – but in principle surmountable – stumbling blocks:

1 *Unlimited territorial access to overflights.* This would demonstrate the commitment of the members to mutual transparency. However, should parties agree on a system of limited aerial access only with restrictions on the sites to be photographed, appropriate procedures could be worked out. Joint sensor operator teams could verify – with the help of

information from navigation satellites – whether imaging sensors are activated only over the locations agreed upon. However, under a scenario of limited territorial access the regime would lose the capability to monitor non-declared but suspect locations.

2 *Agreed resolution of 0.3–0.5 m.* At present several states in the region might object to having their territory photographed at 0.3–0.5 m resolution. This objection will lose its importance over time when other countries in the region obtain high-resolution reconnaissance satellites of their own or gain access to commercial satellite imagery at 0.5–1 m resolution.

3 *Quota system.* States parties decide annually in the Open Skies Consultation Commission on the implementation of their quota entitlements (e.g. which country will observe other states how many times in the coming year). The joint administrative and coordinating effort for such a quota system will be much more complicated in the conflict-ridden Middle East/Gulf region.

4 *Costs.* As a rough guideline an operation unit of some 25 personnel and a budget of about 500,000 Euros would be required for executing ten missions with rented aircraft and for escorting the same number of flights. Compliance with the 0.3 m resolution limit would involve costly test flights by each operating state to determine the flight altitude that corresponds to the resolution limit for each sensor configuration.

To summarize, the stumbling blocks to Open Skies in the East–West context were overcome. It should be possible, given the necessary political will, to overcome them in the Middle East as well with solutions adapted to the needs and capabilities of the region.

Second area: restrictions, reductions, and the context of an MFZ

An MFZ cannot be established overnight. In accordance with the article on Caps and Bans we suggest focusing efforts initially on missiles with a range of at least 70 km which can be equipped with WMD. In the early stages a flight test ban would prove a useful tool to halt or at least to slow down the development of delivery vehicles, since flight testing is essential to building confidence that the system under development will work as intended (Gopalaswamy and Scheffran 2009). Such a ban would be relatively easy to verify. As then-CIA Director William Webster acknowledged in May 1989, the "status of missile development programs is less difficult to track than nuclear weapons development. New missile systems must be tested thoroughly and in the open" (quoted in Lumpe *et al.* 1992: 35). Additional measures to strengthen crisis stability include de-alerting weapon systems by separating missiles from warheads and adequately verifying them through on-site inspections.

As far as the destruction of the missile related production, testing, repair/maintenance, and related storage facilities is concerned, verification is more difficult (Altmann 1993). Restraints on research and development (R&D) are hard to verify, particularly due to the overlap with legitimate space projects, although conversion of military R&D facilities in conjunction with international space cooperation could raise transparency and confidence to acceptable levels. Once a binding MFZ agreement has been implemented, the task of verification would be to prevent violations and breakouts, whether through indigenous capabilities or imports. This means that all tools and facilities will remain inactive, and that no new production and test facilities will be built.

If an MFZ agreement should allow for the production, testing, and deployment of missiles up to a certain range (at least 70 km), the verification measures would have to ensure that no legally permitted missiles or components/technologies (including imports) are diverted to the manufacture and testing of prohibited delivery vehicles. The verification demands increase with a reduction in the permitted range threshold, since delivery systems at low ranges (like the unguided short-range artillery rockets or small UAVs) are hard to detect and track. Control of these systems would be left to later stages in the process, once progress had been achieved on the larger systems. In an environment of trust that is conducive to cooperative security, monitoring of the smaller systems is not excluded because functioning on-site inspections can check any suspected location for the existence and the infrastructure needed to maintain them.

All this requires the permanent verification of critical sites, facilities, and cross-border activities. In the case of the latter, continuous monitoring would be important, since external assistance can advance both space and missile programmes, e.g. through transferred guidance and propulsion systems and their components (Gopalaswamy 2008). A combination of technology and human activities, especially ground-based monitoring and inspections, could provide a solution, as long as this is embedded into a cooperative setting and standards of adequate verification are applied that do not demand the impossible, i.e. absolute or perfect security against all imaginable worst-case scenarios. Such a verification system could be modelled after the INF example to include: on-site inspections at declared missile deployment and destruction sites (as mentioned previously, even the most advanced technologies for the verification of an MFZ are no substitute for human inspections); portal monitoring with x-ray and other detectors at declared missile production sites with unlimited inspector access to the portal monitoring equipment for checking its integrity and operation; and container screening at air, land, and sea entry points of the region in order to check for forbidden imports of missiles or major components (the last monumental task was not part of the INF verification regime).

An additional problem in the region is the imbalance in the use of satellites as NTMs. While Israel is operating surveillance satellites with the required sub-metre resolution, other countries in the Middle East lack comparable capabilities. Nonetheless, the availability and resolution of commercial satellite imagery has improved considerably in recent years (Pabian 2008: 221–49). This would level the playing field somewhat by allowing access to useful information, provided there are no other restrictions on access to advanced satellite imagery, such as the limit of 2 m resolution for US commercial satellites over Israel. Those countries with less resources would benefit from international technical support and assistance, e.g. through a regional monitoring agency. The establishment of such an organization would be truly important, since it could help solve other problems as well. In addition to international cooperation in civilian space programmes involving, for example, the sharing of multilaterally owned space equipment and facilities (satellites, rockets, and ground facilities), a regional (or international) agency which could oversee the gradual destruction of the missile arsenals, could also monitor space launch activities, permit conversion of decommissioned missiles to space launchers, and safeguard missile related facilities (Tarasenko *et al.* 1995: 138–42).

Conclusions

This article has outlined vital elements of a verification system as part of a broader missile control/reduction concept – the premise being that adequate verification is a necessary (although not sufficient) condition for successfully embarking on the incremental process of reducing delivery vehicles with the final goal of an MFZ. We have outlined both the experience gained during the Cold War and in the Middle East itself as points of reference for future activities and have given an overview of the spectrum of verification technologies ranging from optical sensors orbiting in space to a variety of advanced ones that can be carried aboard aircraft and deployed on the ground. In addition, we have pointed to the crucial importance of human inspections in various forms on the ground, and we have elaborated on their utility and limitations. This article has also underscored the importance of cooperative features for any verification system.

We have concluded that adequate verification of the major areas during the reduction process and in the context of ultimately creating an MFZ in the Middle East is possible in principle if technical monitoring devices are supplemented by cooperative human inspections. This will only work, however, in a politically favourable environment (which touches on the issue of domestic acceptability) – a major condition for facilitating arms reductions. The relationship between effectiveness and intrusiveness highlights the crucial verification challenge: the more effective the verification measures, the more intrusive they are and the more they become dependent on the trust and goodwill of the country in question. Space-based surveillance technologies are the least intrusive, but they are hampered by anything under cover or outside their narrow field of view. Ground-based technologies, especially combined with inspections, can observe and detect much more, yet they are highly intrusive, and therefore depend on prior trust – and must of course at the same time be confidence generating.

If the verification experience of the East–West context is any useful guide for the Middle East, then the conditions Allan Krass (1985: 750) has outlined for successful verification (see above) could function in terms of a checklist to help structure and assess expectations for future activities in the conflict region: the first condition of "objective interest in the success and preservation of agreements" is closely associated with the second one, i.e. the need to be "reassured by evidence of … compliance". Both conditions demonstrate the gap between East–West and the Middle East contexts. The challenge is twofold. On the one hand, genuine interest in arms reductions has to be developed in the region. The participating entities, governments and constituencies alike, have to be convinced that the entire incremental arms reduction process towards an MFZ starting with CSBMs meets the security needs of the major players. On the other hand, agreements to be achieved will have to be preserved. Here the record is mixed. While several accords have been implemented, the Nuclear Non-Proliferation Treaty offers a different story. It has been violated and challenged by activities in Iran, Iraq, and, obviously, Syria.

The third condition falls into two parts. The first element – confidence in one's own intelligence capabilities – again shows the differences between the two regional contexts, and it points to the relationship between effectiveness and intrusiveness. While both superpowers could rely on their unilaterally deployed NTMs, the states in the Middle East would have to accept on-site inspections by third party nationals or by UN personnel. Currently the likelihood that they are embraced by the vital actors in the Middle East is low. Once space-based and airborne surveillance technologies are used, they will become likely complements to direct, nationally conducted on-site inspections rather than a substitute for them. As already mentioned, even these limited technologies – mostly American ones – will pose severe problems for political acceptability. This will complicate the process of interpreting collected data and raise the problem of differentiating between legitimate observing and illegitimate spying. The second element of the third condition – a militarily significant advantage cannot be kept secret – addresses the requirements for adequate verification. Indeed, no such system is perfect, and a residual risk remains that countries could build a clandestine technical missile capability by exploiting dual use space launch technology. Krass' fourth condition – treaty ambiguities must be treated "in a calm … manner" on the assumption that the other side has an interest in "correcting any incidents of noncompliance" – points again to forums that do not currently exist and to an interest that has yet to be generated.

At the heart of the entire problem is the limited role of verification which "cannot act as surrogate for trust" but needs in turn "some initial trust". Indeed, paradoxically but not surprisingly, confidence is required to generate confidence. The implications of this are twofold. First, if history is any guide here, it teaches us that the long way of arms reductions towards disarmament gets its momentum in the wake of peace, not vice versa. Second, this insight nevertheless allows a limited but vital role for verification: increasing the readiness of the adversaries and of antagonistic domestic actors to accept mutual on-site inspections might in turn facilitate and endorse the interest in an arms reduction process compatible with security. The question of how to start the verification related process of building confidence and generating interest in it concerns extra-regional actors, independent experts, and governments alike. They therefore should seize the opportunity in the context of the envisaged Middle East Conference process by outlining in greater detail at least two proposals whose implementation would include both non-governmental experts and, above all, officials in the region: first, to conduct seminar-like conferences (like in the 1990s) as a means of building and expanding expertise in the short- and mid-term and, second, as a parallel and complementary activity, to secure support for the idea of a regional monitoring agency (the importance of which we have briefly described). This organization would be another step along the way towards building a constituency for generating a broader culture of verification within a transnational setting.

Notes

1 The terms 'confidence' and 'trust' are used as synonyms.
2 On the need for a non-discriminatory, multilateral, and verifiable treaty banning the production of fissile material and an arms control approach in the Middle East, see Aboul-Enein (2009a). The importance of verification and transparency for nuclear disarmament is discussed by Aboul-Enein (2009b).
3 This holds true even for the UN-based milestone study by Leonard *et al.* (1991: 12).
4 The official definition of the term 'verification' as provided by the U.S. Arms Control and Disarmament Agency in the era of Ronald Reagan reads:

> VERIFICATION: The process of determining, to the extent necessary to adequately safeguard national security, that the other side is complying with an agreement. This process of judging adequacy takes into account the monitoring capabilities of existing and future intelligence collection systems and analysis techniques and the ability of the other side to evade detection if it should attempt to do so. This process also assesses the political and military significance of potential violations and the costs, risks, and gains to a side of cheating.
>
> (ACDA 1983: 367)

This definition ignores the conceptual battle between the traditional standards of adequacy versus effectiveness which the Reagan administration introduced, as it regarded the yardstick of adequacy as insufficient. But "the difference has never been spelled out" by the Republican government (Krepon 1984: 18). This view is corroborated by Scribner *et al.* (1985: 17), who concluded: "The terms adequate and effective arise because ... no arms control agreement is totally verifiable."

5 Harold Brown, former U.S. Secretary of Defense, quoted in Krass (1989: 16); similarly U.S. Secretary of State George Shultz, quoted in ibid. (17).

6 See on this Kubbig (1988: 122–7, 199–203) versus Scribner *et al.* (1985: 33).

7 William Colby, former CIA Director, quoted in Wieß (1989: 12).

8 For more details and sources see Spitzer (2009), Dunay *et al.* (2004).

9 The definition of resolution in Open Skies practice is the ground sampled distance, which is half the ground resolved distance used in photogrammetrics.

References

Aboul-Enein, Sameh (2009a) 'The 2010 NPT Review and the Middle East: Challenges and Opportunities', *Palestine-Israel Journal of Politics, Economics and Culture*, 16(34): 67–76.

Aboul-Enein, Sameh (2009b) 'The Roadmap to Total Nuclear Disarmament', in George Perkovich and James M. Acton (eds) *Abolishing Nuclear Weapons: A Debate*, Washington, DC: Carnegie Endowment for International Peace, 271–85.

ACDA/U.S. Arms Control and Disarmament Agency (1983) *Fiscal Year 1984 Arms Control Impact Statements*, Washington, DC: US Government Printing Office.

Altmann, Jürgen (1993) 'Ballistic Missile Limitations and Their Verification', in Götz Neuneck and Otfried Ischebeck (eds) *Missile Proliferation, Missile Defense, and Arms Control*, Baden-Baden: Nomos, 91–100.

Altmann, Jürgen (2005) 'Acoustic-Seismic Detection of Ballistic-Missile Launches for Cooperative Early Warning of Nuclear Attack', *Science and Global Security*, 13(3): 129–68.

Dunay, Pál (1991) *Verifying Conventional Arms Limitations: The Case of the November 19, 1990 Treaty on Conventional Armed Forces in Europe*, Bochum: UVB-Universitätsverlag Dr N. Brockmeyer.

Dunay, Pál, Krasznai, Márton, Spitzer, Hartwig, Wiemker, Rafael, and Wynne, William (2004) *Open Skies: A Cooperative Approach to Military Transparency and Confidence Building*, Geneva: United Nations Institute for Disarmament Research.

Gopalaswamy, Bharath (2008) 'Is an Iranian Missile Threat Imminent?' *INESAP Information Bulletin*, 28: 20–5.

Gopalaswamy, Bharath (2009) *The Use of Infrasound Technology for Monitoring of Rocket Launches*, London: Verification, Research, Training and Information Centre (Typescript).

Gopalaswamy, Bharath and Scheffran, Jürgen (2009) 'Time for a Missile Test Ban', *Bulletin of the Atomic Scientists*. Online, available at: www.thebulletin.org/web-edition/op-eds/time-missile-test-ban (25 March 2011).

IISS/International Institute for Strategic Studies (2010) *Iran's Ballistic Missile Capabilities: A Net Assessment*, London: IISS.

Krass, Allan S. (1985) 'The Politics of Verification', *World Policy Journal*, 2(4): 731–52.

Krass, Allan S. (1989) *The Verification Revolution*, Cambridge, MA: Union of Concerned Scientists.

Krepon, Michael (1984) *Arms Control: Verification and Compliance*, New York: Foreign Policy Association.

Kubbig, Bernd W. (1988) *Amerikanische Rüstungskontrollpolitik: Die innergesellschaftlichen Kräfteverhältnisse in der ersten Amtszeit Reagans (1981–1985)*, Frankfurt am Main: Campus.

Lederman, Itshak (1994) 'Verification in the Middle East and Europe: Lessons from Past Experience', in Shai Feldman (ed.) *Confidence Building and Verification: Prospects in the Middle East*, Boulder, CO: Westview Press, 137–60.

Leonard, James, Prawitz, Jan, and Sanders, Benjamin (1991) 'Effective and Verifiable Measures Which Would Facilitate the Establishment of a Nuclear-Weapon-Free Zone in the Middle East', Report of the Secretary-General, A/45/435, New York: United Nations Department for Disarmament Affairs.

Lumpe, Lora, Gronlund, Lisbeth, and Wright, David C. (1992) 'Third World Missiles Fall Short', *Bulletin of the Atomic Scientists*, 48(2): 30–7.

Mandell, Brian S. (1987) *The Sinai Experience: Lessons in Multimethod Arms Control Verification and Risk Management*, Ottawa: Department of External Affairs.

Pabian, Frank (2008) 'Commercial Satellite Imagery: Another Tool in the Nonproliferation Verification and Monitoring Toolkit', in James E. Doyle (ed.) *Nuclear Safeguards, Security, and Nonproliferation: Achieving Security with Technology and Policy*, Oxford: Elsevier, 221–49.

Potter, William C. (ed.) (1980) *Verification and SALT: The Challenge of Strategic Deception*, Boulder, CO: Westview Press.

Scheffran, Jürgen (1995) 'Verification of Missile Bans and Monitoring of Space Launches', in Wolfgang Liebert and Jürgen Scheffran (eds) *Against Proliferation: Towards General Disarmament*, Münster: Agenda-Verlag, 156–64.

Scheffran, Jürgen (1997) 'Elimination of Ballistic Missiles: An Important Step towards a Nuclear-Weapon-Free World', in Joseph Rotblat and Michiji Konuma (eds) *Towards a Nuclear-Weapon-Free World: Proceedings of the Forty-Fifth Pugwash Conference on Science and World Affairs*, Singapore: World Scientific, 310–26.

Scribner, Richard A., Ralston, Theodore J., and Metz, William D. (1985) *The Verification Challenge: Problems and Promise of Strategic Nuclear Arms Control Verification*, Boston, MA: Birkhäuser.

Spitzer, Hartwig (2009) *News from Open Skies: A Co-Operative Treaty Maintaining Military Transparency*, London: Verification Research, Training and Information Centre. Online, available at: www.isn.ethz.ch/isn/Digital-Library/Publications/Detail/?ots591=0c54e3b3-1e9c-be1e-2c24-a6a8c7060233&lng=en&id=97949 (25 March 2011).

Steinberg, Gerald M. (1997) *New Verifications Technology and Issues: Space and Airborne Monitoring*, Vienna: International Atomic Energy Agency. Online, available at: http://faculty.biu.ac.il/~steing/arms/verify.htm (25 March 2011).

Tarasenko, Maxim V., Pike, John, and Scheffran, Jürgen (1995) 'International Space Cooperation and Conversion of the Aerospace Complex', in INESAP/International Network of Engineers and Scientists against Proliferation (ed.) *Beyond the NPT: A Nuclear-Weapon-Free World*, New York: INESAP, 138–42.

Wieß, Bettina (1989) *Verifikation und "Compliance Issues": Die amerikanische Diskussion um strategische Rüstungskontrolle (1977–1985) und der INF-Vertrag*, Münster: LIT.

9 The prospects for disarmament

The case of Hezbollah

Judith Palmer Harik and Walid Abu-Dalbouh

The role of rockets and the confrontation between Israel and Hezbollah

Katyusha rockets and guided missiles played a critical part in Hezbollah's (Party of God) tactics during the 2006 War with Israel. As evidence of the continuing volatility of this conflict, these weapons have reportedly not only been replenished in large numbers by the Lebanese organization's partners – Syria and Iran – but there have been indications and reports revealing that surface-to-air missiles (SAMs) have been added to the Party of God's arsenal. In April 2010 Israeli Defense Minister Ehud Barak warned that the transfer of missiles from Syria to Hezbollah "undermines the balance of power and threatens regional stability and calm" (quoted in *YNET*, 4 April 2010). At an earlier date he indicated that SAMs posing a threat to his country's aircraft "are unacceptable to Israel" (quoted in *Haaretz*, 9 August 2008). In the present situation, it is difficult to imagine that either combatant would willingly accept the limitations imposed by either missile disarmament or even a Middle East/Gulf Missile Free Zone (MFZ). Nevertheless, we suggest that changing local, regional, and international circumstances and/or a new approach to the Hezbollah–Israel conflict could create conditions under which participation in an MFZ might be viewed more favourably by the belligerents. Focusing on the Party of God, this chapter explores potential changes in regional politics or attitudes that could induce Hezbollah to reverse its present adamant rejection of any initiative requiring its disarmament.

We should acknowledge at the outset that our analysis would be on firmer footing if the organization's strategic intentions were clearly known. Uncertainty about whether recuperation of Shebaa Farms,[1] or Jerusalem – of course notwithstanding its critical role in fulfilling its regional allies' political interests – is the Party of God's main purpose makes it difficult to point with any certainty at the conditions that could induce its support of Lebanon's participation in an MFZ. We thus approach this problem by examining the goals articulated by Hezbollah's leaders and infer their strategic intent from their behaviour. The intricate task of this assignment is exacerbated by the fact that the Party of God's decisions regarding disarmament or arms limitation and inspection would be conditioned not only by perceptions of its enemy's intent but also by positions taken by Hezbollah's allies Syria and Iran. These actors must decide upon the relative merits of a continuing struggle with Israel, negotiating outstanding particular issues, or pursuing the much sought comprehensive resolution of the Arab–Israeli conflict.

Since Tehran and Damascus are able to influence the Party of God's military agenda by increasing or withdrawing their assistance to the war effort, the interests and policies of these states must also be factored into the present discussion. Before turning to the issue at hand, the article examines Hezbollah's association with Syria and Iran as well as the extent to which both can influence the Party of God's adoption of strategically sensitive policies. This section is followed by a brief review of Hezbollah's emergence and development into a mainstream Lebanese political party, with a resilient fighting force. This paves the way for a discussion of the various possible scenarios under which the Party of God might be induced to relinquish its missiles (or at least accept restrictions). These scenarios are: relying on missile defence (MD), using military force, the possibility of Lebanon regaining its sovereignty, a missile 'no use' agreement, and withdrawal of the support of Hezbollah's allies in the wake of 'land for peace' negotiations.

Hezbollah in regional and Lebanese contexts

The Syrian–Iranian tie

Politically, the Party of God is the product of the long-standing traditional alliance between Syrian President Hafiz al-Asad and Iran's Ayatollah Ruhollah Khomeini struck in 1980.[2] The coalition aimed at changing the strategic balance of power in the Middle East/Gulf. To accomplish this goal both countries apparently focused on responding to three mounting conflicts in the region: first, the confrontation with Israel in and from southern Lebanon; second, the Palestinian resistance to Israeli occupation; and third, the Iran–Iraq War (1980–88). Over the years, Damascus and Tehran have developed and/or supported separate local military allies or surrogates – Hezbollah in Lebanon, Hamas in Palestine, and the insurgents in Iraq during and after Saddam Hussein's rule. While these three joint transnational interventions were executed separately and over different time periods, they nevertheless collectively serve the overall objective of the Syrian–Iranian alliance: confrontation of Israeli occupation and prevention of an American hegemony in the Middle East.

Among the more important particular interests that energized the Damascus–Tehran coalition in 1980 was the United States' effort to weaken the newly formed Islamic Republic by encouraging Saddam

Hussein's military campaign against Iran.[3] Since the 'War of the Cities' effectively isolated Tehran, the Khomeini government sought political influence elsewhere in the region. An entrée into the Middle East conflict was desirable, since it would allow Iran to project its power westward, export the Islamic Revolution, and strike a blow against Israel, commonly called 'the usurper of Holy Jerusalem'. Hafiz al-Asad's goals in the early 1980s were more pressing. He needed to counter perceived Israeli designs on Lebanon so as to protect Syria's flank, while searching for the political leverage needed to regain the Golan Heights, a water-rich area which had been annexed by Israel in the aftermath of the 1967 War.[4] The story of Asad's attempt to control the Lebanon-based Palestine Liberation Organization (PLO) as an indirect way of recouping the Golan Heights is well known and need not be rehearsed here.[5] The relevant point is that this strategy failed, as Damascus' power over the Palestinian commandos never proved strong enough to control the stick they were trying to wield against Israel. The PLO's evacuation from Lebanon during the 1982 Israeli invasion,[6] however, necessitated Asad's search for a more reliable replacement in Lebanon that he could dominate.

Syria turned to Iran for assistance in this endeavour, and Tehran was willing to cooperate, since the government had already marshalled allies from among Iraq's Shiite community during the Iran–Iraq War. Hence the potential for a similar gambit in Lebanon was evident. The authorities were completely paralysed as a result of the ongoing civil war, the Shiite community was already mobilized, and its ally Syria had control over important areas of the country. Moreover, a handful of capable Mullahs who embraced the Khomeini concept of 'activist Islam', a rejection of former Shiite pacifism,[7] were available to command the envisioned surrogate force. Furthermore, several violent military remnants of the 1982 battle of Beirut, some apparently perpetrated by Iran with Syria's consent,[8] provided a climate of resistance guaranteed to speed recruitment of a Shiite force.

Apparently, an agreement was made between Iran and Syria that the latter would have its surrogate force, that this militia would be Shiite, and that it would be a Lebanese version of Iran's Hezbollah, an organization formed during the revolution to fight the Shah's powerful police and military forces by unconventional means. Asad could not hope to finance the long-term conflict with Israel which was guaranteed by the appearance of a Syrian-backed force at the Lebanese–Israeli frontier after 1982. The agreement made with Khomeini, however, solved this problem and produced a natural division of labour between the two actors regarding their respective contributions. Iran would send Revolutionary Guards to train and indoctrinate the recruits of what would become Hezbollah; they would provide the models for command and control as well as troop support systems; and bankroll the project, including soldiers' pay, families' benefits, and weapon delivery.

For its part, Syria was to provide training grounds and the logistics for both cash and arms supplies from Tehran through Damascus to Lebanon's Bekaa Valley where the Party of God's fighting cadres would be formed. At a later date, Syria would also prepare the political ground necessary for Hezbollah's permanent insertion into Lebanese political life through the politico-military influence it had developed during the Civil War period (1975–90).[9] In this alliance, each partner had found itself with an unparalleled opportunity to influence the Middle East peace/war equation from the sidelines. For the Party of God's leaders, it meant that they would be able to conduct a fully supported *Jihad* against their main enemies – directly against Israel, and indirectly against the United States of America, two countries they considered to be the greatest threats to Islam – with a considerable degree of freedom.

While Hezbollah is the product of Syrian–Iranian foreign policy, it would be a mistake to assume that the organization is completely dependent on its benefactors. According to Daniel Byman (2005: 78), "the Lebanese Hizbullah enjoys exceptionally close ties to Tehran and works carefully with Damascus [but] remains its own organization and has not been entirely a proxy of one of its sponsors".[10] This fact, as well as the level of commitment of all three actors to the struggle against Israel and the US, is of great importance for understanding the conditions under which the Party of God might eventually be led to accept arms limitations.

Hezbollah as a militant Lebanese resistance force

In the decades since its formation in 1982, the weak Lebanese government has given the Party of God a free rein to pursue its resistance against Israel. The establishment of a security zone within Lebanon in 1985 handed the organization its resistance credentials and led to a 17 year struggle on the part of the militant group to gain control of the border area. In May 2000 Israel and its ally, the South Lebanese Army – a mixed Maronite Christian/Shiite militia led by former Lebanese Army commanders – withdrew from Lebanon, but Israeli troops retained outposts in the Shebaa Farms area. Their ongoing presence fuels Hezbollah's campaign against its enemy. During the period of Syrian regency in Lebanon, when it was important to emphasize the Party of God's national resistance credentials, the organization transformed itself into a mainstream political party, and it has since consistently won between 11 and 14 seats in the 128-seat chamber. Although the coalition it headed lost to the US-backed 14 March Movement[11] that won 71 seats in the parliamentary elections of June 2009 to the opposition's 57, Hezbollah retained the 11 seats it had won in the 2005 elections (some reckon 13 adding in two of its bloc members) and, for reasons described below, these results are not expected to seriously affect its armament behaviour.

The Party of God's capacity to withstand Israel's assault in 2006 enhanced its credentials while demonstrating the extent of its popular support and its domestic invincibility. Broad social and public services offered to Shiites and others living in Shiite-concentrated areas further cemented the political allegiance to Hezbollah among this economically distressed group. As a result the Party of God's leaders are able to mobilize rapid and massive political and military support when they summon it. The picture presented above is one of a strong, highly organized and well-armed actor that over the years has outflanked efforts by local opponents as well as regional and international enemies to bring it to heel. In this respect, Hezbollah considers the charges expected to be brought against its partisans by the Special Tribunal for Lebanon for their purported role in the assassination of former Prime Minister Rafiq Hariri in February 2005 as simply one more attempt by its adversaries to besmirch its resistance credentials. In a speech carried by al-Manar TV on 16 November 2010, Secretary General Hasan Nasrallah staunchly denied the accusation decrying the "manufactured" evidence presented to the court. He also threatened anyone who attempted to do harm to the resistance or to arrest the men reported to be indicted, saying they "will be cut". Despite these explosive developments, it is nevertheless recognized that Hezbollah has the situation well in control.

In short, it appears that the Party of God retains its political clout as it continues preparations for what its leader Hasan Nasrallah calls "the next and inevitable round of fighting with Israel" (quoted in *Daily Star*, 17 February 2009). But what does Hezbollah want to achieve from the confrontation? What is their final goal? As the Party of God has evolved militarily, has the original goal of the restoration of Shebaa Farms been replaced with a broader agenda?

Divining the Party of God's ultimate goal

Pinpointing Hezbollah's final objectives is made difficult by its leaders' deliberate ambiguity on this subject, stemming from their need to garner broad support from a multi-sect and politically divided population. Lebanon's political system was designed to balance sectarian interests in an effort to ward-off the instability that might arise if any confessional group came to dominate it. As leader of the country's largest sect, the Party of God has an interest in deflating notions that its goal is political hegemony or that it seeks the establishment of an Islamic Republic for Lebanon. Hezbollah has therefore emphasized its national resistance credentials and, except when addressing core Shiite adherents, has tended to de-emphasize goals associated with its religious identity.[12]

The Party of God makes a point of actively courting Christian endorsement of its resistance agenda while at the same time countering Sunni opponents' efforts to characterize its growing influence as part of a generally perceived Shiite threat to Sunni interests in the region. Hezbollah's emphasis on a Lebanese identity and a national agenda – the withdrawal of foreign troops from Lebanese territory – suggests that its objectives in confronting Israel would be satisfied if that condition was met. Yet after departure of Israeli troops and members of its surrogate force from the security zone in May 2000, the Party of God and the Syrian-backed Lebanese government claimed that resistance would continue until all of its territory was evacuated. This was a reference to Shebaa Farms where the Israelis retain a listening post.[13] In other words, as long as national territory remained under Israeli control, Hezbollah's leaders indicated that their organization had no interest in laying down its arms. However, it stands to reason that if Israeli troops evacuated Shebaa Farms it would be more difficult, but not impossible, for the Party of God to justify its present confrontational stance in national terms, since other Lebanese claims on land occupied by Israel exist.

Operating freely and, at the same time, covertly along the frontier, Hezbollah often targeted the Israeli listening post on Shebaa Farms in response to Israeli overflights of Lebanese territory or bombing raids. Its proximity to Israel's border and hence the shorter flight times for its rockets increased the difficulty of defending against them, hence providing the Party of God – and by extension its allies – an important strategic advantage in the continuing struggle with Israel. One special aspect of this conflict is whether Hezbollah continues to pursue a merely Lebanese agenda or whether it evolves into a broader Arab resistance. This self-understanding was publicly announced by Hasan Nasrallah in autumn 2000 as the second Palestinian *Intifada* went into full swing. At that time Nasrallah declared that his organization would "directly interfere" in the uprising while leaving how that would be done ambiguous.[14] Since that time Hezbollah's activities on behalf of Hamas have ranged from opening a carefully calibrated second front against the Israelis during the 2001 military campaign in Jenin, to the reported recent arms transfers to Gaza. The Party of God's television station al-Manar also promotes the Palestinian struggle in both Islamic and Arab nationalist terms, urging emulation of its own feats against the 'occupying forces'.

Do these activities and pronouncements mean that Hezbollah might expand its confrontation with Israel across the Lebanese–Israeli frontier? The contention that "Israel's very existence constitutes an ever-present threat" suggests that taking the battle into Israeli territory is always a possibility (Badran 2009).

Preparations for another confrontation

The fact that both belligerents continue to prepare for future confrontation underscores the need for a rapid resolution to this dangerous conflict. In general it can be inferred from Hezbollah's actions in the 2006

War with Israel and what little is known about its preparations for another battle, that coercion of the Hebrew state rather than any attempt to destroy it seems to be the Party of God's continuing goal. Presumably, this will continue to be carried out by short- and medium-range rocket fire designed to pressure Israel by provoking displacement of the population and targeting strategic assets. However, some experts maintain that since July 2006, Hezbollah has managed not only to recover its military competence, but also to strengthen it far beyond its pre-July capacities (Scheffran 2007). Yossi Kuperwasser (2008: 19) for instance contends that

> Hizbollah's massive force buildup since the end of the Second Lebanon War, which has occurred in part by virtue of the weakness of Security Council resolution 1701, has allowed the organization to amass more rockets and missiles with longer ranges than were in the 2006 arsenal it used against Israel.

Missile saturation by highly mobile cadres as well as hard-to-locate stationary launchers will probably continue to be the backbone of any strategy although this certainly does not exclude other military options. The organization's leaders often refer to "important surprises for Israel" in a future war.[15] The seriousness of this situation is magnified by the fact that the Party of God's effective use of missiles and its modernized guerilla warfare has not been lost on Syria, as can be seen by the restructuring and redeployment of its military along similarly unconventional lines. Additionally, its cadres have probably been equipped with the weapons in Hezbollah's armoury as well as more advanced guided missiles. However, the regime in Damascus has mainly been focused on the domestic unrests that intensified in 2011.[16] It should also be noted that Israel practised war games featuring tank assaults in the early spring of 2009 on the Golan Heights. In this respect, it might be relevant that Hezbollah's use of anti-tank missiles in summer 2006 was considered particularly effective by military analysts. One was led to remark that the Party of God's performance "may have signaled the end of tank warfare".[17] Apparently Israeli military strategists disagree with this assessment.

Further indications that Israel is on a wartime footing is the large-scale exercises of the Israeli Air Force held to simulate war against several foes including Iran, Syria, and Hezbollah from 18–21 May 2009. According to officials the four day exercises involved Israel's entire air arm, regular army personnel, as well as reservists and was meant to prepare for an all-out war, including missile attacks aimed at Israeli cities. Tzachi Hanegbi, who heads the Israeli Parliament's Foreign Affairs and Defense Committee and attended the exercises, told Israel's Army Radio on 22 May 2009 that "[i]t's no secret that the time for decisions is approaching, in all of the world's leaderships". Heightening the tension in this area further, it was announced that Iran had successfully test fired a new rocket, the Sajjil-2 in

November, that it claims to have a range of 1,200 miles while the Israeli war exercises were unfolding (*BBC News Middle East*, 20 May 2009).

Scenarios of Hezbollah's military restraint and disarmament

First scenario: intercepting incoming missiles

To meet a future possible threat from Hezbollah rocket fire, Israel is preparing a multi-layered missile defence umbrella, featuring the Iron Dome MD and Raytheon Company's Vulcan Phalanx system. The latter was developed to defend ships and features a Gatling gun that fires 3,000–4,500 rounds per minute (*Jerusalem Post*, 21 April 2009). Vulcan Phalanx may be used to protect highly strategic sites such as airbases and nuclear facilities from the Party of God's rocket barrages, but the problem of defending cities, towns, and villages from Katyusha short- and medium-range rocket diversification and saturation – the comparatively high costs of MD notwithstanding – has yet to be resolved. The head of the Sha'ar Hanegev Regional Council, Alon Schuster, sent an urgent personal letter to Defense Minister Ehud Barak on this issue on 13 January 2009. Referring to the short flight time of Hamas' Qassam missile, Schuster noted that the "reaction time of the Iron Dome missile defense system is too long to cope with the rockets adding that the system is incapable of protecting Sderot and many of the communities bordering the Gaza Strip" (quoted in Pedatzur 2008). The reliability of missile defence is therefore questionable, although Iron Dome successfully intercepted a Grad rocket in April 2011 (*Haaretz*, 7 April 2011). Towns and villages in northern Galilee that are targeted by Hezbollah's short-range Katyusha rockets face a similar threat.

Having indicated the severe danger to regional stability and the threat to civilian populations on both sides of the Lebanon–Israel frontier, if the Hebrew state and the Party of God continue on another collision course, we shall now examine how Hezbollah might be convinced to submit to missile restrictions or even MFZ regulations. Considering its growing military competence in the conflict with Israel this would appear to be difficult to bring about. The following sections will nevertheless present several hypothetical scenarios which might move the Party of God to forego its weapons.

Second scenario: arms limitation by force

The Party of God comprises the weakest link in the Syria–Iran–Hezbollah alliance and is the one member of the triad whose degradation or elimination by the Israeli military is least likely to escalate into a regional conflict. The relevant question is how this could be accomplished and what would be the cost to the Hebrew state and Lebanon given the results of the 2006

War. A brief review of that conflict is therefore in order. Israel's plan for a quick victory over the Party of God using precise air strikes to eliminate arms depots and strategic points south of the Litani River failed as a result of inadequate intelligence. By the time air raids had attempted to crush Hezbollah's control and command centre in southern Beirut, the Party of God's forces had evacuated the target areas, which meant the civilian residents were the ones mainly affected by the bombing.

Hezbollah fought Israeli ground troops and conventional armour with a combination of conventional and guerrilla tactics as well as using a variety of weapons such as short- and medium-range surface-to-surface rockets and anti-ship missiles (Exum 2006: 6). After weeks without ostensible gains by Israeli ground troops and air force, and under heavy rocket fire by Hezbollah which gradually reached beyond Haifa,[18] Prime Minister Olmert's fall back idea was to establish another security zone in South Lebanon. Nevertheless, this plan was quickly given up, and on 13 August 2006 a ceasefire was declared. An official investigation into Israel's military campaign concluded with a host of criticisms.[19] The relevant point for our discussion, however, is that Israel's use of massive conventional force to roll back the Party of God failed in 2006, as it had done in 1993 and 1996.[20] Intensive Israeli aerial bombardment that reached as far north as Beirut failed to destroy Hezbollah's command and control structure, while, at the same time, it fanned popular support for the resistance.

This leads us to the conclusion that while Israel certainly possesses the military hardware to remove the Party of God from the field of action by force, it remains unknown, given its poor performance in the 2006 War, whether these assets can be used effectively enough to counter the army Hezbollah has now become. In any case, it seems safe to say that if force remains the only option for curbing the Party of God, then future hostilities will likely involve an increase in targeted and collateral damage on both sides while Hezbollah may very well continue to escape defeat. In addition to the use of force against Hezbollah, Syria could also use military means as a tool of dissuading the Party of God from its mission should it be in Damascus' interests to do so. This option is explored below.

Third scenario: Lebanon's sovereignty regained

Hezbollah's capacity to wage war against Israel without government consultation or restraint is another feature of the Middle East crisis. Lebanon lost its sovereignty in 1969 when the Arab League – through the Cairo Accord – granted displaced Palestinian fighters the right to use the country as a base from which to attack Israel. Even before that time and ever since, Lebanon has been a strategic prize to be fought over by Syria and Israel. In that sense the state is still a pawn to foreign interests with the United States standing in for Israel, joined by Saudi Arabia, Egypt, and Jordan in efforts to stymie Damascus' influence on its smaller neighbour.

A Lebanese Sunni militia was formed after the 2006 War with the blessing of Washington and the collaboration of Riyadh, Cairo, and Amman, with the hope of successfully confronting Hezbollah. The militia was, however, disbanded in 2008. Lebanon's armed forces have not only lacked the weapons and training to defend their country, they have also had a history of fractionalization along confessional lines dating back to the lengthy civil war.

Restoring sovereignty is a major goal of US foreign policy in the region and to this effect efforts have been made to strengthen the Lebanese Army with upwards of $500 million. An officers' training programme and delivery of various conventional and second-hand weapons, armoured personnel carriers, and bridge boats was begun in 2007, with France and Britain also contributing. The army operates a few very old Huey helicopters, and has three or four patched up aeroplanes, but it is unlikely to receive or be able to purchase more advanced military assets given the precarious position of the Lebanese Republic in the Middle East conflict. Israeli Defense Minister Ehud Barak has indicated that he is unhappy with the current US military aid to Lebanon saying when "it comes to Lebanon we have not much appreciated the delivery of arms these past months. We fear that this material could wind up in the hands of Hezbollah" (quoted in *Associated Press*, 11 June 2009).

Perhaps a more daunting obstacle to the use of the army for recovering Lebanese sovereignty by disarming the Party of God is that it will not fight against fellow citizens and may balk at any action to curb Hezbollah, given the fact that 65 per cent of the troops are Shiites. To this point the soldiers have acted as a corps of army engineers during confrontations between the Party of God and Israel. During the 2006 War they took defensive action to combat the invasion. Additionally, in May 2008 when Hezbollah and its Lebanese allies disarmed antagonistic militias by force, the army did not interfere. One reason is certainly that it is no match for Hezbollah, but its commanders also view the Party of God's actions against Israel in nationalistic terms, and hence are not loathe lending support when push comes to shove. A joint Party of God/Lebanese Army defence strategy is under discussion, but any design that would deprive Hezbollah of its capacity to call the shots when it comes to fighting Israel or that might breach its communication and logistic systems will assuredly be firmly rejected.

One of the attempts to restore the state's full sovereignty over its own territory was introduced soon after the July War. The UN Security Council unanimously adopted Resolution 1701 of 11 August 2006, which calls for a "full cessation of hostilities" provided under Articles II and III respectively and stresses the need to boost the Lebanese government's control of the southern part of the country. It maintains that upon "full cessation of hostilities, it calls upon the Government of Lebanon and UNIFIL [United Nations Interim Force in Lebanon] ... to deploy their forces together throughout the South". Furthermore Article III emphasizes the

> importance of the extension of the control of the Government of Lebanon over all Lebanese territory … so that there will be no weapons reserved or in use without the consent of the Government of Lebanon and no authority other than that of the Government of Lebanon.
>
> (UNSC 2006)

But Lebanon has a dismal record when it comes to implementing such resolutions. Previous international agreements and regional accords which have tackled the issue of delegating the administration to exercise full sovereignty such as Security Council Resolutions 1559 (2004) and 1680 (2006) as well as relevant provisions of the Taif Accord (1989) have all come to naught. All these legally binding resolutions failed to bring about the state's control over its southern border, since it was not in the Party of God's interest. Reacting to an insinuation that Hezbollah was responsible for Lebanon's loss of sovereignty, Mohammed Fneish, a senior official of the Party who was Energy Minister in the Fouad Siniora cabinet, claimed that

> the state lacked sovereignty because of successive Israeli invasions and occupations, not because of Hezbollah's arms. Had the state assumed its role as a sovereign power and evicted Israel from its territory there would have been no need for the resistance.
>
> (quoted in Saad-Ghorayeb 2006: 4)

In light of the above, it seems safe to say that Lebanon has little chance of regaining its sovereignty as long as the Middle East conflict persists.

Fourth scenario: the 'no use' of missiles option and its conditions

The feasibility of directly engaging Hezbollah in some sort of a 'no use' of missiles agreement bears some discussion. There is precedent for such an accord in the form of the April 1996 understanding. This agreement signed by the United States, France, Israel, Lebanon, and Syria which followed the Hezbollah–Israel clash that spring, bound the belligerents to limit their military operations to the so-called security zone and to spare each other's civilian populations. As part of the accord observers from the signatory states were posted on the Lebanese side of the border. Hezbollah was committed to the understanding and for the first time was formally recognized as a military combatant rather than a terrorist organization. For the Party of God to adopt a 'no use' agreement, however, certain requirements would have to be met.

First of all Syria would have to regard such an accord as a measure worth trying. Then Israel would have to recognize that Hezbollah is its real antagonist and to separate this from its opposition to the Lebanese government; this is something that the Israelis traditionally have been loath to do. But over the past few years, Beirut's foreign policy positions have diverged from Hezbollah's and may continue to do so. Israel would also have to recognize that such an agreement would be enforced without the use of observers on the ground unless they were present on both sides of the border, as is the case on the Golan Heights. Further it is likely that 'no use' would have to apply to all types of missiles, e.g. those launched from land, sea, or air. The key to Hezbollah's acquiescence to such an accord, however, would most likely be Israel's willingness to cease espionage including aerial surveillance and activities by Mossad on Lebanese soil. If a 'no use' agreement could be struck that met these demands, it could provide a reduction of tensions while diplomats pursue their course.

Fifth scenario: withdrawal of Hezbollah's allies' support and its impact

If the Party of God's interest in pressing its confrontation with Israel continues to be buttressed by the foreign policy designs of its allies, what are the conditions under which Iran and Syria might be motivated to reduce or withdraw their support for Hezbollah?

The Iranian case: the option of an Iranian–American deal

The Party of God's ability to engage the more powerful Israeli force with relative impunity to date is an important asset to the Islamic Republic. This is especially the case should Israel attack Iranian nuclear facilities. If Hezbollah were to be drawn into such a conflict by an Israeli preemptive strike, the Hebrew state could be faced with a second military front, in addition to Tehran's response. This scenario makes it very unlikely that the Islamic Republic will waive its support for the Party of God as long as Israel and the US are committed to eliminating Iran's nuclear infrastructure. On the other hand an American–Iranian deal, for example on Iraq's future, could at some point motivate Tehran to reconsider its alliance with Hezbollah. A decision to suspend Iranian financial and military support for the Party of God might also be contingent upon regime change in the Islamic Republic or be taken as part of some sort of capitulation after a military strike against Tehran's nuclear programme.

The Syrian case: the 'land for peace' option

Damascus is unlikely to withdraw support for Hezbollah as long as Israel continues to occupy Arab territory. On the other hand, if successful 'land for peace' negotiations were concluded between Syria, Lebanon, and Israel, the Party of God's resistance would become redundant. Under this scenario, Hezbollah's confrontation with Israel would then threaten Damascus' interests and would likely be met with Syrian-type persuasion.

Yet whether such negotiations are likely in the near future is difficult to assess. Israel regards the strategic depth provided by the Golan Heights a critical element in its security arrangements – even more so since the signing of a defence treaty between Damascus and Tehran in June 2006. For its part, Syria has a problem with concluding a separate peace agreement with its enemy that emanates from its long and strident stance on Palestinian rights. This raises the question whether the Asad regime, especially in view of the domestic upheaval in 2011, could trade its Arab nationalist posture for a narrower Syrian interest without undermining the basis of its domestic support. Of course this problem could be addressed if Damascus distanced itself from Hezbollah and Hamas, while continuing to support the two groups under the table – a situation the Israelis are certainly considering, as they ponder what the outcome of 'land for peace' talks might be.

This scenario may be the most realistic of those we have considered. Syria and Israel have already conducted a number of direct and indirect negotiating rounds. One was initiated as a result of the 1991 Madrid peace conference and another took place in 2000 during Bill Clinton's second presidential term. In 2008 Turkey mediated indirect 'land for peace' talks between the parties. The search for an agreement persists to this day. Moreover, considering the fact that Iran's nuclear ambitions remain Israel's top security concern,[21] successful negotiations with Syria might be pondered as one way to loosen or break up the Damascus–Tehran alliance. There are indications that Israeli decision-makers are seriously considering this option. According to Daniel Levy,[22] in the late 1990s, Prime Minister Benjamin Netanyahu himself conducted secret talks with the Syrians in which he consented to Israeli withdrawal from all Golan territory with some preconditions. The Netanyahu–Asad talks used Ronald Lauderas[23] as mediator. However, Prime Minister Olmert's bureau charged that Netanyahu had entrusted an American businessman to convince the Syrian President to relinquish the Golan Heights even before any negotiations had begun (Levy 2008). President Clinton confirmed the incident in his memoirs saying that in 1996 "Peres wanted me to sign a security treaty with Israel if it gave up the Golan, an idea that was suggested to me later by Netanyahu and would be advanced again by Barak" (Clinton 2004: 883).

The subject of trading land for a guarantee from Damascus to break with Tehran and Hezbollah also came up while Ehud Olmert was still Prime Minister. He reportedly informed Syria via a third party of his awareness that a peace agreement "means returning the Golan to Syrian sovereignty" but he also indicated that he wished to know if Damascus, in return, would "dismantle, in stages, its alliance with Iran, Hezbollah and Palestinian terrorist organizations" (quoted in *Yediot Aharonot*, 7 June 2007). By making these demands a precondition for negotiations it may well be that Israeli leaders hope to deflect attention from the issue of occupation. Despite the difficulties Syria might have in fulfilling these demands, as long as the regime remains in power, the pragmatic approach which the Asad administration has been adopting to attain its national security interests can be expected to allow for some compromise on its part, challenging the Party of God's role. Then Damascus would have relied on Hezbollah to both pressure Israel for a return of the Golan Heights and as a means of undermining the development of any opposition movement in Lebanon (Rabil 2007). If in exchange for making peace with Israel the regime in Syria could fulfil its long-standing national securing interests, i.e. recovering the Golan Heights, then Hezbollah's continued military activities could be viewed by Damascus as a serious hindrance to its strategic and political interests.

Again, this scenario may be the most viable of the options we have considered for achieving Hezbollah's disarmament, as Israel continues to press its campaign against Iran's nuclear programme. However, whether Damascus will agree to end its support for the Party of God before finalizing a treaty with Israel and – also very importantly – before any concrete steps are taken to resolve outstanding Palestinian issues, remains questionable. Naturally, a resolution to the Israeli–Palestinian conflict would probably mitigate Syrian apprehension over becoming embroiled in continuing violence in that theatre despite a 'land for peace' deal with Israel. In that case it would be doubtful as to whether arming Hezbollah would be in Damascus' interest. That said it might be assumed that if final status agreements were managed in a manner acceptable to Israel and all Palestinian parties, it would be more difficult for Hezbollah and its partners to interfere.

Conclusions

The very essence of strategic studies requires scrutinizing the role of military power in producing political outcomes (Buzan 1987). In this case, it was shown that the military power Israel employed to destroy the Party of God over a lengthy period of years and applied even more massively against Hezbollah in 2006, was unable to procure its objectives. Moreover, the political fallout for Israel was more serious after this conflict than previously since crippling the Party of God's military capabilities constitutes an important step towards undermining Iranian–Syrian influence in the region. Likewise, the Lebanon War in the summer of 2006 also exposed Israel's vulnerability to Hezbollah's short- and medium-range rockets of varying levels of sophistication. In general, post war assessments indicate that the Hebrew state faces mounting challenges in elaborating offensive as well as defensive military tactics against the Party of God. The setback suggests that unlike in the case of its wars with Arab states, military power alone cannot serve Israel's foreign policy interests in its confrontation with Hezbollah. In addition, the reliability of missile defence is still questionable. This state of affairs should encourage its policy-makers to seek alternatives to military force to counter threats emanating from South Lebanon.

One measure would be to provide aid to strengthen the weak Lebanese government. Yet the increasing discord within the government is exacerbated by foreign and local attempts to stir up Sunni–Shiite hostility as a means of countering Hezbollah's political dominance. For its part, the poorly trained and armed Lebanese Army cooperates in intelligence and military operations with Hezbollah, which is the greater force. This situation combined with sectarian divisions among the troops raises the question whether building up Lebanese troop strength and providing the army with modern weaponry may be self-defeating as a means of establishing an effective counterforce to the Party of God. The 'no use' option might reduce tensions between the belligerents, while negotiations over 'land for peace' were taking place, Hezbollah's preconditions for participation in such an agreement are likely to be very difficult for Israel to accept. Finally, it was pointed out that the continued stalemate in the 'land for peace' talks provides a further incentive for the Party of God to retain its arms and pursue its resistance course. For their part, Syrian and Iranian strategists probably believe that, lacking serious efforts to find a political solution to this crisis, they have more to gain from continuing to play the Hezbollah foreign policy card.

Reaching such a political agreement however, will not be easy. Although Damascus has shown strong interest in negotiations to resolve the Golan Heights issue, the regime may face criticism for abandoning the Arab cause to pursue narrower interests. Under the prevailing conditions of instability, Israel's interest lies in maintaining the strategic depth that occupation of the mountainous Golan area provides. The Hebrew state has also demanded that important issues be resolved such as the abandonment of Hezbollah and its security pact with Tehran by Damascus, before handing back occupied Syrian and Lebanese lands. It is nevertheless doubtful that the current regime in Damascus would take this action before an acceptable arrangement with Israel is finalized. Still, working towards a political solution of the outstanding issues between Lebanon, Syria, and Israel is certainly a more productive way of encouraging the Party of God to lay down its arms than continuing to rely upon force. If 'land for peace' negotiations between the parties were to succeed, and there were positive signs that strides were being made to resolve the Palestinian issue, it would be in Hezbollah's interest to turn its full attention to the challenges of leading Lebanon's Shiite community. While the Party of God would assuredly be an active participant in discussions over Lebanon's role in an MFZ, that decision would ultimately depend on the positions taken by the governments of Lebanon, Syria, and Israel at the appropriate time.

Notes

1 This is a rugged water-rich territory that constitutes about 2 per cent of Lebanon's national territory. The area was not included in the 2000 UN-drawn withdrawal line as in 1948, when Syrian troops were there and a different UN regime was established to cover it. Deeds to properties located in Shebaa Farms, however, are registered in the courts of Sidon, Lebanon.

2 For background information on this alliance and information about revolutionary Iran and Hafiz al-Asad's Syria, see Agha (1996), Ehteshami and Hinnebusch (1997), Esposito (1990), Lesch (2005), Seale (1988), van Dam (1996).

3 On the first Gulf War, see Chubin and Tripp (1988) as well as Hiro (1989).

4 For background information on the Golan Heights issue, see Lesch (2005: 20–36).

5 For a history of this rough relationship, see Brynen (1990), Byman (2005: 117–54); regarding the PLO's role in Lebanon's disintegration, see El Khazen (2000), Sayigh (1997).

6 An explanation of the dynamics and outcomes of the 1982 Israeli invasion can be found in Schiff and Yaari (1984).

7 For information on Khomeini's ideology and its impact in Lebanon, see Brumberg (1997: 16–82), Dabashi (1993). For works on Hezbollah, see Kramer (1990, 1993), Palmer Harik (2004), Ranstorp (1994), Saad-Ghorayeb (2002).

8 Among the grave incidents that followed the Israeli invasion were the massacre perpetrated by Christian militiamen in the Sabra and Shatila Palestinian camps in September 1982, the car bombing of the U.S. Embassy in Beirut, and the Marine Barracks explosion at Beirut International Airport in October 1983.

9 There is a vast literature on the Lebanese Civil War, see for instance Hudson (1985), Khalidi (1979).

10 Sheikh Naim Qassem (2005: 242), Hezbollah's second in-command, describes his organization's relationship with Syria as follows: "At this historic milestone, Lebanon faces one of two alternatives: a Syrian versus Israeli alliance, two mutually exclusive choices... we consider the relationship with Syria the cornerstone for facing major *regional* obligations" (emphasis added).

11 This mass movement was formed in 2005 after the assassination of former Prime Minister Rafik Hariri. Headed by his son Saadeddin, it comprises anti-Syrian Maronite Christian and Sunni factions. The Druze pillar of this coalition headed by Walid Jumblat has reconsidered its membership in the movement as a result of what appears to be considerable Syrian pressure.

12 In an open letter published in *al-Safir* daily newspaper on 24 February 1985, Hezbollah introduced itself as an enemy of Israel, Lebanese working with that country, and the United States. A special note to Christians assured this community that there was no intention of pressuring it to accept Islam or to change the Lebanese political system.

13 The UN position on Shebaa Farms was that this area had to be treated differently from the rest of the security zone since it came under UN jurisdiction initiated by agreement with Syria and Israel to disengage in 1973. Syrian troops had been operating freely in this unmarked border area claimed by Lebanon. The property deeds for this territory are all registered in Sidon, Lebanon.

14 Hasan Nasrallah, Speech, 29 September 2001 to commemorate the first anniversary of the second *Intifada*. For more information, see Palmer Harik (2004: 160–2).

15 While these tactics and strategies may not create an existential threat for Israel at this time, they have in fact led to a serious discussion in US and Israeli military circles as to the effectiveness of conventional military warfare in

countering the kind of threats now posed by sub-state actors to the Hebrew state's security, see for example Biddle and Friedman (2008).

16 Of additional interest as far as Hezbollah's partner Syria is concerned, is the fact that Damascus carried out joint military exercises with Turkey for the first time ever during the last week of April 2009. According to Ehud Barak this development was "worrying" although he said that he believed the strategic ties uniting Israel and Turkey would prevail (*Al-Khaleej Times*, 29 April 2009). Yet in May 2010, tensions between the two countries heightened dramatically when a Turkish flotilla of boats that intended to challenge the Israeli blockade of Gaza was halted by Israeli commandos. In the ensuing fracas nine Turks lost their lives. Enraged over this incident, Turkish Prime Minister Recep Tayyip Erdogan threatened to break diplomatic relations with Israel for what he labelled an act of "state terrorism". Jewish leaders interviewed in Istanbul about the possible disintegration of Turkish–Israeli relations indicated that "the flap is more about Erdogan's dislike of the Natanyahu government than his view of Israel" (quoted in Tavernise and Slackman 2010).

17 H. John Poole, an expert on asymmetrical warfare and highly mobile small unit infantry, made this remark to Judith Palmer Harik in June 2008 in Fairfield, VA. Andrew Exum (2006), a former U.S. Army officer, was equally impressed by Hezbollah's anti-tank tactics.

18 The Israeli police reported 4,228 rocket impacts in Israel during the war (Rubin 2006); see also the article of Dennis M. Gormley *et al.*

19 See *New York Times*, 30 January 2008, for a summary of the final report of the Winograd Commission, a group chaired by Israeli retired judge Eliyahu Winograd and charged by the Israeli government to investigate and draw lessons from the war. The report harshly criticized Israeli decision-makers and pointed out tactical, operational, and logistic failures on the part of the military.

20 For an analysis of Hezbollah's battlefield performance, see Exum (2006) and *Jane's Intelligence Review* (1 November 2006: 20–7).

21 During his first visit to Washington in May 2009, Israeli Prime Minister Benjamin Netanyahu exerted great efforts to convince the Obama administration to place Iran's nuclear programme first in importance before the Israeli–Palestinian conflict. Addressing a meeting of the American Israel Public Affairs Committee he stated "the only way to make Israeli-Palestinian peace possible is to prevent Iran from acquiring a nuclear capability and throw it back on its heels" (quoted in Hadar 2009).

22 In the Barak administration, he worked in the Prime Minister's office as special advisor and head of the Jerusalem Affairs unit under Minister Haim Ramon and previously served on the negotiating team to the Oslo B Agreement from May to September 1995 under Prime Minister Rabin.

23 Lauderas became the President of the World Jewish Congress on 10 June 2007. Levy refers to the handwritten date on the original Netanyahu–Lauderas Israel–Syria peace treaty as Saturday 29 August 1998.

References

Agha, Hussein (1996) 'The Syrian–Iranian Axis in Lebanon', in Rosemary Hollis and Nadim Shehadi (eds) *Lebanon on Hold: Implications for Middle East Peace*, London: Royal Institute for International Affairs, 24–30.

Badran, Tony (2009) 'Politics Won't Tame Hizbullah', *Jerusalem Post*, 26 May.

Biddle, Stephen and Friedman, Jeffrey A. (2008) *The 2006 Lebanon Campaign and the Future of Warfare: Implications for Army and Defense Policy*, Carlisle, PA: U.S. Army War College.

Brumberg, Daniel (1997) 'Khomeini's Legacy: Islamic Rule and Islamic Social Justice', in R. Scott Appleby (ed.) *Spokesmen for the Despised: Fundamentalist Leaders of the Middle East*, Chicago, IL: University of Chicago Press, 16–82.

Brynen, Rex (1990) *Sanctuary and Survival: The PLO in Lebanon*, Boulder, CO: Westview Press.

Buzan, Barry (1987) *An Introduction to Strategic Studies: Military Technology and International Relations*, London: Macmillan.

Byman, Daniel (2005) *Deadly Connections: States that Sponsor Terrorism*, Cambridge: Cambridge University Press.

Chubin, Shahram and Tripp, Charles (1988) *Iran and Iraq at War*, London: I.B. Tauris.

Clinton, Bill (2004) *My Life*, New York: Alfred A. Knopf.

Dabashi, Hamid (1993) *Theology and Discontent: The Ideological Foundation of the Islamic Revolution in Lebanon*, New York: New York University Press.

Ehteshami, Anoushiravan and Hinnebusch, Raymond A. (1997) *Syria and Iran: Middle Powers in a Penetrated Regional System*, London: Routledge.

El Khazen, Farid (2000) *The Breakdown of the State in Lebanon: 1967–1976*, London: I.B. Tauris.

Esposito, John L. (ed.) (1990) *The Iranian Revolution: Its Global Impact*, Gainesville, FL: University Press of Florida.

Exum, Andrew (2006) *Hizballah at War: A Military Assessment*, Washington, DC: Washington Institute for Near East Policy. Online, available at: www.washingtoninstitute.org/pubPDFs/PolicyFocus63.pdf (25 March 2011).

Hadar, Leon (2009) 'Misreading the Map: The Road to Jerusalem Does Not Lead through Tehran', *Foreign Affairs*, 15 May. Online, available at: www.foreignaffairs.com/articles/65095/leon-hadar/misreading-the-map (25 March 2011).

Hiro, Dilip (1989) *The Longest War: The Iran-Iraq Military Conflict*, London: Grafton.

Hudson, Michael C. (1985) 'The Breakdown of Democracy in Lebanon', *Journal of International Affairs*, 38(2): 277–82.

Khalidi, Walid (1979) *Conflict and Violence in Lebanon: Confrontation in the Middle East*, Cambridge, MA: Center for International Affairs, Harvard University.

Kramer, Martin (1990) 'The Moral Logic of Hizballah', in Walter Reich (ed.) *The Origins of Terrorism: Psychologies, Theologies, States of Mind*, Cambridge: Cambridge University Press, 131–57.

Kramer, Martin (1993) 'Hezbollah: The Calculus of Jihad', in Martin E. Marty and R. Scott Appleby (eds) *Fundamentalisms and the State: Remaking Polities, Economies, and Militance*, Chicago, IL: University of Chicago Press, 539–56.

Kuperwasser, Yossi (2008) 'The Next War with Hizbollah: Should Lebanon be the Target?' *Strategic Assessment*, 11(2): 19–28.

Lesch, David W. (2005) *The New Lion of Damascus: Bashar al-Asad and Modern Syria*, New Haven, CT: Yale University Press.

Levy, Daniel (2008) 'Netanyahu's Syria Credibility Problem', *Prospects for Peace Initiative*, 29 April. Online, available at: www.prospectsforpeace.com/2008/04/ (25 March 2011).

Qassem, Naim (2005) *Hizbullah: The Story from within*, London: Saqi.

Palmer Harik, Judith (2004) *Hezbollah: The Changing Face of Terrorism*, London: I.B. Tauris.

Pedatzur, Reuven and *Haaretz* Correspondent (2008) 'Iron Dome System Found to Be Helpless against Qassams', *Haaretz*, 22 February. Online, available at: www.haaretz.com/hasen/spages/956859.html (25 March 2011).

Rabil, Robert G. (2007) 'Has Hezbollah's Rise Come at Syria's Expense?' *Middle East Quarterly*, 14(4): 43–51.

Ranstorp, Magnus (1994) 'Hizbollah's Command Leadership: Its Structure, Decision-Making and Relationship with Iranian Clergy and Institution', *Terrorism and Political Violence*, 6(3): 303–39.

Rubin, Uzi (2006) 'Hizballah's Rocket Campaign against Northern Israel: A Preliminary Report, *Jerusalem Issue Brief*, 6(10). Online, available at: www.jcpa.org/brief/brief006–10.htm (25 March 2011).

Saad-Ghorayeb, Amal (2002) *Hizbu'llah: Politics and Religion*, London: Pluto Press.

Saad-Ghorayeb, Amal (2006) *Hizbollah's Outlook in the Current Conflict: Part Two: Accommodating Diplomacy and Preparing for the Postwar Context*, Washington, DC: Carnegie Endowment for International Peace. Online, available at: www.carnegieendowment.org/files/ghorayeb_part2_final.pdf (25 March 2011).

Sayigh, Yezid (1997) *Armed Struggle and the Search for State: The Palestinian National Movement, 1949–1993*, Oxford: Clarendon Press.

Scheffran, Jürgen (2007) 'Missiles in Conflict: The Issue of Missiles in All Its Complexity in United Nations Institute for Disarmament Research Missile Control?', *Disarmament Forum*, Special Volume, No. 1. Online, available at: www.unidir.org/bdd/fiche-periodique.php?ref_periodique=1020-7287-2007-1-en#contents (25 March 2011).

Schiff, Zeev and Yaari, Ehud (eds) (1984) *Israel's Lebanon War*, New York: Simon & Schuster.

Seale, Patrick (1988) *Asad: The Struggle for the Middle East*, Berkeley, CA: University of California Press.

Tavernise, Sabrina and Slackman, Michael (2010) 'Turkey Goes from Pliable Ally to Thorn for U.S.', *New York Times*, 8 June.

UNSC/United Nations Security Council (2006) 'Resolution 1701: The Situation in the Middle East', New York: United Nations. Online, available at: http://daccess-dds-ny.un.org/doc/UNDOC/GEN/N06/465/03/PDF/N0646503.pdf?OpenElement (15 February 2011).

van Dam, Nikolaos (1996) *The Struggle for Power in Syria: Politics and Society under Asad and the Ba'th Party*, London: I.B. Tauris.

10 Designing disarmament strategies

The case of Hamas

Margret Johannsen, Ghassan Khatib, and Anat Kurz

The legacy of the Gaza War

The three week Gaza War in the winter of 2008/09 not only inflicted death and destruction on the inhabitants of a densely populated territory causing many civilian casualties (UNHRC 2009), it also shipwrecked the Israeli–Palestinian peace process. Israel's military offensive into Gaza[1] managed to reduce the attacks of unguided rocket and mortar fire against its homeland, which compared to Palestinian suicide bombings of the past, however, seemed an almost negligible physical threat. Still, the most recent tactic employed in the history of the Palestinian national struggle may well turn into a crucial threat and curb efforts to advance peace. As long as the Gaza crisis remains unresolved, it is difficult to imagine either of the combatants willingly accepting the limitations a Middle East/Gulf Missile Free Zone (MFZ) might impose. In this contribution, we nevertheless suggest that incremental steps can be taken to move towards a resolution of the conflict, based on a new approach of dealing with Hamas, and that this could create conditions under which each side might view participation in an MFZ more favourably.

This article investigates some steps which might induce Hamas to reassess the value of its rocket arsenal, including the creation of a norm-building process that would encourage its members to forego the use of these weapons, view them as obsolete, and finally remove them from their armoury. First we present a brief overview of the origin and historical development of Hamas, including its reorientation within and adjustment to Palestinian politics since the second *Intifada*; next we address the threat to Israel posed by Hamas' rocket arsenal as well as various measures which may be taken to counter the threat. While in accord with our assignment for this volume we focus on Hamas, changes in its policies regarding the acquisition and use of rockets is clearly part of the overall dynamic of Israeli–Palestinian relations and dynamics.

Hamas: between armed resistance and politics

Origin: the Muslim Brotherhood

Hamas[2] was established in response to the outbreak of the first *Intifada* in December 1987.[3] It emerged out of the Society of the Muslim Brothers which was founded in Egypt in 1928 with the twin goals of re-Islamizing Egyptian society and liberating the country from colonial rule. During the Arab revolt in 1936, the Egyptian Brotherhood became involved in Palestinian politics (El-Awaisi 1991), founding in 1945 a Palestinian offshoot, which quickly gained a following in Palestine. In 1948 its branches were severed as a result of the first Israeli–Arab War, which brought the Gaza Strip under Egyptian and the West Bank under Jordanian control. Until the War of 1967, when both territories fell under Israeli occupation, the Palestinian Brotherhood had been more or less marginalized, whereas the secular nationalists had flourished in exile. Under Israeli occupation, the Brotherhood reemerged in the mid 1970s. Due to the relative strength of the nationalists vis-à-vis Islamist groups and echoing the Egyptian Brotherhood's quietist agenda, the group initially focused on institution-building as well as social, educational, and welfare activities, especially work in refugee camps and poor urban quarters, turning the Gaza Strip into its stronghold. By the late 1970s, the Palestinian Brotherhood had established itself as a respected actor within civil society, being able to successfully compete for representation in professional, student, and trade union organizations – more so in the Gaza Strip, less so in the West Bank. In the first half of the 1980s, the Gaza branch under the guidance of its then spiritual leader Sheikh Ahmad Yassin established a military arm and began collecting weapons and organizing military training. Early in 1988, the Brotherhood joined the *Intifada* under the name of Hamas.

This group developed into a formidable political rival to the Palestinian nationalists in the course of the *Intifada.* Unlike the Palestine Liberation Organization (PLO) it rejected eventual partition of Mandatory Palestine as a basis for resolving the conflict with Israel. Hamas' charter, published on 18 August 1988, couches their position in religious terms. In Article 11 it states:

> The Islamic Resistance Movement believes that the land of Palestine is an Islamic Waqf consecrated for future Moslem generations until Judgement Day. It, or any part of it, should not be squandered: it, or any part of it, should not be given up.
>
> (Hamas 1988)

According to the document, the stated goal is to liberate all of Palestine, extending from the River Jordan to the Mediterranean Sea, from Israeli occupation.[4] By giving this political goal a religious justification, it has 'Islamasized' the Palestinian conflict, but it has also " 'Palestinianized' the universal claim of Islam and given the movement a national-religious-political profile" (Klein 1997: 112–13), limiting its military operations to Israel, the West Bank, and the Gaza Strip. In 1991, Hamas formally established the Izz ad-Din al-Qassam Brigades as its military wing, named for an influential militant Islamic preacher who was killed in action against British forces in 1935. After the signing of the Oslo Declaration in 1993, the Brigades implemented Hamas' intentions to sabotage progress towards any negotiated Israeli–Palestinian agreement or conciliation. Hamas was partly motivated by a rigid, religiously inspired ideology, and partly because it "did not believe that the political solution agreed between Israel and Fatah would resolve the conflict between Palestinians and Israelis or Muslims and Jews", especially since it did not address the basic issues of the conflict, aside from mentioning them for final status negotiations. Viewing its relationship with Israel in zero-sum terms, Hamas preferred military resistance to dialogue and negotiations, which were not believed likely to secure the Palestinians' legitimate rights (Ghanem 2008: 473–4). In 1994, Hamas began a series of attacks both against Israeli soldiers and against civilians. During the second *Intifada*, the Qassam Brigades tried to outdo other Palestinian groups in militancy, while occasionally cooperating with them in armed operations. While by 2004 most of the Brigades' West Bank cells had been destroyed by the Israeli Army, in the Gaza Strip they retained a forceful presence (see below).

Structure and decision-making

Hamas comprises three wings engaged, respectively, in social, political, and military activities. Opinions differ widely about how to characterize the organization: Shaul Mishal (2003: 572) suggests Hamas "is essentially a social and political movement, providing extensive community services and responding constantly to political reality through bargaining and power brokering". By contrast, Matthew Levitt (2006: 33) claims "virtually every Hamas political and social activity... inextricably bound up with its terrorist mission" (see also Satloff 2006: 5). In principle, the political leadership oversees all of the organization's activities. Control of the military by the political wing is highly significant, as the main concerns of the two can differ considerably. Jeroen Gunning (2004: 236) identifies the military wing as focusing on operational efficiency and secrecy, whereas the political arm is predominantly concerned with maintaining popularity, legitimacy, and visibility. The political leadership consists of an external branch in Damascus and local leaders in the West Bank and the Gaza Strip.

A consultative council is considered the organization's overarching political and decision-making body and it supervises Hamas' activities through a number of committees. The council includes members from the Gaza Strip, the West Bank, and foreign representatives as well as inmates in Israeli prisons. The external leadership exercises control over much of the military wing's funding. Conventional wisdom has it that it is more hawkish than the internal leadership, which has to cope with the situation on the ground and hence tends to be more pragmatic. If this assessment is correct, it bodes ill for a scenario in which Hamas forsakes the use and possession of its rockets. However, the assessment seems at odds with repeated indications by the Damascus-based chairman of the political leadership, Khalid Mishal, that the movement could settle with a Palestinian state within the boundaries of 1967 and with his proposal of a truce (*Hudna*) of ten years.

External assistance

Hamas enjoys a great deal of support and recognition in the Arab and Islamic world. Most of the money raised on behalf of Hamas comes from outside the Palestinian territories, where the organization can rely on the generosity of the Muslim Brotherhood organizations in more than 80 countries worldwide, on donations from the pious Saudi Arabian middle class, and from Iran, which views Hamas as instrumental in antagonizing Israel. The 1979 Iranian Revolution was a source of inspiration to Hamas, although suspicion of Tehran's ambitions to export a Shiite revolution precluded any close relationship until in 2006, when it turned to the Islamic Republic for funding after the freeze on the transfer of money to Hamas through the global banking system. According to Israeli, US, and British sources, Iran supports the Gaza government financially and supplies its military wing with weapons, including rockets with a longer range than the home-made Qassams (*Jane's Defence Weekly*, 28 January 2009: 7, and 8 April 2009: 6). According to Israeli intelligence experts, however, Syria has had a counterbalancing effect, such that Hamas has not become an Iranian surrogate (McGeough 2009: 367). Syria's support is manifold ranging from harbouring Hamas' political bureau and approving weapon deliveries for the Gaza Strip, to tolerating Hezbollah's military training of Hamas fighters (Sharp *et al.* 2006: 33).

The picture of outside support for Hamas would not be complete, however, without mentioning the diplomatic efforts of a number of regional states. Saudi Arabia and Egypt as well as Qatar have been making efforts to mend the rift between Hamas and Fatah; Turkey has offered to assume responsibility for inter-factional Palestinian talks, and Egypt has mediated between Hamas and Israel on sensitive issues such as a ceasefire arrangement and a prisoner exchange.

Ideology revisited

Rhetorically, Hamas is still committed to its vision of an Islamic Palestine from the Jordan River to the Mediterranean Sea. However, the Hamas charter which presents this vision is rarely cited or referred to by the organization's leadership or spokesmen, and senior officials have even criticized the text as inadequate and misleading. In the aftermath of the 11 September 2001 terror attacks, Hamas commissioned a draft for a new charter designed to counter efforts to identify all Islamic movements and organizations with al Qaida. However, the project was put on hold after Hamas' electoral victory in 2006 lest it be interpreted as bowing to outside pressure (Tamimi 2007). Since 2003, when then-President Yasser Arafat's hold on power was weakening and the newly appointed Prime Minister Mahmoud Abbas opted for a dialogue with Hamas, the organization has been undergoing a process of transformation from a rebel movement to a political party, whose popularity is proving to be an asset in local and national elections.

Whether electoral politics will eventually encourage Hamas to forsake the radical vision of its founding document is a matter of speculation. On the one hand, the charter's ideological commitment could be viewed as an organizational imperative, the absence of which might invite internal schisms and fragmentation. On the other hand, the political costs of violent operations have increased incentives to negotiate a long-term truce. Possibly, this process could lead Hamas to redefine resistance by adopting a non-violent alternative to their current course. Agreeing to abstain from the use of violence for a limited period of time could be instrumental in bringing about a strategic shift of this kind (Gunning 2009: 236–7).

The rocket threat

Under the Israeli occupation, the massive loss of land and the rapid growth of the settler population increasingly alarmed the Palestinians. The international community, which officially opposes Israeli settlements in the occupied territories as illegal and obstructive to peace, has nonetheless failed to stop them (B'Tselem 2010). Faced with the gaining momentum of Israel's settlements in the occupied territories, Palestinians have seen themselves left with no other choice than to fight back. Yezid Sayigh put it most eloquently:

> Armed struggle provided the necessary mobilizing theme for the Palestinians, and their instrument of liberation. ... In all cases, it was the defining dynamic that drove the reconstruction and reorganization of Palestinian national politics, and that allowed the search for state to proceed.
>
> (Sayigh 1997: 23)

Resistance continued on and off in different forms using violent as well as non-violent means. In the mid 1990s, the most violent form of Palestinian militancy was the use of suicide bombings.[5] Ten years later, these so-called "martyr operations" have been practically replaced by rocket and mortar grenade fire by Hamas and other militant factions emanating from the Gaza Strip. Most of these weapons are manufactured locally and have a range of up to 10 km. The Qassams are inaccurate and mostly fall harmlessly into uninhabited areas – a citizen of Sderot compared the chances of hit or miss to a lottery (Bouveret 2009: 82). A minority of the weapons, in particular Katyusha-type Grad rockets which have a longer range than the Qassams, are imported. Compared to suicide attacks or Israel's military operations in the Gaza Strip, the number of casualties resulting from rocket attacks is low. From 2001 until 2008, some 8,000 rockets and mortar bombs were launched at Israeli territory. Altogether 16 Israelis were killed (Israel Ministry of Foreign Affairs 2008). Sderot, situated only 1 km from the Gaza Strip, has suffered most from the attacks. During the Gaza War, however, rockets did reach as far as 40 km beyond the border, landing close to Beer Sheva, a city with a population of about 200,000 inhabitants.

Notwithstanding their limited value as war-fighting weapons,[6] the rockets have had a devastating effect on the sense of security of a quarter of a million residents of the western Negev. Their psychological value consists in being an 'equalizer' to Israeli attacks, which have long deprived the Palestinian population of basic security. This could explain why the population in the Gaza Strip views military operations against Israeli targets much more favourably than the Palestinians in the West Bank, while the overall ratio of Palestinians who believe they are useful in achieving national goals went down from 50.8 per cent in January 2009 to 25.4 per cent in April 2011 (JMCC 2011). Until now, rockets fired by Palestinians at Israeli territory have been launched solely from the Gaza Strip and most of Israel remains out of the range of these ultra short-range weapons. However, without eliminating this threat, the Palestinian armed groups can be expected to continue to improve the range, accuracy, and lethality of these weapons so that the security and well-being of an increasing portion of Israel's population is at stake (Weiss 2008). Moreover, if rockets were ever launched from the West Bank, they would pose a threat not only to the Jewish settlements, but also to large Israeli cities and the industrialized areas in the densely populated coastal region, including the main international airport in Tel Aviv.

Since the Gaza War of 2008/09, rocket attacks against Israel have tapered off. Apparently, Hamas is geared to solidify its rule in the Strip, including its recovery from the devastations of the War, and does not want to run the risk of another major military confrontation. However, according to Israeli intelligence sources, Hamas is continuing its weapon smuggling operations (Lis 2009), suggesting that the prevailing calm does not stem from a principled decision but rather from tactical considerations.

Measures to counter the rocket threat

Missile defence, early warning, and fortification

Israeli defensive measures against rocket attacks include early warning systems as well as the fortification of public buildings and private homes in towns within close proximity of the Gaza Strip. In addition, various systems designed to destroy rockets in flight are under development. After the experimental tactical high energy laser Nautilus alias Skyguard was shelved, the Israeli government in 2007 selected the missile defence system Iron Dome which is designed to detect incoming short-range rockets and fire missiles to destroy them mid-air. Alongside production and procurement of the Iron Dome interceptor, Israel is also developing a ground-based laser system, which is advertized as the future in short-range missile interception (Katz 2008). It is doubtful whether these systems will eliminate the threat to Israel posed by the Qassam rockets and comparable weapons. Reuven Pedatzur argues that Iron Dome is not effective against ultra short-range rockets and therefore cannot protect Sderot, since "to intercept a rocket using Iron Dome requires at least 30 seconds. This is the time it takes a Qassam to travel six kilometres" (Pedatzur and *Haaretz* Correspondent 2008). In addition to the technological challenges, Iron Dome also faces severe financial constraints. The estimated cost of the interceptor is between $30,000 and $100,000. By contrast, the cost of making a Qassam rocket is estimated at between $200 and $400.[7] Yiftah Shapir (2007) points out that "the Palestinians could draw Israel into an endless cycle of escalating expenses by producing more and more cheap and simple rockets and forcing Israel to build more and more expensive interceptors". Given the high cost of the interceptor and the ease with which primitive rockets can be manufactured, Israel could lose an arms race sooner or later.

Containment and roll back by military operations

In line with its offensive defence doctrine (Levite 1990; Arens 2007), Israel's armed forces have repeatedly launched military operations in the Gaza Strip. They have ranged from short-term incursions of small forces to the fully fledged war from late 2008 to early 2009. Although there was no political consensus among Israeli leaders about the purpose of the Gaza War, the Israeli Army specified a number of goals of the operation:

> To strike a direct and hard blow against Hamas while increasing the deterrent strength of the IDF [Israeli Defense Forces], in order to bring about an improved and more stable security situation for residents of southern Israel in the long term.
>
> (Cordesman 2009: 36)

The goal of striking "a direct and hard blow" at Hamas included targeting its paramilitary infrastructure and the political leadership. Still, the operation failed to seriously impair Hamas' military capabilities (Catignani 2009). The war left behind more than 1,400 Palestinians killed in the densely populated territory. The report issued by the United Nations Fact Finding Mission on the Gaza War headed by Justice Richard Goldstone found both Israel and Hamas accountable for war crimes. In light of the imbalance between the two sides, however, Israel was more harshly criticized for using excessive force and employing tactics that resulted in the deaths of hundreds of civilians and caused massive damage to the civil infrastructure in the Gaza Strip (UNHRC 2009).

Containment and roll back by economic sanctions

Israel's adoption of a "unilateral strategy" (Sharon 2003), including the complete removal of settlements from the Gaza Strip and military ground withdrawal while maintaining control over the Gaza borders, did not halt the rocket fire. Israel enforced a crippling blockade on the Strip while intensifying settlement expansion in the West Bank, including East Jerusalem, thus lending credibility to Hamas' claims that the occupation of Palestinian territories continued unabated and legitimized continued armed resistance. Politically, the Israeli unilateral approach exacted a heavy price from the Palestinians, severely restricting the movement of Palestinian persons and goods through its checkpoint system and creating the West Bank separation wall, leading to economic deterioration and a dramatic rise in unemployment and poverty.[8] These developments, along with the failure to get the peace process back on track, contributed to the radicalization of Palestinian society, culminating in the votes for Hamas in the elections to the second Palestinian Legislative Council in January 2006. Hamas' victory, facilitated by election procedures, strengthened Israel's objection to further withdrawals from the West Bank and its insistence on a declared end to violence.

Following the change of government in the Palestinian Authority (PA), the ensuing policy of boycott by Israel and the West of the new Palestinian government not only drove the PA bankrupt, but also heated up the political contest between Fatah and Hamas, the former unwilling to share power and the latter unable to pay the price of international acceptance, i.e. formally recognizing Israel's existence, unconditionally renouncing violence, and confirming all past agreements between Israel and the PLO. On 12 June 2007, fierce fighting between Hamas militias and Fatah-led PA security forces ended with Hamas' military takeover of the Gaza Strip. Thus, the geographic separation between the Gaza Strip and the West Bank was aggravated by the political division between the two Palestinian territories, with the Gaza Strip controlled by Hamas and the West Bank controlled by the PA chaired by President Abbas from Fatah.

Responding to the schism, Israel and key figures in the donor community resumed and further increased the direct transfer of financial resources to the new government in the West Bank, while the economic blockade of Gaza was further tightened. The aim of this approach, commonly referred to as the 'West Bank First' strategy, was to create two drastically different realities in the two Palestinian territories, whereby the West Bank prospers and Gaza stagnates. The underlying assumption of this approach was that despair in Gaza would turn its population against Hamas and towards joining their fellow Palestinians on the West Bank in order to have a share in the latter's economic good fortune (Samhouri 2007). The West Bank model, however, was undermined by the reality of occupation there, including the continued confiscation of land, expansion of settlements, restrictions on movements, and the detention of thousands of Palestinians in Israeli prisons (UNRWA 2011). By contrast, Gaza was afflicted by massive poverty but free of direct Israeli occupation. Given that the Palestinian population is essentially interested in ending the occupation, the West Bank was not significantly better off than Gaza. In addition, negotiations failed to achieve the goals agreed upon at the Annapolis conference (see below). To date the 'West Bank First' strategy, a novel approach to the Palestinian political realities, has not yielded the desired political results. Although it was designed to isolate Hamas and undermine its power while strengthening the PA, it instead induced Hamas to sabotage the strategy by eroding the Palestinian Authority's political standing.

Faced with continued rocket fire from the Gaza Strip, which Israel largely saw as evidence that an evacuation of the Palestinian territory was too risky, it was reluctant to negotiate drastic reductions of its military presence and activities in the West Bank. As a result, Israel denied the Abbas government the opportunity to demonstrate the superiority of negotiations to armed resistance for delivering on the Palestinians' national aspirations. Instead of encouraging unity,[9] the rocket fire exacerbated the intra-Palestinian split. After the Gaza War, the marked decrease of attacks was attributed to Israeli Prime Minister Benjamin Netanyahu's follow-up of the 'West Bank First' strategy, by offering to the PA 'economic peace' (Ahren 2008), while at the same time applying a policy of the hard hand vis-à-vis the Gaza Strip. Israel's loosening of restrictions on civilian goods entering Gaza (Netanyahu 2010), announced in the aftermath of the Gaza flotilla incident on 31 May 2010, arguably mitigated the daily hardship facing the residents of the Strip. However, the easing of the boycott was merely the result of intensive international pressure on Israel to lift the closure. It neither indicated a reversal of Israeli policy to isolate the Hamas-ruled Gaza Strip, nor did it affect Hamas' cost–benefit calculus regarding the build-up of its military infrastructure. Its interest in maintaining relative calm, while at the same time replenishing and improving its rocket arsenal, merely demonstrated that Hamas was seeking to consolidate its hold on the Strip. Indeed, it was notably successful in fighting the lawlessness of the pre-2007 era (Sayigh 2010) thereby partly compensating for its failure to stem popular dissatisfaction on the economic front (PCPSR 2010).

A dual strategy of carrots and sticks

Within the framework of the Annapolis process launched on 27 November 2007, there were two negotiating tracks being pursued in the conflict between Israel and the PA. One focused on conflict management, that is, on following the guidelines stipulated in the first stage of the Road Map, while the other on an effort to reach agreements on final status issues pertaining to refugees, Jerusalem, borders, and settlements in order to find a resolution to the conflict. However, against the backdrop of the security tension in the Gaza sphere, Israel asked for more solid evidence of the PA's ability to deal with belligerent factions before drastically reducing its military presence and activity in the West Bank. On the other hand, PA spokesmen insisted that its ability to enlist public support against the militant opposition, specifically Hamas, would remain limited as long as Israel did not reduce its presence in the area.[10] The contents of the discussions on the core issues have remained secret. Given the discrepancies in the positions of the two sides and their policy that "nothing is agreed until everything is agreed" (UNSG 2010), by the end of 2007, when according to the Annapolis formula an agreement was to be completed, the only accomplishment that could be cited was the parties' readiness to continue the dialogue.

By late 2008 the talks had been suspended. Israel entered a pre-election period, and the confrontation between Israel and Hamas in the Gaza sphere culminated in an all-out war. The centre and right-wing parties Likud and Yisrael Beitenu posted strong election gains. The Israeli elections were widely interpreted as a direct consequence of the Gaza War. European and American efforts to push for resumption of the talks were frustrated by the realities on the ground. Due to the deterioration of the overall political situation, particularly caused by the three week Gaza War and the lack of progress in the negotiations between Israel and the PA, the EU postponed the upgrading of its relations with Israel within the framework of the European Neighbourhood Policy.

Resumption of the upgrading process seems to be linked to progress on the diplomatic front. Regarding US efforts, the peace process was originally high on the agenda of the Obama administration. The President himself declared his unequivocal commitment to a permanent settlement on the basis of the principles endorsed at Annapolis (Obama 2009). However, due to controversies over a freeze on ongoing Israeli settlement construction, direct Israeli–Palestinian talks did not begin until September 2010 only to be suspended by Abbas four weeks later, following Netanyahu's reluctance to extend a moratorium imposing limits on West Bank settlement construction. Renewal of the talks alone, however, cannot guarantee real progress towards a settlement. This is mainly because the existing tensions between Israel and Hamas – even if a renewed ceasefire is formulated and kept – has reduced Israel's already limited readiness to accept territorial concessions and their direct and long-term security consequences. Still, concrete progress in the Israel–PA political negotiations could contribute to reversing the trend towards radicalization in Palestinian internal politics, strengthening the peace camp led by President Abbas, and facilitating the implementation of an agreement – assuming a deal can be reached given the fundamental differences between the two sides on core issues.

Hamas' persistent refusal to recognize Israel precludes the latter from engaging the organization in political dialogue. Moreover, for almost four years, the rift between Fatah and Hamas seemed far from being healed. While both agreed on the Mecca Agreement to form a national unity government based on the results of the 2006 elections, they disagreed on the particulars of power-sharing and, well into 2011, Hamas persistently refused to allow the Fatah-led PA to resume control in the Gaza Strip; they also failed to agree on a common political platform. Hamas argued for a re-enactment of the Mecca Agreement whereas Fatah favoured recognition of Israel and explicit acceptance of the Arab Peace Initiative. In addition, in light of the ongoing Israeli security control of the West Bank, if and when an agreement on a national unity government is implemented, Fatah's ability to guarantee Hamas a share in governing the West Bank (so as to reciprocate Fatah's share in controlling Gaza) remains contingent upon Israel's acceptance of a power-sharing arrangement between the two actors. After Fatah and Hamas declared in May 2011 their intention to overcome the split and form a unity government which was to organize general elections, the difficulties of implementing the unity agreement, in addition to the political and security situation in the Israeli–Palestinian arena, still suggests that even prospects for formulating a 'shelf agreement' between Israel and the PA remain rather slim.

Under these circumstances, it appears that the best Israel and the PA can expect is continuation of the dialogue. However, for Fatah, such process can be a double-edged sword. On the one hand, it legitimizes political and security coordination with Israel, and enhances its regional and international legitimacy compensating for its institutional weakness and precarious standing in the Palestinian domestic sphere (PCPSR 2009b). On the other hand, a prolonged process without tangible results in conjunction with the continued expansion of settlements would undermine the credibility of the peace camp led by Fatah and President Abbas, especially since they have tried hard to fulfil the Palestinian obligations on reforms, economic stability, and effective security spelled out in the first phase of the Road Map. As for Israel, cooperation with international efforts to renew the dialogue, particularly with the US, the EU, and also Egypt, taking into account the Military Council's declared willingness to honour the old regime's agreements with Israel, will answer questions about its willingness to move towards a compromise. However, without the hope of advancing a political settlement, it will be difficult for Israel and the PA to overcome the obstacles posed by the first stage of the Road Map. Therefore, measures to improve the economic situation and foster good governance in both the West Bank and the Gaza Strip, in addition to significantly reducing the daily friction with Israel, could help establish continuity in dialogue – even if in the immediate future it does not bring the political process to the final implementation stage.

Cooperative approaches to coping with the rocket threat

The promise of engaging Hamas

Lack of empirical evidence makes generalizations about the relative value of coercive or non-coercive inducements for initiating de-escalation risky (Kriesberg 1992: 83–4). However, neither the defensive nor the offensive measures presented so far have seriously hampered Hamas' military build-up or the improvements in its operational capabilities, nor have they diminished the potential for another escalation of rocket fire. Since not war, improved technology, or boycotts offer satisfactory long-term answers to the rocket problem, we recommend seeking alternatives to the punitive approach that has so far dominated the policy vis-à-vis Hamas. In fact, Israel, the US, and the EU still largely view Hamas through the lens of the war on terror proclaimed in the wake of 11 September 2001. Devising alternatives would require revising this paradigm and the Manichean world view underlying it, including its ill-considered 'one size fits all' approach. Indeed, whereas al Qaida's apocalyptic agenda is non-negotiable, Hamas' national-religious stance does not in principle rule out compromise. Therefore an attempt should be made towards an inclusive strategy which engages Hamas so as to encourage its transformation from a rebel movement to a responsible political actor. Whether Hamas is likely to undergo such transformation is a matter of controversy.[11] The following suggestions are based on the two premises: first, Hamas will not disappear, either voluntarily or under coercion; second, it will not give up its goal of ending foreign occupation. A strategy of engaging the organization should therefore be given the benefit of the doubt; this would include providing incentives to Hamas to become less militant and more politically accountable, such that the possession and use of rockets would eventually become a liability.

This transformation will be aided by supporting Palestinian national reconciliation and unity which can be done through signalling, under certain conditions, namely endorsement by Hamas of the two-state solution in line with the Arab League's Peace Initiative of 2002 and 2007, acceptance of the pending election's outcome, irrespective of who carries the vote. Engaging Hamas, however, is not a technical issue. The important thing is the message to be conveyed by a unity government: if its policies are governed by Hamas' platform – as it is today – prospects for advancing peace will remain slight. Engaging Hamas does not mean turning a blind eye to the dynamics of the struggle in general and the rocket attacks in particular. The assumption that Israel can be defeated by a strategy of attrition still resonates particularly among the armed wings of the resistance movements that view the withdrawals from South Lebanon in 2000 and the Gaza Strip in 2005 as proof that Israel can be defeated militarily. The perception that the rockets create a "Balance of Fear" (Myre 2006) obviously satisfies a deeply ingrained need among the population of the occupied territories to see the enemy suffer as they have.

Military operations that exact a high price from the enemy contribute to the image of an organization willing to stand up against great odds and, in an atmosphere of disillusionment towards ongoing negotiations, can be useful in the competition with other Palestinian factions for popular consent. Military 'tit for tat' tactics can be employed to undermine the credibility of rivals that have opted for negotiations with the enemy. In the absence of a long-term solution, the dynamics of violence can be revived easily under the proper circumstances. Engaging Hamas is suggested here as a solution by default, given that other strategies have not been effective, but it is also made with the expectation that it could be instrumental in reducing and eventually eliminating the rocket threat as well as in containing and finally dismantling Hamas' veto power, thereby facilitating negotiations to resolve the seemingly intractable conflict.

Palestinian statehood – Israeli security: two sides of the same coin

Even after eventually mending the Palestinians' political split, which for a number of years complicated meaningful negotiations, the basic problem is that a permanent peace requires mitigating the deeply rooted and fundamental causes of mutual hostility, hatred, and violence. This could be achieved in part by promoting a negotiated settlement between the parties. This would

> result in the emergence of an independent, democratic, and viable Palestinian state living side by side in peace and security with Israel.... The settlement will resolve the Israel–Palestinian conflict, and end the occupation that began in 1967, based on the foundations of the Madrid Conference, the principle of land for peace, UNSCRs 242, 338 and 1397, agreements previously reached by the parties, and the initiative of Saudi Crown Prince Abdullah – endorsed by the Beirut Arab League Summit.
>
> (UN 2003)

This will include a just, agreed-upon solution to the problem of Palestinian refugees (Morris 2004). In fact, the idea of a two-state solution to the conflict had earned public and political support, both from Israel and the Palestinians, long before the endorsement of the Road Map. Israel has come to recognize compromise as a solution to security and political challenges. The PLO, under Fatah leadership, chose the diplomatic process as a way of promoting national aspirations and establishing its leadership. The two-state solution has also been widely supported in the international arena. However, differences regarding core issues – mainly the borders between the two states (including the future of the Israeli settlements in the West Bank), the future of Jerusalem, and a solution to the problem of the Palestinian refugees – have blocked progress towards realization of this vision. Freedom and sovereign statehood for Palestinians and security for Israel are two sides of the same coin. This is important because the recent peace-making attempts face the dilemma of deciding on which should be a prerequisite for the other – ending the occupation or achieving security.

The constructive approach would be to progress simultaneously on both fronts. That would allow the positive developments in the one area to reinforce the other. By contrast, if one or the other is given priority, neither will be achieved. The Quartet-sponsored Road Map is a good example of the approach suggested here.[12] In conclusion, the two-state solution is a necessary condition for a lasting peace. Progress towards that end would be facilitated by guaranteeing the unity and integrity of the Palestinian territory. While this goal can in principle be achieved unilaterally – in the sense that a lasting intra-Palestinian reconciliation is an internal matter, and ceasing to preempt the outcome of negotiations on the borders of a future Palestinian state is the responsibility of the government of Israel – such moves could be taken with less domestic risk to the parties in power if violence could be overcome. External actors committed to a two-state settlement can support such moves by encouraging the steps underway towards a new power-sharing arrangement between the rival factions in the Palestinian polity.

Stemming the tide of radicalization: incremental steps towards threat reduction

Conditions for rocket control

In principle, disarming the Palestinian militias could be a stepping stone in this process. As numerous cases show,[13] disarmament of paramilitaries can be either the result of or the precondition for ending a conflict; in reality it is best understood as a contribution to security and stability in post-conflict environments so that recovery and development can begin. Disarmament of militias can be achieved forcefully or by mutual agreement; in reality it happens more often than not as the result of coercion coupled with incentives. It can be achieved by collecting and destroying the militia's weapons, by putting them under the control of a legitimate authority, by reliably witnessing that they have been disabled by the paramilitaries themselves, or by integrating armed militias into the regular military or police forces. Disarmament is essentially a politically driven process requiring genuine political will, the cooperation of all parties involved and a political climate conducive to success. It is hard to conceive of such conditions being generated unless the root causes of the conflict are addressed. Arguably, given the current sorry state of the peace process, a formal agreement to end the conflict and remove the rocket threat appears to be an unlikely possibility. Still, a negotiated long-term settlement can hardly be achieved 'under fire'. So a pragmatic policy may be a matter of taking incremental steps to reduce the violence in Israeli–Palestinian relations, thereby making meaningful negotiations possible.

The prospects for a 'no use' agreement

Barring disarmament, a pledge by Hamas not to use its rockets would be a substantial confidence-building measure and would facilitate negotiations on final issues. In principle and expecting that the other side will eventually reciprocate,[14] such a step could be issued unilaterally.[15] Short of a settlement of the conflict, however, and taking into consideration the present level of animosity and danger of escalation, it is inconceivable that Hamas (or Israel, for that matter) would commit unilaterally to such a norm. Instead, a truce could serve as a functional equivalent. Obviously, it would not only prohibit the firing of rockets or missiles, but would also have to include other military means within the overall asymmetric equation, such as Israeli combat aircraft and Palestinian suicide bombings. A long-term truce would go a long way towards a 'no use' agreement.

The concept of a ceasefire has repeatedly been promulgated by Hamas (Milton-Edwards and Crooke 2004: 299; Mishal 2003: 577–9). A month after it assumed control of the Gaza Strip, Hamas' de facto Prime Minister Ismail Haniyeh put forth an offer for a ten year truce with Israel, which Hamas leaders have kept on the table as a way to end the hostilities. The timing of this announcement was very significant because it signalled the organization's willingness to accept coexistence with Israel following Hamas' takeover of the Gaza Strip. Its efforts in presenting itself as willing and capable of enforcing law and order in the area under its control were not lost on international observers. After meeting with Hamas' top leader in Damascus, former U.S. President Jimmy Carter confirmed that "the Islamic group Hamas was willing to accept Israel as a 'neighbour next door' " (quoted in McGregor-Wood 2008). The Israeli daily *Haaretz* quoted Khalid Mishal, chairman of the political bureau, saying: "We have offered a truce if Israel withdraws to the 1967 borders, a truce of 10 years as proof of recognition" (21 April 2008).

While Hamas was responsible for most of the rocket fire from Gaza into Israel, its ceasefire message was consistent with the claim that it is the only Palestinian party capable of keeping its border quiet. Indeed, in the aftermath of the Gaza War, Mahmoud Al-Zahar, Hamas' hawkish senior official in Gaza, told a Saudi newspaper that his organization would move against any armed group that attempted to violate the ceasefire with Israel, including members of Hamas and Islamic Jihad. He also warned that any Palestinian group deviating from the national consensus was harming the Palestinian people and could no longer be considered a part of the resistance, but rather as someone who seeks to weaken it. He added: "Hamas has recently arrested gunmen who fired missiles at the Karni crossing as well as at trucks carrying goods that passed through the crossings" (quoted in *Haaretz*, 28 June 2008). Hamas' strict policy of adhering to a ceasefire with Israel – which, however, did not include an end to the smuggling of weapons and explosives – was confirmed by the other factions that had tried to violate it (Salim 2008).

Hamas was willing to go even further, indicating that under certain conditions it could accept Palestinian statehood alongside Israel. In fact, Hamas leaders have for a number of years been consistent in calling for a truce that includes ending hostilities with Israel in return for ending the occupation and allowing the establishment of a Palestinian state in the West Bank, including East Jerusalem and the Gaza Strip. As Haniyeh pointed out in April 2008,

> [o]n the issue of the state and the 1967 borders, this is not a new position; Hamas founder Sheikh Ahmad Yassin said in 1988 "End of occupation in the 1967 borders and a long truce with the occupation can be granted".
>
> (quoted in the Palestinian Information Center 2008)

Clearly, such a position is not identical with the PLO's and PA's commitment to a lasting, comprehensive peace. Still, in the year since winning the majority of Legislative Council seats, and particularly in the platform of the short-lived national unity government of 2007, Hamas has moved much closer towards a position that arguably no longer disqualifies it from international legitimacy. The platform stipulates the following:

> The government will work with our Arab brothers and with the international community to terminate the occupation and to return our people's legitimate rights, the first of which is the establishment of an independent Palestinian state with full sovereignty over all the lands occupied in 1967 and with Jerusalem as its capital. That will enable us to build a firm, consolidated foundation for peace, security and prosperity throughout the region and [for] generations to come.
>
> (PNA 2007)

Another article of the same platform states: "The government will honour the legitimate international decisions [i.e. decisions made by the UN and international agreements] and agreements signed by the Palestine Liberation Organization" (ibid.). However, in addition to undermining Palestinian aspirations for independence and statehood, the split between Hamas in Gaza and Fatah in the West Bank caused a competition by each party to create a monopoly of force in their respective areas of control. Furthermore, the internal split and Hamas' control over Gaza were a pretext for Israel to limit economic and security interactions, and, above all, confine political dealings to contacts with the PA. But continuing to render Hamas irrelevant to the political process undermines the prospects for advancing peace, and diminishes the prospects of implementing Israeli–Palestinian understandings that might eventually be reached. Against the backdrop of political stagnation, Israel also continued its expansion of the illegal settlements in the West Bank, including East Jerusalem, and maintained its closure policy in both the West Bank and the Gaza Strip, thus taking advantage of the lull in international pressure and the perceived silent acquiescence of the US while President George W. Bush's second term was coming to an end. This situation encouraged militancy and possible violent escalation by the Palestinians. So, in order to defuse the tension and primarily to bring about a more conducive environment for advancing a long-term truce, Israel and the international community – as well as the PA – should bolster Palestinian unity by engaging Hamas as an integral party to a legitimate national government.

Arguably, 'no use' of rockets under the terms of a truce may prepare Hamas and its constituency for relinquishing its emphasis on a military course of action and arms build-up. This approach presupposes that Hamas is not fundamentally opposed to ceasing armed operations and that, under the appropriate circumstances, it can be expected to develop a pragmatic interest in holding its fire. This would make it, in terms of conflict theory, a 'limited spoiler', that is, an actor whose cost–benefit calculus is influenced both by sanctions and by incentives from the outside as well as by domestic considerations (Gunning 2007). Indeed, in the past, Hamas was able to bring the campaign of suicide attacks to a halt (Al-Zahar 2008). Military pressure might have also contributed to this operational change. The same could be said about the de facto ceasefire that was sustained by Hamas following the Gaza War. If Hamas is indeed not principally oriented towards violence, it can also be expected to decrease its militancy not necessarily by military and economic pressure, but rather in response to a favourable shift in its environment.

Hamas, as a 'limited spoiler', can be expected to respond positively if its demands, primarily acceptance by Fatah and external actors of the principle of power-sharing as well as the results of future elections in the PA, are addressed. Other environmental changes, first and foremost lifting the blockade on the Gaza Strip, could also help revise Hamas' calculations of costs and benefits of escalatory resistance operations on the one hand and ceasefire arrangements on the other. To be sure, intra-Palestinian violence and a failure to overcome the political division of the Palestinian territories are not conducive to the transformation of Hamas – and, for that matter, of the overall Palestinian political system whose factions have not yet parted with the narratives of their 'heroic' past (Münkler 2005: 109) and are still maintaining their military wings to be prepared for the worst. As long as Hamas is not assured that it will be accepted as a legitimate force in the Palestinian state-to-be, it will not shed its features as a resistance movement reserving the right to resume the armed struggle.

Establishing a truce

Apparently, neither the relative calm in the Gaza region nor Hamas' offer of a ten year truce is sufficient to alleviate Israel's perception of the threat. As long as it has not departed from its vision of an Islamic state in all of Palestine, extending from the Jordan River to the Mediterranean Sea, and refuses to commit to a final status settlement, the truce offer actually reinforces Israel's fears that the lull will be used for Palestinian military entrenchment and build-up both in the Gaza Strip and in the West Bank. Israel's rejection of a long-term truce, serving as a substitute for a declared end of the conflict, does not rule out, however, coming to terms with Hamas on a short-term basis, if this serves both parties' temporary interests. In this regard, the six month truce that was declared by Israel and Hamas in June 2008 deserves special attention. Although it collapsed after five months of tense calm, it is worth noting that since the resumption of violent confrontations between Israel and the Palestinians in 2000, it has been the most successful bilateral, jointly agreed ceasefire to date. However, in retrospect and taking the strategic interests of both actors into consideration, it was doomed to be short lived. Israel and Hamas had agreed to the lull, hoping that it would fend off criticism in their respective arenas of their inability to secure normal life on both sides of the Gaza Strip border.

The truce was also designed by the two parties to promote totally conflicting strategic interests. Israel sought to prevent the security tension in the Gaza sphere from obstructing the foot-dragging yet continuous political process that, as part of the effort to promote a comprehensive settlement, also aimed at delaying the advance of Hamas towards the Palestinian political helm. For its part, Hamas hoped the lull would ease the consolidation of its hold on Gaza, essentially at the expense of the Fatah-led PA. In addition, the ceasefire was not designed as part of dialogue aiming at establishing trust and advancing a political settlement but as a goal unto itself. As such, it lacked jointly accepted mechanisms for dealing with violations and crises. In all other respects, the dynamics of the ceasefire echoed the conflicting expectations and demands that had developed between Israel and the Palestinians in the years since the Oslo process and which have thwarted repeated attempts to promote a political breakthrough. Hamas made ending the rocket fire contingent upon Israel loosening political and economic pressure, particularly the opening of the border crossings, while Israel made easing the pressure contingent on a complete end to the rocket fire and a halt to weapon smuggling into the Gaza Strip. Thus there were opposing views about why the confrontation persisted: rocket fire and counterstrikes have been interpreted alternatively as a response to a provocation or a provocation requiring a response.

For several months, this truce was largely adhered to, but mutual violations showed it was only temporary. The border crossings remained largely closed on both the Israeli and Egyptian sides, Israeli military operations in the West Bank against Hamas institutions and representatives continued, and weapon smuggling through tunnels went on at a much greater pace than before. The possibility of renewing the ceasefire agreement, let alone extending it to include the West Bank, did not materialize.[16] The conflict erupted into war once more in late December 2008. After indirect ceasefire negotiations hosted by Egypt during the last week of the war, both sides unilaterally declared an end to the hostilities. The subsequent retreat of the Israeli Army was facilitated by understandings reached with the US and Egypt, and welcomed by the EU, concerning an internationally coordinated effort to stop weapon smuggling. However, the calm has remained fragile partly because no agreement was reached about how to reconcile the security interests of both sides so that Israel continued the blockade on Gaza, preventing material needed for reconstruction and economic development from reaching the Strip, and partly because an exchange of prisoners that would allow for the release of abducted Israeli soldier Gilad Shalit took over five years to materialize, effectively feeding popular resentment in Israel against accommodating Hamas in other respects.

The Olmert and Netanyahu governments – the one that waged the Gaza War and the other one that was established following the February 2009 elections – embraced a policy based on the assessment that the threat of military action would be sufficient to deter Hamas and other belligerent factions from firing. Yet, although in the several months that followed the Gaza War rocket fire decreased significantly, it did not stop completely, indicating that without a meaningful incentive such as lifting the blockade and lightening of the burden borne by the residents of the Gaza Strip, a lasting, comprehensive ceasefire would be impossible to restore. Still, a renewed agreement – focusing on Hamas halting the rocket fire in exchange for Israel ending its military activity in the Gaza Strip and gradually lifting the closure on the area – is quite plausible since the basic interests that led Israel and Hamas to agree to a truce in June 2008 and following the Gaza War, are still in play. But a future agreement will not prevent another escalation unless the pitfalls that precipitated the collapse of the previous one are avoided, and lessons that emerged while it was formally in force are implemented. The task both actors are facing is to focus on a coordinated management of the conflict, if they wish to promote a short-term if not mid-term security arrangement. The following steps should be taken:

- Israel and Hamas should try and avoid the vicious cycle of mutual accusations. Unilateral gestures on the part of Israel, particularly further easing the restrictions on the movement of persons and goods, could decrease public support for militants bent on provoking escalation. The declared intention by Hamas to halt rocket and mortar fire, accompanied by assiduous efforts to impose restraint on other militant factions, can be expected to encourage the easing of the Israeli pressure.
- Unilateral accommodating moves will be constructive if they are supported by formal security coordination. Hence regulations for preventing violations and subsequent escalation should be jointly formulated and endorsed. Drawing up terms for a lull ought to be internationally mediated.
- A ceasefire formulation should be supplemented by an internationally coordinated monitoring apparatus.
- Furthermore, reversing the radicalization of public opinion, in both Israel and Palestine respectively, requires constructive international efforts to move the peace process forward. This will restore hope among the Palestinians and Israelis of a possible negotiated end to the occupation and the realization of peace and security.

Hamas' ruling status in the Gaza Strip has won considerable external acknowledgement. This is reflected in the demand, which was also persistently voiced by Israel, that Hamas give its consent to a lull, respect it, and prevent its breach by other factions. International bodies, both Western and Arab, involved in the rehabilitation and reconstruction of the civilian infrastructure in the Gaza Strip following the recent confrontation, will find it impossible to avoid working with officials of the organization. Probably, the contacts will weaken the boycott against Hamas, without the latter complying with formal demands set by the Quartet as prerequisites for engaging it in a political dialogue. Given that the international quarantine of Hamas is already showing signs of erosion, it will not be long before Israel finds itself isolated in its struggle against the organization. Unless Israel coordinates efforts to establish a lull with Hamas, it will face a constant threat of escalation – an inevitable consequence of an acute political conflict and a prolonged economic crisis in the Gaza Strip – while lacking international legitimacy for continuing the embargo and blockade on Hamas as well as explicit or implicit support for military operations. For Israel, this would mean more serious security and diplomatic challenges than those they currently face in the conflict with Hamas. Certainly, the above-mentioned steps will not secure but a precarious and at best a mid-term lull. Even if fully translated into practice, they are unlikely to safeguard a long-term truce, let alone a political settlement of the conflict between Israel and Hamas. Such an agreement will have to be promoted in the context of a comprehensive resolution of the Israeli–Palestinian conflict.

After the Gaza War, however, prudent policies in dealing with the threat have become more difficult and at the same time more urgent: more difficult, because the war has strengthened the hawkish positions both on the Israeli and on the Palestinian sides (PCPSR 2009a), and more urgent, because it has underscored a dilemma that not only Israel, but also external actors like the US and the EU are facing. As long as Israeli towns are threatened by rockets, neither the Americans nor the Europeans, both committed to a two-state solution to the Israeli–Palestinian conflict, will spend the political capital to pressure Israel into taking moves designed to advance a comprehensive peace agreement, including military withdrawal from the West Bank, evacuation of settlements there, and commitment to ending the occupation in the framework of a negotiated political and territorial agreement.

Absorption of violent actors into national security structures

No matter what the terms of an eventual ceasefire, the commitment not to use the weapons at one's disposal would be solely a declaration of intent on the condition of reciprocity. Facing what it sees as Western and Israeli hostility and without establishing a minimum of trust in the working relations between Hamas and Fatah, Hamas will remain armed during negotiations. An unconditional retreat in the face of external pressure is not an option because it would cause the movement to lose both legitimacy and authority. However, strengthening the pragmatic forces within Hamas cannot be achieved by isolation and boycott. With their backs to the wall, the organization will close ranks and entrench instead of rising to the challenges of diplomacy and compromise. As long as Hamas is threatened with annihilation, it will not forsake weapons it sees as an 'equalizer'. Therefore, within the framework of a truce, gradual demilitarization, which would include curbing rearmament and modernization of Hamas' rocket arsenal, may be the more promising path to follow. One way of implementing this option would be to absorb Hamas militants and their weaponry into a national security framework as part of a comprehensive reform of the Palestinian security sector. As yet, the reform measures are limited to the West Bank and remain politicized, being employed by the PA against the opposition. As demonstrated by the meagre results of the past efforts, including US and EU support (Asseburg 2009: 40), a comprehensive reform needs to be non-partisan in nature and requires Hamas–Fatah institutional coordination and cooperation. As part of a larger national security structure Hamas may eventually share control of its rocket arsenal. Only 'nationalized' rockets will be negotiated away.

Such a process can be facilitated by enhancing economic development in Gaza and by capping Hamas' support from regional actors. Given the difficult economic situation that Gazans are facing, rockets, in addition to the growing weapon smuggling industry resulting from the blockade and benefiting mainly Hamas supporters, have proved an important source of income within the semi-formal sector of the economy (ICG 2007: 7). Reconstruction and economic development of the Gaza Strip together with a resumption of externally supported infrastructure projects would reduce the economic incentives to produce, smuggle, or fire rockets. This, however, is only possible if the border crossings are opened and border controls function smoothly. This could be achieved by implementing the Access and Movement Agreement of 2005 (EP 2005). After its takeover, Hamas, as part of efforts to impose law and order, has targeted Gaza's 'tunnel economy' and, by taxing its goods, turned it into a source of government revenue. If willing, Hamas can ensure that the implementation of the agreement does not lead to an increase in weapon smuggling. Its cooperation is needed for a return of the EU Border Assistance Mission at the Rafah crossing point as well as for closing and sealing the tunnels and preventing the establishment of new smuggle routes (Asseburg 2009: 43).

Finally, an easing of regional tensions would go a long way towards mitigating the danger of increasing violence, including escalations in the Gaza sphere. More specifically, the easing of American–Iranian tensions might weaken Tehran's, and consequently Damascus', motivation to support Hamas, hence reducing the potential for violence. In addition, settling the Israeli–Palestinian and the Israeli–Syrian conflicts could dry up Damascus' and Tehran's support for Hamas' military posture. A concrete breakthrough towards a comprehensive resolution of the Arab–Israeli conflict would also make it increasingly difficult for Iran to use Hamas as a bargaining chip against the US and Israel. Consequently, it would improve chances for sustainable Palestinian reconciliation and an Israeli–Palestinian peace.

Conclusions and implications

In scrutinizing the implications of asymmetric military power in producing political outcomes, it was shown that Israel cannot defeat Hamas militarily without incurring high political costs in the region and beyond. The same goes for economic coercion, which has been widely condemned as a form of collective punishment. The option of missile defence faces technological and financial constraints and its reliability is unproven. On the assumption that the 'land for peace' formula is still valid, we suggest that engaging Hamas as part of the Palestinian political system and conducting meaningful negotiations on final status issues would go a long way to significantly promote the elimination of the rocket threat posed by Hamas and other militant Palestinian factions. A Palestinian state alongside Israel could also be part of a future regional arms control regime. Such an arrangement would fit in well with efforts to negotiate and establish a regional Missile Free Zone. It would enable Hamas to sharpen its profile as a conservative party with an ideology based on a religious outlook. Whether a Palestinian state would eventually be governed according to secular or religious values (or a mix of the two, for that matter) would be solely for the Palestinians to decide. If learning from past cases of de-escalating conflict is possible, adherence to the rules of antagonistic cooperation (Schmid 1993: 158) or peaceful coexistence (*BBC News*, 25 December 2004) is all that can be demanded of a Palestinian state in its neighbourly relations.

Without suggesting to preempt the outcome of permanent status negotiations, which, amongst others, will cover the issue of security (UNGA and UNSC 1993: Art. V), it can be safely assumed that, before the conclusion of a peace treaty, the rockets need to be decommissioned so as to facilitate Israel's withdrawal from the West Bank. Without foreign military presence replacing its army, however, Israel is likely to remain reluctant to do so due to security concerns, and unable to do so because of public opinion at home. To be sure, international troops can only be instrumental in securing progress towards a final status agreement and peace if both parties agree to their deployment. Israeli objection to the deployment of foreign troops in the territories, mainly due to the concern that their presence would limit its freedom of action in case of perceived security threats, can be expected to prevent materialization of the idea. At the same time, Palestinian objection to the presence of international forces in the territories, particularly due to the perception of foreign troops as yet another form of occupation would also prevent their deployment there. Without an agreement between the parties to the conflict, the idea of a foreign military presence is a non-starter. However, with an accord, it might be a constructive component of the solution (Hauswedell *et al.* 2009).

Notes

1 It is called "Operation Cast Lead" in Israel and "the Israeli war on Gaza" by the Palestinians.

2 Hamas is the acronym of Harakat al-Muqawwama al-Islamiyya, meaning Islamic Resistance Movement.

3 For background on the *Intifada* and Hamas' role in the uprising, see Alimi (2006), Aronson (1990), Dajani (1994), King (2007), Nassar and Heacock (1990).

4 See also Articles 6, 9, 15, 32 of the Hamas Covenant (Hamas 1988).

5 For an in-depth analysis of the roots and rationales of suicide bombing in the context of the Israeli–Palestinian conflict, see Ahmed (2005).
6 For this reason, they are also perceived as "political" or "symbolic" weapons, see Karp (2009: 387).
7 See Ben-David (2008), Ravid *et al.* (2007), *World Tribune* (2008).
8 For further information, see World Bank (2005).
9 In this light, Karp's (2009: 387) assessment that the rockets "only have to work well enough to encourage Palestinian unity and keep pressure on Israel for political concessions" does not capture the dynamics within the inner-Palestinian rivalry.
10 See, for example, a letter of complaint against Israel sent by Prime Minister Salam Fayyad to the EU (AIC 2008).
11 As representative for one school of thought that on account of religious dogmas deems Hamas unlikely to undergo such transformation, see Levitt (2006). For the opposite school of thought that emphasizes the political flexibility of the organization as a potential for transformation, see Gunning (2009). For a brief overview of the debate on whether Hamas is an immutable enemy of peace or whether it is feasible to include Hamas in a peace settlement, see Gunning (2004: 233–4).
12 The Road Map stipulates:

> In Phase I, the Palestinians immediately undertake an unconditional cessation of violence ... Palestinians and Israelis resume security co-operation ... Israel withdraws from Palestinian areas occupied from September 28, 2000 ... Israel also freezes all settlement activity, consistent with the Mitchell report.
>
> (UN 2003)

13 For a number of case studies regarding disarmament as part of a comprehensive approach, see United Nations Disarmament, Demobilization and Reintegration Resource Centre. Online, available at: www.unddr.org/whatisddr.php.
14 See the concept of "graduated reciprocated initiatives in tension reduction" (GRIT) (Osgood 1962). For the application of the concept as part of an overall de-escalation strategy towards achieving long-term goals in protracted conflicts, see Kriesberg (2002: 206, 216).
15 For the utility of unilateral moves, see Kriesberg (1992: 219).
16 For the collapse of the ceasefire, see IISS (2009: 236–7).

References

Ahmed, Hisham H. (2005) 'Palestinian Resistance and "Suicide Bombing": Causes and Consequences', in Tore Bjørgo (ed.) *Root Causes of Terrorism: Myth, Reality and Ways Forward*, London: Routledge, 87–102.

Ahren, Raphael (2008) 'Netanyahu: Economics, Not Politics, Is the Key to Peace', *Haaretz*, 20 November. Online, available at: www.haaretz.com/hasen/spages/1038970.html (25 March 2011).

AIC/Alternative Information Center (2008) 'Palestinian Prime Minister Salam Fayyad Writes to EU Leadership regarding the Potential Upgrade of EU-Israel Relations', Jerusalem: AIC. Online, available at: www.alternativenews.org/english/index.php/topics/news/1212-palestinian-prime-minister-salam-fayyad-writes-to-eu-leadership-regarding-the-potential-upgrade-of-eu-israel-relations (25 March 2011).

Al Zahar, Mahmoud (2007) 'Statement in Interview: Rocket Barrage of Sderot Is Hamas Strategy', in Israel Ministry of Foreign Affairs (2008) *The Hamas Terror War against Israel*, Jerusalem: Israel Ministry of Foreign Affairs. Online, available at: www.mfa.gov.il/MFA/Terrorism-+Obstacle+to+Peace/Hamas+war+against+Israel/Missile+fire+from+Gaza+on+Israeli+civilian+targets+Aug+2007.htm (25 March 2011).

Alimi, Eitan Y. (2006) *Israeli Politics and the First Palestinian Intifada: Political Opportunities, Framing Processes and Contentious Politics*, London: Routledge.

Arens, Moshe (2007) 'Defense or Offense?' *Haaretz*, 5 October. Online, available at: www.haaretz.com/hasen/spages/882543.html (25 March 2011).

Aronson, Geoffrey (1990) *Israel, Palestinians and the Intifada: Creating Facts on the West Bank*, London: Kegan Paul International.

Asseburg, Muriel (2009) *European Conflict Management in the Middle East: Toward a More Effective Approach*, Berlin: Stiftung Wissenschaft und Politik.

B'Tselem (2010) 'Land Expropriation and Settlements'. Online, available at: www.btselem.org/English/Settlements (25 March 2011).

BBC News (2004) 'Pope Delivers Sombre Message', 25 December. Online, available at: http://news.bbc.co.uk/2/hi/europe/4124961.stm (25 March 2011).

Ben-David, Alon (2008) 'Iron Dome Advances to Meet Qassam Threat', *Jane's Information Group*, Bracknell: IHS Global Limited. Online, available at: www.janes.com/news/defence/systems/jdw/jdw080318_2_n.shtml (25 March 2011).

Bouveret, Patrice (2009) 'Les Roquettes du Hamas', in Patrice Bouveret, Pascal Fenaux, Caroline Pailhe, and Cédric Poitevin (eds) *Qui Arme Israël et le Hamas? La Paix Pass(é)e par les Armes?* Brussels: Groupe de Recherche et d'Information sur la Paix et la Sécurité, 82–5.

Catignani, Sergio (2009) 'Variation on the Theme: Israel's Operation Cast Lead and the Gaza Strip Missile Conundrum', *Royal United Services Institute Journal*, 154(4): 66–73.

Cordesman, Anthony H. (2009) *The "Gaza War": A Strategic Analysis*, Washington, DC: Center for Strategic and International Studies.

Dajani, Souad R. (1994) *Eyes Without Country: Searching for a Palestinian Strategy of Liberation*, Philadelphia, PA: Temple University Press.

El-Awaisi, Abd Al-Fattah (1991) 'The Conceptual Approach of the Egyptian Muslim Brothers towards the Palestine Question, 1928–1949', *Journal of Islamic Studies*, 2(2): 225–44.

EP/European Parliament (2005) 'Agreed Documents on Movement and Access from and to Gaza', Brussels: EP. Online, available at: www.europarl.europa.eu/meetdocs/2004_2009/documents/dv/200/200902/20090204UNSCO_AMA_EN.pdf (25 March 2011).

Ghanem, As'ad (2008) 'Palestinian Politics after Arafat: The Predicament of the "Palestinian National Movement"', *Journal of Developing Societies*, 24(4): 465–87.

Gunning, Jeroen (2004) 'Peace with Hamas? The Transforming Potential of Political Participation', *International Affairs*, 80(2): 233–55.

Gunning, Jeroen (2007) 'Hamas: Socialization and the Logic of Compromise', in Marianne Heiberg, Brendan O'Leary, and John Tirman (eds) *Terror, Insurgency, and the State: Ending Protracted Conflicts*, Philadelphia, PA: University of Pennsylvania Press, 123–54.

Gunning, Jeroen (2009) *Hamas in Politics: Democracy, Religion, Violence*, New York: Columbia University Press.

Hamas (1988) 'The Covenant of the Islamic Resistance Movement', Translation. Online, available at: http://avalon.law.yale.edu/20th_century/hamas.asp (25 March 2011).

Hauswedell, Corinna, Johannsen, Margret, and Nolan, Paul (eds) (2009) *Demilitarizing Conflicts: Learning Lessons in Northern Ireland, Palestine and Israel*, Rehburg-Loccum: Evangelische Akademie Loccum.

ICG/International Crisis Group (2007) 'Inside Gaza: The Challenge of Clans and Families'. Online, available at: www.crisisgroup.org/~/media/Files/Middle%20East%20North%20Africa/Israel%20Palestine/71_inside_gaza___the_challenge_of_clans_and_families.ashx (25 March 2011).

Israel Ministry of Foreign Affairs (2008) *Suicide and Other Bombing Attacks in Israel since the Declaration of Principles (Sept 1993)*, Jerusalem: Israel Ministry of Foreign Affairs. Online, available at: www.mfa.gov.il/MFA/Terrorism-+Obstacle+to+Peace/Palestinian+terror+since+2000/Suicide+and+Other+Bombing+Attacks+in+Israel+Since (25 March 2011).

IISS/International Institute for Strategic Studies (2009) *Strategic Survey 2009: The Annual Review of World Affairs*, London: Routledge.

JMCC/Jerusalem Media and Communication Centre (2011) 'Governance and Reconciliation', Poll No. 73, April. Online, available at: www.jmcc.org/documentsandmaps.aspx?id=827 (9 June 2011).

Karp, Aaron (2009) 'The Changing Ownership of War: States, Insurgencies and Technology', *Contemporary Security Policy*, 30(2): 375–94.

Katz, Yaakov (2008) 'IDF Starts Using New Laser System near Gaza', *Jerusalem Post*, 28 July. Online, available at: www.jpost.com/servlet/Satellite?cid=1215331128662&pagename=JPost/JPArticle/ShowFull (15 February 2011).

King, Mary E. (2007) *A Quiet Revolution: The First Palestinian Intifada and Nonviolent Resistance*, New York: Nation Books.

Klein, Menachem (1997) 'Competing Brothers: The Web of Hamas–PLO Relations', in Bruce Maddy-Wietzman and Efraim Inbar (eds) *Religious Radicalism in the Middle East*, London: Frank Cass & Co., 111–32.

Kriesberg, Louis (1992) *International Conflict Resolution: The U.S.-USSR and Middle East Cases*, New Haven, CT: Yale University Press.

Kriesberg, Louis (2002) *Constructive Conflicts: From Escalation to Resolution*, Lanham, MD: Rowman & Littlefield.

Levite, Ariel (1990) *Offense and Defense in Israeli Military Doctrine*, Boulder, CO: Westview Press.

Levitt, Matthew (2006) *Hamas: Politics, Charity, and Terrorism in the Service of Jihad*, New Haven, CT: Yale University Press.

Lis, Jonathan (2009) 'Sources: Iran Gave Hamas Missile That Can Hit Tel Aviv', *Haaretz*, 3 November. Online, available at: http://haaretz.com/hasen/spages/1125561.html (25 March 2011).

McGeough, Paul (2009) *Kill Khalid: The Failed Mossad Assassination of Khalid Mishal and the Rise of Hamas*, New York: New Press.

McGregor-Wood, Simon (2008) 'Carter-Hamas Talks May Bear Fruit', *ABC News*, 21 April. Online, available at: http://abcnews.go.com/International/story?id=4693208&page=1 (25 March 2011).

Milton-Edwards, Beverly and Crooke, Alastair (2004) 'Waving, not Drowning: Strategic Dimensions of Ceasefires and Islamic Movements', *Security Dialogue*, 35(3): 295–310.

Mishal, Shaul (2003) 'The Pragmatic Dimension of the Palestinian Hamas: A Network Perspective', *Armed Forces and Society*, 29(4): 569–89.

Morris, Benny (2004) *The Birth of the Palestine Refugee Problem Revisited*, Cambridge: Cambridge University Press.

Münkler, Herfried (2005) *The New Wars*, Cambridge: Polity Press.

Myre, Greg (2006) 'Rockets Create a "Balance of Fear" with Israel, Gaza Residents Say', *New York Times*, 8 July. Online, available at: www.nytimes.com/2006/07/09/world/middleeast/09rockets.html?_r=1&pagewanted=print&oref=slogin (25 March 2011).

Nassar, Jamal R. and Heacock, Roger (eds) (1990) *Intifada: Palestine at the Crossroads*, New York: Praeger.

Netanyahu, Benjamin (2010) 'Office Statement Following the Israeli Security Cabinet Meeting', 20 June. Online, available at: www.mfa.gov.il/MFA/Government/Communiques/2010/Prime_Minister_Office_statement_20-Jun-2010.htm (25 March 2011).

Obama, Barack (2009) 'Remarks by the President on a New Beginning', Speech Held at Cairo University, Egypt, 4 June. Online, available at: www.whitehouse.gov/the-press-office/remarks-president-cairo-university-6-04-09 (25 March 2011).

Osgood, Charles E. (1962) *An Alternative to War and Surrender*, Urbana, IL: University of Illinois Press.

Palestinian Information Center (2008) 'Speech of Ismail Haniyeh on 23 April', Arabic. Online, available at: www.palestine-info.info/ar/default.aspx?xyz=U6Qq7k%2bcOd87MDI46m9rUxJEpMO%2bi1s7ywuWtgFXMRCLwX/tIXaRy/aQnq-Z7bZDKM9kRMfzcDUrG44hI186aQE9OZCmomn%2bYtCYslbloD6ggsMGykZ2llZte8dUmv657nXvC3hc1fyc= (15 February 2011).

PCPSR/Palestinian Center for Policy and Survey Research (2009a) 'Joint Israeli–Palestinian Poll, March 2009: In the Aftermath of Israel's Military Operation in Gaza and the February Elections, Israelis and Palestinians Share Hawkish Positions and Gloomy Expectations Regarding the Peace Process'. Online, available at: www.pcpsr.org/survey/polls/2009/p31ejoint.html (25 March 2011).

PCPSR/Palestinian Center for Policy and Survey Research (2009b) 'Palestinian Public Opinion Poll, No. 34', December. Online, available at: www.pcpsr.org/survey/polls/2009/p34e.html (25 March 2011).

PCPSR/Palestinian Center for Policy and Survey Research (2010) 'Palestinian Public Opinion Poll, No. 37'. Online, available at: www.pcpsr.org/survey/polls/2010/p37e.html#domestic (25 March 2011).

Pedatzur, Reuven and *Haaretz* Correspondent (2008) 'Iron Dome System Found to Be Helpless against Qassams', *Haaretz*, 22 February. Online, available at: www.haaretz.com/hasen/spages/956859.html (25 March 2011).

PNA/Palestinian National Authority (2007) 'The Program of the National Unity Government'. Online, available at: www.almubadara.org/new/edetails.php?id=2520 (25 March 2011).

Ravid, Barak, Azoulay, Yuval, and Grinberg, Mijal (2007) 'Security Cabinet Okays Funding for "Iron Dome" Rocket Defense System', *Haaretz*, 23 December. Online, available at: www.haaretz.com/hasen/spages/937429.html (15 February 2011).

Salim, Abu (2008) 'Hamas Arrest of Al-Aqsa Projectile Launchers Disrupts National Unity', *Ma'an News Agency*, 11 July. Online, available at: www.maannews.net/en/index.php?opr=ShowDetails&ID=30479 (15 February 2011).

Samhouri, Mohammed (2007) *The "West Bank First" Strategy: A Political-Economy Critical Assessment*, Waltham, MA: Crown Center for Middle East Studies, Brandeis University.

Satloff, Robert (2006) 'A Primer on Hamas: Origins, Tactics, Strategy and Response', in Robert Satloff (ed.) *Hamas Triumphant: Implications for Security, Politics, Economy, and Strategy*, Washington, DC: Washington Institute for Near East Policy, 5–9.

Sayigh, Yezid (1997) *Armed Struggle and the Search for State: The Palestinian National Movement, 1949–1993*, Oxford: Clarendon Press.

Sayigh, Yezid (2010) *Hamas Rule in Gaza: Three Years On*, Waltham, MA: Crown Center for Middle East Studies, Brandeis University.

Schmid, Claudia (1993) *Der Israel-Palästina-Konflikt und die Bedeutung des Vorderen Orients als sicherheitspolitische Region nach dem Ost-West-Konflikt*, Baden-Baden: Nomos.

Shapir, Yiftah (2007) 'Anti-Rocket Defense: A Waste of Taxpayers' Money?' Online, available at: www.inss.org.il/publications.php?cat=21&incat=&read=146 (25 March 2011).

Sharon, Ariel (2003) 'Address by Prime Minister Ariel Sharon at the Fourth Herzliya Conference', 18 December. Online, available at: www.mfa.gov.il/MFA/Government/Speeches+by+Israeli+leaders/2003/Address+by+PM+Ariel+Sharon+at+the+Fourth+Herzliya.htm (25 March 2011).

Sharp, Jeremy M. (Coordinator), Blanchard, Christopher, Katzman, Kenneth, Migdalovitz, Carol, Gallis, Paul, Rennack, Dianne, Rollins, John, Bowman, Steve, and Veillette, Connie (2006) *Lebanon: The Israel-Hamas-Hezbollah Conflict*, Washington, DC: Congressional Research Service.

Tamimi, Azzam (2007) *Hamas: A History from Within*, Northampton, MA: Olive Branch Press.

UN/United Nations (2003) *A Performance-Based Roadmap to a Permanent Two-State Solution to the Israeli-Palestinian Conflict*, New York: UN. Online, available at: www.un.org/media/main/roadmap122002.pdf (25 March 2011).

UNGA/United Nations General Assembly and UNSC/United Nations Security Council (1993) 'Letter Dated 8 October 1993 from the Permanent Representatives of the Russian Federation and the United States of America to the United Nations Addressed to the Secretary-General', New York: United Nations. Online, available at: http://unispal.un.org/unispal.nsf/0/71DC8C9D96D2F0FF85256117007CB6CA (25 March 2011).

UNHRC/United Nations Human Rights Council (2009) 'Human Rights in Palestine and Other Occupied Arab Territories', Report of the United Nations Fact Finding Mission on the Gaza Conflict, New York: United Nations. Online, available at: http://www2.ohchr.org/english/bodies/hrcouncil/specialsession/9/docs/UNFFMGC_Report.pdf (25 March 2011).

UNRWA/United Nations Relief and Works Agency for Palestine Refugees in the Near East (2011) 'Labour Market Briefing West Bank Second Half 2010'. Online, available at: www.unrwa.org/userfiles/201106082849.pdf (11 June 2011).

UNSG/United Nations Secretary-General (2010) 'Transcript of Press Conference by Secretary-General Ban Ki-Moon at United Nations Headquarters', New York: United Nations. Online, available at: http://unispal.un.org/UNISPAL.NSF/0/781AA16C576B5F4E85257523004D7A3B (25 March 2011).

Weiss, Margaret (2008) *Weapon of Terror: Development and Impact of the Qassam Rocket*, Washington, DC: Washington Institute of Near East Policy.

World Bank (2005) *Papers on Movement and Access: Contributions to a Negotiated Solution*, Washington, DC: World Bank. Online, available at: http://web.worldbank.org/WBSITE/EXTERNAL/COUNTRIES/MENAEXT/WESTBANKGAZAEXTN/0,,contentMDK:20745897~menuPK:294370~pagePK:1497618~piPK:217854~theSitePK:294365,00.html (25 March 2011).

World Tribune (2008) 'Rockets on the Cheap: Hamas Can Produce Kassam-Class Missiles for $400 Each', 25 November. Online, available at: www.worldtribune.com/worldtribune/WTARC/2008/me_*Hamas*0739_11_25.asp (25 March 2011).

Part III

Three milestones towards a Missile Free Zone

11 The promise of military transparency

Building on East–West experiences and the UN Register of Conventional Arms

Sven-Eric Fikenscher, Patrick Flamm, Lina Golob, Bernd W. Kubbig, and Christian Weidlich

Given the precarious situation in the Middle East/Gulf, the establishment of a Missile Free Zone (MFZ) will only be possible in the long run. In addition to the fulfilment of the conditions already analysed in this volume, an effective approach to arms control and eventual disarmament must therefore be incremental. As a result, it is important to find ways to progressively reduce and finally overcome the security dilemma which has prevented substantial arms control progress. As the past has shown confidence- and security-building measures (CSBMs) can be a promising starting point for facilitating regional stability (see the article of Emily B. Landau and Dalia Dassa Kaye). Such a CSBM process could eventually lead to the disclosure of existing military capabilities as the European experience has demonstrated. In this article we analyse the prospects for the above-mentioned trust-building[1] process and work out proposals for facilitating it. A special focus will be on the conditions for advancing CSBMs and the setting of a transparency agreement with missiles being at its core.

We start by evaluating the (conceptual) merits of the Conference on Security and Co-operation in Europe (CSCE). In this respect far-reaching CSBMs with transparency being an element of rising importance were adopted; especially the various Vienna Documents of the Negotiations on Confidence- and Security-Building Measures which require the disclosure of military assets constitute an unmatched success of regional trust-building. That is why we explore the applicability of the CSCE process to the conflict realities in the Middle East/Gulf. On this basis a regional framework for generating trust is developed which puts an emphasis on the Arab Peace Initiative (API). In order to facilitate military transparency in the Middle East we focus on the United Nations Register of Conventional Arms (UNROCA) which was accepted by almost all states in the region. In order to live up to its ambitious goal the register needs to be reformed and put into the overall military context of the Middle East/Gulf, requiring more demanding reporting standards as well as a special focus on missiles and air capabilities.

Transparency: anatomy of the (European) concept

Europe stands out as the region with the most successful trust-building record. Confidence-building was instrumental in overcoming the zero-sum conceptions of national security which were dominant in the first decades of the East–West conflict. Military openness/transparency was central to the confidence-building measures (CBMs) or CSBMs, respectively, that were initiated in the context of the CSCE. The 1975 Helsinki Accords between East and West included a first set of CBMs which aimed at "reducing the dangers of armed conflict and of misunderstanding or miscalculation of military activities which could give rise to apprehension, particularly in a situation where states lack clear and timely information" (quoted in Lachowski 2004: 11). This definition demonstrates the importance of openness/transparency for confidence- (and security-)building measures. Through the disclosure of military data, which provides the adversary with "clear and timely information", these initiatives reduced misperceptions and the risk of armed conflict. On the basis of this general understanding the CBMs of the Helsinki Final Act were conceptualized (Darilek 1994: 13).

More than a decade later, negotiators achieved a breakthrough in the wake of the 1984–86 Stockholm Conference by agreeing on the next round of more ambitious trust-building initiatives, termed confidence- and security-building measures. The final declaration outlines four criteria for CSBMs: they are to be politically binding, militarily significant, verifiable, when possible, and to extend from the Atlantic Ocean to the Ural Mountains (Lachowski 2004: 12–13). This new approach was seen as sufficient to justify the new term 'confidence- *and security*-building measures' and includes a dimension of trust-building largely ignored by the Helsinki Accords: the danger of cheating and the use of verification tools in order to guarantee compliance; previous monitoring activities were voluntary (Krepon 1994: 1). The logic behind this novel approach is fairly simple: any step towards confidence-building can be undermined by an actor's unwillingness to comply with the terms of the negotiated agreement or simply its adversaries' suspicion that key information is being concealed. So trust-building is linked with mechanisms for successful compliance control. This raises the question of how to create effective verification measures (see the article of Jürgen Scheffran *et al.*).

The literature on confidence-building includes various definitions of the terms 'CBMs' and 'CSBMs', which despite some critical voices, are largely similar to the understanding of those notions worked out in Helsinki and Stockholm.[2] We shall therefore stick to the terminology of the CSCE process with the notable exception that we use the terms 'confidence-building' and 'confidence- and security-building' interchangeably if they refer to measures in the security sector that aim at enhancing political and military stability – whether they are verifiable or not. The decision to view only far-reaching initiatives as security enhancing and to refer to all other measures only as confidence-building does not seem to be warranted. In times of crisis even minor steps such as the exchange of sensitive information about troop movements outlined in the Helsinki Accords can fundamentally alter the security environment of the actors involved.

The range of CSBMs depends on the definition of militarily significant data. This leads to the question: what information is to be considered "clear and timely" and therefore needs to be part of any major CSBM initiative? In the case of the CSCE process the negotiators from both blocs mainly focused on advance notification of major troop manoeuvres until the end of the East–West conflict.[3] The 1990 Vienna Document went one step further, changing the focus from crisis stability to arms race stability. The accord obliged CSCE members to disclose in-depth information about their defence postures such as force levels, command structure, defence expenditures, and especially military holdings (battle tanks, helicopters, armoured combat vehicles, anti-tank missile launchers, artillery, and armoured vehicle-launched bridges). The same year the decision was made to set up a so-called Conflict Prevention Centre which was charged with tracking the information provided by the contract parties (Lachowski 2004: 23, 48; Spector and Dean 1994: 140). Subsequent Vienna Documents in 1992, 1994, and 1999 extended the definitions of the weapon categories at stake, eventually including certain warships and artillery (Lachowski 2004: 48, 106). This understanding of transparency is virtually all-encompassing inasmuch as it involves the disclosure of military arsenals.

Its unmatched success justifies considering the CSCE process, especially the disclosure of sensitive military data, a role model for overcoming zero-sum conceptions of national security in other regions. However, it remains to be seen whether this approach can indeed serve as a guiding line in other parts of the world. A satisfying answer requires the analysis of the conditions for success which facilitated the agreements on and the compliance with trust-building initiatives in Europe as well as of the conflict formations and CSBM experiences in the region at stake, in our case the Middle East/Gulf.

The conditions of success in the European context

Patience was one of the key conditions of success in reaching C(S)BM agreements in Europe during the East–West conflict. The first Soviet proposal for a European security conference dates back to 1954. Hence it took almost 20 years of preparation until the process got started in 1973, concluding with the Helsinki Final Act two years later. The entire official "CSCE-CSBM road" (Goetze 1989: 140) required a comparable amount of time and had various – sometimes rocky – phases: from Helsinki to Geneva

(1973–75), from Belgrade (1977–78) to Madrid (1980–83), and from Stockholm (1984–86) to the development of the provisions for the Vienna Documents (Lachowski 2004: 46–84; Richter 1994: 31–3). The negotiations were time consuming because of the political conditions under which the talks were convened.

After having been hostile to the idea of engagement for more than two decades, NATO's 1967 Harmel report changed the security environment for the better. This document paved the way for détente by arguing that engagement with the East was no longer incompatible with Western security interests. Another important step in this regard was the German policy of rapprochement with the Soviet Union and its Eastern European allies ("Entspannungspolitik"). The decision of the government of Chancellor Willy Brandt to abandon the confrontational stance towards the Warsaw Pact and to pursue a policy of détente led to the renunciation of force agreements with Moscow and Warsaw (Fisher 1994: 32). These steps helped initiate negotiations between East and West.

As a result, the scholarly literature has concluded that the implementation of CSBMs, especially in times of crisis, depends on the willingness of the actors involved to take risks at the international and domestic level. The Brandt administration risked being rebuffed by the Warsaw Pact countries and losing voters' approval – its conservative counterparts launched an aggressive campaign against the government's policy of engagement. The dangers faced by Mikhail Gorbachev in the mid 1980s were even greater. He attempted to convince Western powers that he was serious about 'glasnost' in a series of public declarations, including his admission that the Krasnoyarsk radar constituted a violation of the Anti-Ballistic Missile Treaty (Krepon 1994: 7). At a time of East–West tensions in the wake of an ongoing arms race, Gorbachev's frankness could have proven costly for him, since he had to expect opposition from the US and maybe even a *coup d'état* orchestrated by Soviet hardliners.

During the preparatory period for the CSCE, détente was generally accepted in East and West. While this approach forbade the use of force to eliminate the adversaries or their political structures, it did not necessarily imply agreement with them. This shared understanding enabled an approach to bargaining that aimed at achieving an overall balance of interests/priorities between East and West in various policy areas (Fisher 1994: 31). The Western states demanded above all the acceptance of basic human rights from the members of the Warsaw Pact which in return wanted the recognition of the territorial status quo and access to Western technology and capital (Krepon 2000: 7; Meyer 2007: 19). Under these circumstances three baskets were created dealing with security, environmental and economic issues as well as human rights, allowing a 'tit for tat' approach to the negotiations. In fact the Helsinki Final Act is based on taking the priorities of both sides into account: it includes agreements on European boundaries and on human rights. In addition, the establishment of a conference forum guaranteed that dialogue was maintained even in times of renewed tensions. Moreover, the political stability in both blocs, led by two superpowers that guaranteed a certain degree of coherence, assured the adversaries that a change in government would not endanger the agreed-upon accords. The fact that the border disputes were solved by the Helsinki Accords – comparatively soon within the CSCE process – also contributed to the latter's success.

The applicability of the CSCE process to the Middle East/ Gulf

Regional conflict realities

In the Middle East/Gulf talks would have to be initiated under less favourable circumstances. The Middle East has not undergone the century-long process of state-building and recognition of borders comparable to the process leading to the Peace of Westphalia in Europe. State boundaries have been mainly drawn by external powers and in many cases are still contested by neighbouring actors. A second difference to Europe and an obstacle to effective regional confidence-building is the complexity of Middle Eastern conflict dynamics. Unlike the confrontation between East and West which was bipolar, consisting of two alliances with their hegemonic structure, and whose central element was nuclear deterrence, the Middle East is not characterized by a single overarching conflict formation. Instead, there are overlapping disputes ranging from territorial, rule and security related topics to identity related ones. Currently there are two paramount conflicts, the traditional Arab–Israeli dispute and the Iran-centred one.

The Arab–Israeli conflict which includes specific questions such as the dispute between Israelis and the Palestinians as well as unresolved border issues between Israel, Lebanon (Shebaa Farms), and Syria (Golan Heights), is mainly about territory and resources. The conflict between Palestinians and Israelis in addition involves the issues of self-determination, identity, and human rights on the part of the Palestinians. The Iran-centred dispute includes the long-standing rivalry for hegemony mainly between Tehran and Riyadh as well as Cairo (with the nuclear conflict about the ambitions of the Islamic Republic looming large), Iranian support for (mainly Shiite) actors in regional state and sub-state conflicts, and the staunch anti-Israeli rhetoric and policies of the regime in Tehran.[4] The lack of diplomatic relations between Israel and most Arab states, not to mention the Islamic Republic, and the fact that civil society contacts between these countries are practically non-existent further contributes to the troublesome conflict realities. To sum up, essential conditions of success that proved to be vital in the East–West context are absent in the conflict-ridden Middle East.

A framework for generating trust

However, the reconciliatory approach of the Brandt and especially the Gorbachev administrations has demonstrated that tremendous gridlocks can be overcome, if actors are willing to take the necessary risks, accept year-long negotiations, and conflicting interests/priorities can be balanced as was the case in Helsinki. In the Middle East the peace accord between Israel and Egypt that took the demands of both sides into account – peaceful bilateral relations (Israel) and a return of the Sinai (Egypt) – is one of the few shining examples of such cooperation. The willingness of the Egyptian government of Anwar al-Sadat who had the courage to withstand domestic pressure and reached out to Israel eventually paved the way for the conclusion of the treaty. In the absence of direct communication channels such unilateral steps by one government or a coalition of states are central for changing the mutual perceptions of Israel, most Arab states, and Iran as central threats to each other's security (Steinberg 2004: 265). Nowadays, given the multifaceted conflicts in the Middle East, especially between Israel and Iran, a Graduated and Reciprocated Initiatives in Tension Reduction (GRIT) process (Osgood 1962) would be a promising first step towards successful regional CSBMs.

In this connection the Arab Peace Initiative comes to mind. This proposal was worked out by the Arab League under Saudi leadership in 2002 and renewed in 2007. It proposes peaceful and normalized relations with the Jewish state in exchange for an Israeli withdrawal to its 1967 boundaries, the establishment of an independent Palestinian state, and granting to refugees the right to return to their homelands in accordance with UN General Assembly Resolution 194 (Rifkind 2008). Although Israel did not actually respond to the API, it is the only initiative based on the 'land for peace' formula to which all Arab governments have agreed and remains the most promising starting point for transforming Israeli–Arab relations. As the radical forces within the Arab world – many of them have large groups of supporters – vehemently reject any conciliatory approach, the API already lives up to the condition of risk taking.

Nevertheless, considerable obstacles still have to be overcome. Since the API does not include the offer to start negotiations with Israel, the Jewish state is confronted with a 'take it or leave it' situation: accepting all basic demands or rejecting the proposal entirely. Currently there is only a slight chance that Israel will accept all three requirements. UN General Assembly Resolution 194 is in particular heavily contested. The Jewish state rejects the Arab interpretation of the resolution as granting all refugees the right to return and instead insists that this right depends on practicability. So the question is how the parties might be convinced to accept lengthy negotiations that finally lead to acceptable outcomes for all of them. In this connection a still more limited compromise could provide a trust-building process with the badly needed momentum. Such an agreement might include opening up so-called 'hotlines' between the diplomatic services and the Arab commitment to foster civil society exchange in exchange for ending settlement constructions in the West Bank, Israel's return to the 1967 borders, and compensation for refugees. Such steps would at least promote communication between the adversaries and might over time contribute to changing their mutually hostile perceptions.

This would resemble the 'basket approach' of the CSCE negotiations, since a compromise in one area of special importance for the Arab states – the situation of the Palestinians – would go hand in hand with the realization of CSBMs (i.e. 'hotline' agreements) which are at the core of Israel's security policy. Both issues could be negotiated in separate baskets. This implies that the talks would have to be set up differently from how they were within the CSCE, since human rights is not at the centre of any regional party's interests/priorities. The approach of the Arms Control and Regional Security (ACRS) working group which was convened from 1992–95 with the participation of 14 regional actors cannot serve as a role model either, because regional security was largely subordinated to the conclusion of bi- and multilateral peace agreements, while the refugee issue was treated separately (see the article of Emily B. Landau and Dalia Dassa Kaye).

One major challenge will be to find a way to integrate Iran into such talks because the Islamic Republic officially denies Israel's right to exist. Tehran has constantly refused to reach out to Israel and considers rapprochement with its ideological counterpart illegitimate.[5] Even former President Mohammad Khatami's "dialogue among civilizations" which modestly enhanced the Islamic Republic's relations with the United States and the Arab world excluded Israel. The only major Iranian attempt at regional trust-building, largely in form of non-aggression accords and security pacts, has been limited to the Gulf (Junnola 2000: 86–8). Under the current conditions setting up a conference modelled after the CSCE or ACRS would almost certainly lead to Iran's refusal to participate. Until Tehran changes its official stance on Israel, confidence- and security-building initiatives between the two states will have to be realized either in secrecy or in a conference setting with half-open adjoining rooms, allowing Tehran's delegation to avoid sitting at a common table with Israelis (Meyer 2007: 23, 25).

Nevertheless, on the basis of facilitating cooperation where a balance of (security) interests is given, one can find some leeway for region-wide cooperative measures. One core issue on which the most important actors should be able to agree relatively easily is on fighting al Qaida's influence in the Middle East. There is an overall interest in rolling back the terrorist organization, at least since the radical Sunni group has even attacked Saudi targets. This provides a promising starting point for a trust-building process. The exchange of information on the organization's activities in the region might be a helpful first step. And there is precedent for the positive effects of anti-terrorism activities: since 1975 an agreement between the Israeli Mossad and the Jordanian Mukhabarat intelligence agencies allowed both states to coordinate their fight against underground militias (Krepon 2000: 3) and helped to initiate a stabilization process in bilateral relations that finally paved the ground for the Peace Treaty of 1994. In general terms such an accord, if successful, assures each participant of the adversaries' general willingness to cooperate. This could in turn enhance the political climate and might initiate the badly needed GRIT process. The information exchange could be organized by a newly formed security centre under the management of a neutral actor such as the Organization for Security and Co-operation in Europe, which already manages the Conflict Prevention Centre. This might allow the Islamic Republic to share its intelligence without having to openly participate in joint sessions with Israelis.

The Middle Eastern/Gulf record on CSBMs: a history to build on?

The importance of a regional framework for confidence-building notwithstanding, the question of how to conceptualize far-reaching transparency measures like the Vienna Documents remains. The idea of fostering trust and political stabilization via the establishment of CSBMs is not alien to the Middle East/Gulf, although such measures have been rarely used. Nevertheless, three groundbreaking agreements (Sinai I and Sinai II between Israel and Egypt as well as the Golan Heights Agreement between the Jewish state and Syria) were negotiated in the aftermath of the October War in 1973 (see the article of Jürgen Scheffran *et al.*). Having been negotiated immediately after a major war, without any previous trust-building initiatives to build on and including verification measures, these agreements were – unlike the CSBMs initiated in Europe during the Cold War – not even the result of gradual development. However, they primarily aimed at monitoring troop movements to decrease the danger of a surprise attack and were therefore based on a rather limited understanding of transparency that is comparable to the pre-Vienna phase of the CSCE process.

In addition they were conceived in bilateral settings, so that only some of the problems that hinder regional arrangements had to be overcome, making it much easier to reach an accord. In the ACRS working group – for which the CSCE experience served as a role model – 14 Middle Eastern actors reached agreement on maritime CSBMs, information exchange on and the pre-notification of military exercises, the set up of a communications network as well as the creation of three Regional Security Centres, although no initiative was realized. Again transparency was understood in a much more limited way than in the context of the Vienna Documents. Instead of requiring access to information about the parties' conventional military capabilities, the participants only agreed on prior notification of military exercises involving more than 4,000 troops or 110 tanks (Junnola 2000: 75), sharing information on military personnel, unclassified security documents, as well as military training and education. The participants did not find consensus on verification measures, a situation comparable to the first generation of European trust-building initiatives (Yaffee 2001: 15; see also the article of Emily B. Landau and Dalia Dassa Kaye).

Measures of military transparency: the UN Register of Conventional Arms

The limited success of CSBMs in the Middle East raises the question of whether the Vienna Documents might serve as a role model for military transparency. However, the accords only cover anti-tank missile launchers and hence only a portion of the missile sector. This major weakness of the agreements disqualifies them from being the ideal guide for transparency measures which are meant to be a cornerstone for the establishment of a Missile Free Zone in the Middle East/Gulf. In this regard the UNROCA, also rooted in an ambitious understanding of transparency, provides a much better basis for promoting greater regional security. The UN Register covers seven categories of military goods: battle tanks, armoured combat vehicles, large-calibre artillery systems, combat aircraft, attack helicopters, warships, missiles and missile launchers. These categories constitute a standardized reporting form all states are asked to fill in. In Chapter VII the register specifically includes

> [g]uided or unguided rockets, ballistic or cruise missiles capable of delivering a warhead or weapon of destruction to a range of at least 25 kilometres, and means designed or modified specifically for launching such missiles or rockets, if not covered by categories I through VI. … this category: (a) Also includes remotely-piloted vehicles with the characteristics for missiles as defined above; (b) Does not include ground-to-air missiles.
>
> (quoted in Chalmers and Greene 1995: 23)

Since 2003 the register also includes man-portable air defence systems, called MANPADS (UNGA 2009: 11). Additionally, Category III, covers "[g]uns, howitzers, artillery pieces, combining the characteristics of a gun or a howitzer, mortars or multiple-launch rocket systems, capable of engaging surface targets by delivering primarily indirect fire, with a calibre of 100mm and above" (quoted in Chalmers and Greene 1995: 23). In 2003 the threshold was lowered to 75 mm (UNGA 2009: 11). In addition, the register was in principle supported by almost all Middle Eastern/Gulf states. The Resolution "Transparency in Armaments" (UNGAR 46/36 L) that established the UNROCA was passed with 150 members of the General Assembly voting in favour, Iraq and Cuba abstaining, and 12 countries (including Syria) declining to participate (Wezeman 2003: 5). Since the ousting of the regime of Saddam Hussein the initiative no longer faces fundamental opposition in the region.

The need to reform the register

Taking the specifics of the Middle/Gulf into account

Despite the decision of the countries in the Middle East to accept the norm of military transparency they rarely live up to it in practice. Only Israel, Jordan, until 1998 Iran, and occasionally Qatar, disclosed their arms transfers as the register requires (Perlo-Freeman 2009: 6, Wezeman 2003: 12). One of the main challenges is to convince the Arab states to report the requested data. The vast majority of them, led by Egypt, are troubled by the lack of attention paid by the UNROCA to existing weapon stockpiles and indigenous production, two areas in which Israel is greatly ahead of the Arab world (Feldman 1994: 273). Countries are only requested to report imports and exports. This is especially problematic in the case of Cairo which, largely thanks to American aid, belongs to the main regional recipients of weapon technology. The reporting form includes a question on "Background Information" covering military arsenals, procurement from national production, as well as the policies and regulations relating to arms transfers, but the participating states are only *invited* to provide such information (Chalmers and Greene 1995: 79). As a consequence many countries that fill in the United Nations' standardized form do not provide information on military stockpiles.

Another controversial issue regards the UNROCA's focus on conventional weapons, which further shifts the burden of reporting to the Arab states since Israel's nuclear capabilities are not captured by the register. The Arab League has therefore maintained that ignoring weapons of mass destruction (WMD) is discriminatory (Wezeman 2003: 12). Egypt tried to promote nuclear arms control in the early 1990s, when the UNROCA was negotiated and concretized. After the 1994 working group of governmental representatives, tasked with analysing and reforming the register, had failed to make progress on these issues, Cairo decided not to participate in any further development of the UNROCA (Chalmers and Greene 1997: 17). Even Israel's commitment to transparency in the missile sector is unclear. This assessment may be surprising given its support for CSBMs and its proposal to control and eventually reduce ballistic missiles during the ACRS negotiations (Landau 2006: 52). Nonetheless, one obstacle remains: complete disclosure of its missile capacities will affect Israel's strategy of nuclear deterrence, since its adversaries would have access to information about its delivery systems (see the chapter of Dinshaw Mistry and Mark Smith). Nevertheless, the decision of the Jewish state to fill in the UN form, including Category VII, is an encouraging sign.

A register for the Middle East

The above-mentioned areas of continuing dispute suggest that the UNROCA needs to be expanded and reformed to require all regional states to disclose their military capabilities and to achieve complete transparency. Both aims could be pursued within a Register of Conventional Arms in the Middle East (ROCAME). In view of differing needs for military openness, Resolution 46/36 L invites all states to establish (sub-) regional transparency agreements. The Organization of American States has already heeded this call and initiated an accord with more demanding reporting standards (Wezeman 2003: 21). In the Middle East a similar process would – most importantly – have to include the mandatory disclosure of information about the whole array of military capabilities targeted by the to-be-convened initiative. This should minimally require reporting all missile systems, whether recently procured from domestic production or imported, or as part of their long-term weapon holdings. Sticking to a too narrow focus on arms transfers undermines regional compliance and transparency.

Regarding the issue of military asymmetries in the conventional realm, one has to conclude that the way the UNROCA is currently modelled benefits Israel. A missile-focused ROCAME would as well. But despite their dismal reporting record and their missile-based deterrence strategies (see the chapter of Bernd W. Kubbig *et al.*), the Arab states have an obvious interest in transparency as far as certain weapon systems are concerned. This is especially true when it comes to disclosures about the Israeli Air Force, which is considered by the adversaries of the Jewish state to pose the gravest threat to their security (Feldman 1994: 273). Possible air strikes of Israel against Iran's nuclear facilities are certainly concerns of the Islamic Republic, too. As a result, missile capabilities cannot be the only conventional weapons to be part of a reformed and detailed data exchange measure on the basis of the UN Register. Israel's qualitative edge in weapon technology must also be taken into account. Since its well-equipped air force is particularly troublesome for its adversaries, expanded reporting standards that significantly go beyond the disclosure of military transfers, would have to apply to fighter jets as well, in order to get the Arab countries and Iran on board.

However, even a reformed ROCAME with a special focus on missiles and aircraft would surely lead to renewed calls of the Arab League to include the reporting of WMD. Unfortunately, openness in the conventional sector cannot be equated with openness in the nuclear realm. Such demands, while ethically laudable and worth pursuing on the global level (Müller and Frank 1997), ignore the fact that Israel pursues an opaque nuclear programme (see the chapters of Uri Bar-Joseph and of Gawdat Bahgat *et al.*). Its strategy of ambiguity would be dramatically affected by a far-reaching nuclear data exchange measure. Given the nearly unconditioned support for the policy of opaqueness in political and military circles, it is highly unlikely that the Israeli leadership would accept any nuclear transparency agreement in the early stages of an arms control process. Additionally – and most importantly – retaliating measures by the Arab world and Tehran can be expected, if they are confronted with the unambiguous fact that Israel has nuclear weapons.

The Arab League has declared that its members will collectively withdraw from the Nuclear Non-Proliferation Treaty should Israel admit that it has nuclear weapons. Against this backdrop, the inclusion of WMD in the envisaged ROCAME would be feasible only when the security dilemma has been diminished by effective trust-building measures such as the ratification of the Comprehensive Test Ban Treaty and the implementation of a Nuclear Test Free Zone (see the chapter of Gawdat Bahgat *et al.*). These steps would signal a willingness to initiate nuclear arms control measures. Combined with a more ambitious agenda, including the commitment to implement additional arms control initiatives in the nuclear realm at a later stage, both steps might satisfy the Arab states and convince them to disclose their missile holdings.

The question whether transparency in the missile realm is acceptable for Israel depends on its calculation of the risks. An opaque nuclear deterrence is not necessarily more effective than an open one which would allow adversaries to obtain information about its delivery systems. Indeed exchanging data on missile capabilities may even strengthen deterrence since it would leave no doubt about Israel's ability to deliver a nuclear attack. And as long as the number of the warheads and their locations do not have to be disclosed, Israel could live up to the mantra not to be the first to "introduce" nuclear weapons in the region. Nevertheless, the Jewish state can be expected to demand that stability-enhancing measures which ease the tensions between itself and its Arab neighbours are initiated and that the conflict about Iran's nuclear programme is tackled before addressing the issue of opacity.

Even if the above-mentioned obstacles were overcome one challenge would persist: the current reporting system only requires countries to list their suppliers and the numbers of arms received, as well as the general category of weapon, but not to disclose the type of arms (Wezeman 2003: 5, 8). Therefore it would still be possible that participating states circumvent meaningful transparency by providing only vague data. Although the standardized form includes a "Remarks" column in which governments can provide descriptions of and comments on the transferred military goods, reporting only highly aggregated information has been a core problem of the transparency agreement. Interestingly the missile category has been especially affected by this practice (Chalmers and Greene 1994: 76). This does not violate the UN Register, but undermines its central aim, namely to enhance stability and to build confidence by strengthening transparency in the security realm, since the military value and the strategic importance of the exported or imported missiles and missile launchers cannot be derived solely from their numbers. Disclosing missile and rocket transfers may therefore even contribute to existing misperceptions or create new ones (Laurance and Wulf 1994: 41) – at least as long as no additional information on the transferred items is provided. Therefore a meaningful register for the Middle East would have to mandate the disclosure of all relevant information including the type of weapon, its characteristics, and year of production.

With regard to the reporting of weapon deals, the UNROCA suffers from another weakness which might undermine the establishment of a fruitful ROCAME: the register does not offer a concrete definition of the term 'arms transfer'. Resolution 46/36 L only mentions that items should be disclosed which are "imported into or exported from [a state's] territory". It also mentions that "exports and imports represent, including its annex, all forms of arms transfers under terms of grant, credit, barter or cash" (quoted in Laurance *et al.* 1993: 11). So it is unclear whether the data the UN is asking for is based on orders or on deliveries (ibid.). The US, for instance, takes the view that the resolution does not cover the lease of military goods and as a result has refused to report weapon deals of this kind (Chalmers and Greene 1997: 29). Since the participating states apply different standards for providing information about the transfer of military goods, imports are sometimes listed by the recipient but are not reported as exports by the supplier, or vice versa.

Due to the failure of the Conference on Disarmament as well as the 1994 Panel of Governmental Experts and subsequent working groups to reach a consensus on a more precise wording, a task with which both bodies were entrusted, no agreement has been worked out in this regard. A convincing solution could be modelled on the transparency accord of the Organization of American States which requires the disclosure of any new weapon within 90 days of its incorporation in the armed forces. This would apply to imported arms as well as procurement from domestic production and therefore contribute to the crucial objectives of regional compliance (by heeding the calls of the Arab states) and greater transparency of the envisaged ROCAME.

An all-encompassing data exchange on missiles *and* rockets would require the participation of Hamas and Hezbollah, who have primarily relied upon Qassam and Katyusha arsenals – some of which are covered by UNROCA's definition of Category VII – in their conflicts with Israel. But their inclusion would complicate the entire confidence-building process, since bringing in these actors would probably lead to a change in the trust-building agenda. For Hamas and Hezbollah, missiles and rockets do not so much constitute a deterrent, but these offensive weapons are frequently used to exert pressure on Israel or to retaliate against it (see the chapters of Margret Johannsen *et al.* as well as of Judith Palmer Harik and Walid Abu-Dalbouh). So the issue of the actual *use* of missiles and rockets would suddenly loom large and due to Israeli demands for immediate non-use and eventually complete disarmament, the question of abolishing weapons would be on the agenda of talks originally designed to focus only on disclosing capabilities. Additionally, it is highly doubtful whether Israel would accept Hamas and Hezbollah as participants in a multilateral framework or vice versa since such a development would be a signal of mutual acceptance. This would add substantial problems to the already ambitious goal of missile transparency, making the inclusion of both actors an unwise step under current circumstances. However, the situation might change if there were positive developments in the conflicts of both parties with Israel.

A pie in the sky? Facilitating confidence-building and transparency in the missile sector

The European experience teaches that CSBMs can only be achieved gradually by actors committed to a time consuming process facilitated by continued dialogue, the willingness of core actors to take risks, agreed-upon borders, and a 'tit for tat' approach to negotiations. In the Middle East almost none of these conditions are currently present, although the Arab Peace Initiative with which leading Arab states have both risked being rebuffed by Israel and domestic dissent, may turn out to be a window of opportunity. On its basis a new approach which takes the interests of the states involved into account is possible. Such a concept should, as we suggest, be composed of talks to be convened on the rights of the Palestinians and simultaneously on setting up modest CSBMs, which would put the priorities of both Israel and the Arab world on the agenda. Tacit cooperation in the fight against terrorism might facilitate a trust-building process and allow for the inclusion of Iran.

One of the central achievements of the European experience was the implementation of far-reaching military transparency measures. A similar success would constitute a milestone for a future disarmament concept in the Middle East/Gulf. Including the missile sector in such an initiative could pave the way for restrictions that might eventually lead to an MFZ, since disclosing capabilities can be expected to contribute to confidence-building. At the same time such a step would set the agenda for arms control/disarmament talks by providing all actors with information on the military holdings of their adversaries. We conclude that the UNROCA which requests states to report arms transfers would be the most promising starting point for realizing such an endeavour. The UNROCA includes missiles as a weapon category and has in principle been accepted by almost all countries in the Middle East. The fact that most of them did not fill in the standardized form shows that the register needs to be reformed in order to ensure compliance. Additionally its focus on arms transfers at the expense of indigenous production and military stockpiles limits meaningful transparency. Overcoming both shortcomings requires mandatory disclosures on arms transfers and procurement of domestic production as well as military holdings in the missile and aircraft sector. Information should be provided on the basis of solid data, not just ambiguous numbers. In addition, a clear-cut definition of what constitutes an import/export of a weapon is warranted. Besides, the first steps in nuclear arms control have to be taken in parallel. Currently, the participation of Hamas and Hezbollah would severely complicate the agenda, since it would probably mean that restrictions on the use of missiles were on the table. Therefore encouraging an agreement between the state actors seems to be more promising in the near- and mid-term future.

Notes

1 The terms 'trust' and 'confidence' are used interchangeably.
2 For overviews and differing suggestions, see Krepon (1994: 6–7), Richter (1994: 5).
3 Following the agreements of 1986, 1990, and 1992 the initiation of all ground and air manoeuvres involving more than 9,000 troops must be disclosed. In addition, activities involving at least 13,000 men and women in uniform and 3,500 amphibious and parachute military personnel have to be monitored (Spector and Dean 1994: 138).
4 See on this Asseburg and Steinberg (2007), Meyer (2007), Schimmelfennig (1991).
5 Despite this *official* position leading politicians of the Islamic Republic have announced that they would accept a two-state solution, if the Palestinians themselves do (see the chapter of Uri Bar-Joseph; Hadian-Jazy 2004: 47).

References

Asseburg, Muriel and Steinberg, Guido (2007) 'Konfliktdynamik im Nahen und Mittleren Osten', *Aus Politik und Zeitgeschichte*, 19: 6–12.

Chalmers, Malcolm and Greene, Owen (1994) 'Further Development of the Register Reporting System', in Malcolm Chalmers, Owen Greene, Edward J. Laurance, and Herbert Wulf (eds) *Developing the UN Register of Conventional Arms*, Trowbridge: Redwood Books, 51–86.

Chalmers, Malcolm and Greene, Owen (1995) *Taking Stock: The UN Register after Two Years*, Trowbridge: Redwood Books.

Chalmers, Malcom and Greene, Owen (1997) 'The Register in Its First Four Years', in Malcom Chalmers, Mitsuro Donowaki, and Owen Greene (eds) *Developing Arms Transparency: The Future of the United Nations Register*, Trowbridge: Redwood Books, 9–38.

Darilek, Richard E. (1994) 'East–West Confidence-Building: Defusing the Cold War in Europe', in Michael Krepon (ed.) with the assistance of Jill Junnola, Dominique M. McCoy, Lisa Owens, Matthew C.J. Rudolph, and Amit Sevak, *A Handbook of Confidence-Building Measures for Regional Security*, Washington, DC: Henry L. Stimson Center, 11–23.

Feldman, Shai (1994) *Nuclear Weapons and Arms Control in the Middle East*, Cambridge, MA: MIT Press.

Fisher, Cathleen S. (1994) 'The Preconditions of Confidence-Building: Lessons from the European Experience', in Michael Krepon (ed.) with the assistance of

Jill Junnola, Dominique M. McCoy, Lisa Owens, Matthew C. J. Rudolph, and Amit Sevak, *A Handbook of Confidence-Building Measures for Regional Security*, Washington, DC: Henry L. Stimson Center, 25–39.

Goetze, Bernd A. (1989) 'Verification of Confidence and Security Building Measures: Evolution and Future Prospects', in Jürgen Altmann and Joseph Rotblat (eds) *Verification of Arms Reductions: Nuclear, Conventional and Chemical*, Berlin: Springer, 140–50.

Hadian-Jazy, Nasser (2004) 'Statement', in 108/1 US Congress, Senate, Committee on Foreign Relations, *Iran: Security Threats and U.S. Policy*, Hearing on 28 October 2003, Washington, DC: U.S. Government Printing Office, 40–53.

Junnola, Jill R. (2000) 'Confidence-Building Measures in the Middle East: Developments in the Arab-Israeli Peace Process, the Arabian Peninsula, and the Persian Gulf', in Michael Krepon, Michael Newbill, Khurshid Khoja, and Jenny S. Drezin (eds) *Global Confidence Building: New Tools for Troubled Regions*, London: Macmillan, 71–100.

Krepon, Michael (1994) 'The Decade for Confidence-Building Measures', in Michael Krepon (ed.) with the assistance of Jill Junnola, Dominique M. McCoy, Lisa Owens, Matthew C. J. Rudolph, and Amit Sevak, *A Handbook of Confidence-Building Measures for Regional Security*, Washington, DC: Henry L. Stimson Center, 1–10.

Krepon, Michael (2000) 'Conflict Avoidance, Confidence Building, and Peacemaking', in Michael Krepon, Michael Newbill, Khurshid Khoja, and Jenny S. Drezin (eds), *Global Confidence Building: New Tools for Troubled Regions*, London: Macmillan, 1–13.

Lachowski, Zdzislaw (2004) *Confidence- and Security-Building Measures in the New Europe*, Oxford: Stockholm International Peace Research Institute.

Laurance, Edward J. and Wulf, Herbert (1994) 'Lessons from the First Year', in Malcolm Chalmers, Owen Greene, Edward J. Laurance, and Herbert Wulf (eds) *Developing the UN Register of Conventional Arms*, Trowbridge: Redwood Books, 37–50.

Laurance, Edward J., Wezeman, Siemon T., and Wulf, Herbert (1993) *Arms Watch: SIPRI Report on the First Year of the UN Register of Conventional Arms*, Oxford: Stockholm International Peace Research Institute.

Landau, Emily B. (2006) *Arms Control in the Middle East: Cooperative Security Dialogue and Regional Constraints*, Eastbourne: Sussex Academic Press.

Meyer, Berthold (2007) 'Sub-Regional or Multilateral? New Approaches to Conflict Management in the Middle Eastern Region under Scrutiny', Research Paper Drafted at the Request of Friedrich-Naumann-Stifung für die Freiheit. Online, available at: www.la.fnst-freiheit.org/uploads/896/Meyer-Nahost-deutsch.pdf (15 February 2011).

Müller, Harald and Frank, Katja (1997) 'A Nuclear Weapons Register: Concepts, Issues, and Opportunities', in Malcolm Chalmers, Mitsuro Donowaki, and Owen Greene (eds) *Developing Arms Transparency: The Future of the UN Register*, Trowbridge: Redwood Books, 233–55.

Osgood, Charles (1962) *An Alternative to War and Surrender*, Chicago, IL: University of Illinois Press.

Perlo-Freeman, Sam (2009) *Arms Transfers to the Middle East*, Stockholm: Stockholm International Peace Research Institute.

Richter, Andrew (1994) *Reconsidering Confidence and Security Building Measures: A Critical Analysis*, Toronto: Centre for International and Strategic Studies.

Rifkind, Gabrielle (2008) 'The Arab Peace Initiative: Why Now? One Year On from Annapolis – The Need for a Regional Frame for Negotiations', Report following an Oxford Research Group Meeting, 15–17 October, London: Oxford Research Group.

Schimmelfennig, Frank (1991) *Konferenzdiplomatie als regionale Friedensstrategie: Läßt sich das KSZE-Modell auf den Vorderen Orient übertragen?* Hamburg: Institut für Friedensforschung und Sicherheitspolitik.

Spector, Leonard S. and Dean, Jonathan (1994) 'Cooperative Security: Assessing the Tools of the Trade', in Janne E. Nolan (ed.) *Global Engagement: Cooperation and Security in the 21st Century*, Washington, DC: Brookings Institution, 131–73.

Steinberg, Gerald M. (2004) 'The Centrality of Confidence Building Measures: Lessons from the Middle East', in David Carment and Albrecht Schnabel (eds) *Conflict Resolution: Rhetoric and Reality*, New York: Lexington Books, 259–85.

UNGA/United Nations General Assembly (2009) *Continuing Operation of the United Nations Register of Conventional Arms and Its Further Development*, A/64/296, New York: United Nations.

Wezeman, Siemon T. (2003) *The Future of the United Nations Register of Conventional Arms*, Stockholm: Stockholm International Peace Research Institute. Online, available at: http://books.sipri.org/files/PP/SIPRIPP04.pdf (25 March 2011).

Yaffe, Michael (2001) 'Promoting Arms Control and Regional Security in the Middle East', *Disarmament Forum*, 2: 9–25.

12 The Missile Technology Control Regime, the Hague Code of Conduct, and missile proliferation

Dinshaw Mistry and Mark Smith

Two main multilateral security instruments address missile proliferation which has been an international security concern for many years. One is the Missile Technology Control Regime (MTCR), established in 1987. The MTCR seeks to curb proliferation by denying regional powers the technology to build missiles. The second is the Hague Code of Conduct Against Ballistic Missile Proliferation (HCOC), which was opened for signature in 2002. The Hague Code is a transparency- and confidence-building mechanism. It calls upon states to make annual declarations about their missile and space rocket policies and to give advance notice of testing. What are the key aspects of these instruments? How well have they helped curb missile proliferation in the Middle East/Gulf? These questions are addressed in this article. We first discuss the background to the MTCR before we turn to its impact and that of other supply-side approaches to non-proliferation in the Middle East. Next, we analyse the HCOC and its application to the region. Finally, we conclude by examining current Middle East/Gulf missile programmes. The key questions here are: which state actors still retain missiles, despite the existing supply- and demand-side approaches to non-proliferation? Are new instruments required to tackle missile proliferation in the region?[1]

The MTCR and supply-side strategies

The background to the regime

The MTCR was established by the Group of 7 in 1987. It subsequently expanded to include 34 formal members.[2] A few more states (China, Israel, India, and Romania) formally adhere to the treaty but are not official members. The MTCR aims to curb proliferation through a supply-side approach that seeks to deny regional powers complete missiles and relevant technology. In terms of complete systems, the MTCR initially covered missiles capable of delivering a 500 kg payload (the estimated lightest weight of an actor's nuclear warhead) to a distance beyond 300 km. By mid 1993, the regime addressed all missiles intended for delivery of weapons of mass destruction (WMD). This extended scope covered missiles such as China's M-11, which is declared to have ranges less than 300 km but can carry payloads lighter than 500 kg (e.g. biological and chemical payloads) over greater distances. The MTCR deals with both ballistic and cruise missiles, and its plenary in 2002 began to address some shortcomings regarding unmanned aerial vehicles (UAVs) and cruise missiles (Gormley 2008: 152–4). At this plenary, members established a new definition for the range of these systems, based on the most fuel-efficient flight profile, thereby covering a wider set of cruise missiles and UAVs.

Regarding missile technology, the MTCR Annex contains lists of items whose transfer is banned. They address missile-relevant technology such as chemicals for propulsion systems, specialized steel for bodies and airframes, structural composites, instrumentation and navigation technology, as well as test and manufacturing equipment (MTCR 2008). This Annex has been expanded to keep up with technological advances and political events. For example, from 2002 onward, regime members sought to limit terrorists from accessing MTCR-controlled items and UAVs. To address this issue, at their 2005 plenary, contracting states added controls on UAVs equipped with aerosol dispensers. In this session, the controls on items such as accelerometers and liquid propellants were also expanded. In 2006, members further tightened restrictions on cruise missile propulsion, guidance, and control systems; and the 2007 plenary again updated the control lists.

The impact of the MTCR

The MTCR achieved substantial success in restricting the transfer of missiles and related technology (CIA 2008).[3] In its first two decades, all major suppliers except North Korea halted their transfer of complete missiles, and their supply of related technology was also restricted. This resulted in modest success towards achieving the regime's ultimate objective of preventing aspirants from building and acquiring missiles, but progress has varied by region and over time. Since the 1980s when the MTCR was established, 14 states outside the treaty's framework have attempted to build missiles. Nine either gave up their ambitions, were restrained to building cruise and short-range ballistic missiles such as 300 km range Scud-Bs and 600 km range Scud-Cs, or built space launch vehicles. These restraints, although partly due to the MTCR's technological obstacles, were also induced by political considerations.

In general, the denial of foreign technology was sufficient to thwart or considerably delay missile development in the technologically weakest countries, such as Libya. For six other powers – Argentina, Brazil, Egypt, South Africa, South Korea, and Taiwan – the MTCR's barriers generally hindered, raised the costs of, and delayed their missile programmes. During the time provided by the regime-induced delays, two political factors restrained their missile projects: US diplomacy as well as American political and economic pressure influenced all the above-mentioned states. The easing of regional tensions that reduced the security related demand for weapons had an additional impact on the programmes of Argentina, Brazil, and South Africa. The MTCR's barriers could not prevent five other states – Israel, India, North Korea, Pakistan, and Iran – from building and testing medium-range ballistic missiles. These countries have short-range missiles and 1,000–2,500 km range single stage and two stage medium-range capabilities. In addition, India has tested the 3,500 km range Agni-3, and North Korea tested an approximately 6,000–8,000 km range Taepodong-2 which could have a greater range with a third stage.[4]

In summary, it is hard to quantify the success of the MTCR. One measure of achievement is that nine out of 14 aspirants have not tested medium-range missiles. But such success cannot be entirely, or in some cases substantially, attributed to the regime. Accordingly, one study examining both the political and MTCR-induced restraints on 14 countries' programmes, concluded the MTCR achieved a mid-level success rate by 1997 and a mid-to-low success rate by 2002 (Mistry 2003). Although the current state of research does not permit a reliable assessment regarding the developments since then, our analysis below shows that at least in the Middle East the missile projects have steadily progressed.

Supply-side issues in the Middle East/Gulf

The MTCR has established a multilateral norm against missile technology transfers. This was reinforced in 2004 by UN Security Council Resolution 1540, which calls upon all states to "take and enforce effective measures to establish domestic controls to prevent the proliferation of nuclear, chemical, or biological weapons and their means of delivery" (UNSC 2004). Although there is no direct link between the MTCR and Resolution 1540, and the latter deals primarily with WMD, it also covers WMD delivery systems (i.e. missiles), and thereby complements the MTCR. In addition, the United Nations Security Council adopted Resolutions 1695, 1718, and 1874 concerning North Korea (in response to its July 2006 missile tests, and the October 2006 and May 2009 nuclear tests), and Resolutions 1696, 1737, 1747, 1803, 1835, and 1929 concerning Iran's nuclear programmes but also covering its missiles. Four of these resolutions called on countries to prevent these states from procuring missiles and related technologies and to obstruct financial transactions related to their projects; they also imposed technology transfer restrictions and other sanctions on missile related entities.

To what extent have rules against missile technology transfer been adopted in the Middle East/Gulf? Three aspects of this issue are examined below. First, to what degree do parties outside the region transfer missiles and related technology to the Middle East? Second, to what extent do states in the Middle East/Gulf transfer missiles and technology within and outside the region? Third, what other supply-side approaches, such as the Proliferation Security Initiative (PSI), curb missile proliferation in the Middle East?

External suppliers and their missile technology transfers to the Middle East/Gulf

The behaviour of some main missile suppliers – Russia and China – and others, such as the US, has been analysed in the article on external actors in this volume. Briefly summarized, Moscow and Beijing ended their supply of complete missiles as defined in this chapter to regional powers by the early 1990s, and substantially reduced their transfer of missile related technology. As Russian and Chinese transfers of complete missiles ended, Middle Eastern states turned to North Korea.[5] Pyongyang supplied an estimated 400 Scud-B and -C missiles to Tehran and Damascus in the late 1980s and early 1990s. It then exported a smaller quantity of Scuds, or components such as engines, to Egypt, Syria, and Yemen. North Korea has also delivered 1,000–1,300 km range Nodong missiles or components to Iran since the 1990s, from which Tehran's Shahab-3 missiles have been derived (Bermudez Jr 1999; IISS 2010: 19–26). During the early 2000s, Pyongyang's missile exports to Yemen probably ceased.

North Korea continued to transfer missile technology to Iran and Syria in the 2000s, although other states blocked some of them, as noted in the section on the PSI, below. At that time the Islamic Republic may have acquired a new Nodong-B from North Korea, which could form the basis for its Shahab-3B. Some reports note that a missile tested in January 2006 was the Nodong-B missile, which had a 2,500–3,000 km range and was based on technology from the Russian SS-N-6/R-27 submarine-launched missiles. US officials noted that Iran acquired 18 such missiles (or their components) from Kim Jong-Il's regime in 2006 (*Jane's Defence Weekly*, 16 May 2007: 17). In December 2009, US officials again referred to this transfer, informing their Russian counterparts that North Korea had sent 19 of these missiles to Iran (*New York Times*, 29 November 2010). However, Theodore Postol (2009) notes that Tehran developed its own new variant of the Shahab-3, the Shahab-3M or Ghadr-1, by extending the fuel tank of the original Shahab-3. He adds that the Islamic Republic used the upper stage of the SS-N-6/R-27 as the second stage of its Safir satellite-launching rocket, whose first stage is derived by considerably extending the Shahab-3. North Korean officials reportedly witnessed Iran's launch of this rocket in February 2009, and a team from the Islamic Republic possibly observed the April 2009 launch of a Taepodong-2 in the East Asian country (Goodenough 2009).

Other reports indicate additional transfers from North Korea to the Middle East. In November 2007, US officials informed China about the transfer of missile jet vanes from North Korea to Beijing, from where they were to be sent on an Iran Air flight to Tehran (*New York Times*, 28 November 2010). A further article indicates that Pyongyang collaborated with the Islamic Republic in developing a new nosecone for some of Iran's Shahab-3 missiles. In October 2010, North Korea displayed a variant of its Nodong missile with a triconic nosecone, and a similar nosecone had previously been seen on an Iranian medium-range missile (Pollack 2010). Still, a study by a panel of UN experts stated that North Korea continued to transfer missile technology to Iran and Syria in the late 2000s. The panel noted that

> evidence provided in these reports [various non-governmental analyses and government reports] indicates that the DPRK [Democratic People's Republic of Korea] has continued to provide missiles, components, and technology to certain countries including Iran and Syria since the imposition of these measures [i.e. since UN Security Council Resolution 1718 of 2006].
>
> (Report to the Security Council 2010: 18)

To summarize, since the MTCR was created, Russia and China have reduced their missile-relevant transfers to the Middle East, whereas a third external supplier – North Korea – has not. Such North Korean exports are a hurdle for implementing missile import restrictions into the Middle East. They may have to be addressed through a missile dialogue with North Korea, whereby Pyongyang is given political and economic incentives to halt its missile technology transfers.

Regional states as missile proliferators

Middle Eastern countries have not been formally invited to join the MTCR, but they have informally followed practices against missile technology transfers. At the same time, some of them have undertaken minor transfers of related technology or short-range missile systems, both within and outside the region. Israel was influenced by the United States to halt missile related supplies and comply with the MTCR. In the 1970s, Israel had considered cooperation on a 100–300 km range cruise or ballistic missile project with Iran. In the 1980s, it transferred anti-ship cruise missile technology to Taiwan and was involved in ballistic missile transactions with South Africa (Parsi 2007: 74–8; Shichor 1998: 71, 76). After talks with Washington, Israel implemented the MTCR's guidelines from December 1991, a year earlier than it had wanted to. In subsequent years, Israel largely adhered to its missile export restraints, although it did consider selling Arrow missile defence interceptors to India.[6] Egypt supplied two Scuds to North Korea (either in 1976 or 1979–81). In the 1980s, Cairo participated in the Condor medium-range missile project with Argentina and Iraq; it was halted by MTCR restrictions and American diplomatic pressure (Navias 1992: 120).

Syria was involved in small missile-relevant transfers. It supplied, or gave access to, the 120 km range Russian-made SS-21 missile to North Korea in 1994 or 1996. Pyongyang has since used reverse engineering to produce them. Damascus reportedly also supplied, or considered delivering, Scuds to Sudan in 2004 (MissileThreat.com 2009). Another newspaper article noted that, in 2010, Israeli officials stated that Syria had transferred Scud missiles to Hezbollah, although the timing and details of the arms deal remained unclear (*New York Times*, 14 April 2010). Tehran has undertaken limited missile related deliveries after its programme matured in the late 1990s and 2000s. In the mid and late 1990s, Iranian technicians allegedly were assisting Libya with a liquid fuel al-Fatah missile, although Tripoli eventually acquired more substantial assistance from North Korea rather than from the Islamic Republic (NTI 2010). Iran possibly supplied or considered supplying Scuds to the Congo in 1999 (*Washington Times*, 22 November 1999). In addition, the Islamic Republic delivered the C-802 anti-ship cruise missile to the Lebanese group Hezbollah, which used it to hit an Israeli vessel in 2006. Further, Tehran may have collaborated with Damascus in a project on missile warheads.[7]

To summarize, Middle Eastern/Gulf states have been minor suppliers of missile related technology within (and in some cases outside) the region. Although minor, these transfers illustrate the potential of Middle Eastern/Gulf states to become more substantial sources of missile proliferation, both within and outside the region, as they develop additional missile capabilities.

Other supply-side approaches – the Proliferation Security Initiative and UN resolutions

In 2003, 11 countries established the PSI which seeks to interdict ships and aircraft carrying missiles and WMD to states of concern (Etzioni 2009: 7–11; Joseph 2004). By 2009, some 90 countries had stated that they would participate in the PSI, although not all had formally joined. The PSI, and the interdiction of missile cargoes by states outside the initiative, sometimes acting under the guidelines of UN resolutions, has had some impact on missile proliferation in the Middle East/Gulf. Two transfers took place before the PSI was established. A North Korean shipment of Scuds to Yemen was interdicted in 2002, and a vessel carrying missile technology to Libya was halted in an Indian port in 1999. Thereafter, PSI participants (and actors such as Beijing and New Delhi that did not formally join) interdicted missile technology transfers to Iran at least five times between 2004 and 2008.

The first intercept in this period was noted by then-Secretary of State Condoleezza Rice in May 2005, when she mentioned that the initiative's members had undertaken 11 actions during the previous year and had blocked missile technology transfers to Tehran. In June 2005, at the request of Washington, China and a Central Asian nation denied overflight permission to a plane flying from the Islamic Republic to North Korea (*New York Times*, 24 October 2005). In November 2006, acting under UN Security Council Resolution 1696, an unspecified country stopped an Asian company's transfer of chromium nickel steel plates (useful in missile components) to Iran (Boese 2008a). In April 2007, a shipment of sodium perchlorate (useful for solid fuel propellant) to the Islamic Republic was detoured to an Asian port, whose government sent the consignment back to the country of origin. This shipment was intended for an Iranian entity barred from receiving missile technologies by UN Security Council Resolution 1737. In August 2008, India denied overflight permission to a North Korean plane delivering a suspected missile cargo to the Islamic Republic (*Wall Street Journal*, 1 November 2008).

In 2007, PSI participants and other states also interdicted two missile technology transfers to Syria. In February that year, equipment bound for Damascus that could be used to test ballistic missiles was seized by port authorities in an unspecified country – four initiative member countries worked to halt this equipment (*Telegraph*, 29 May 2008). In June 2007, a Syrian plane flying to North Korea was denied overflight rights and could not proceed to Pyongyang, and the states involved may have acted under UN Security Council Resolution 1737.

To summarize, in the 2000s, the PSI and similar efforts by other states blocked some missile-relevant transactions with Tehran and Damascus. As a result, the PSI likely hindered the pace and sophistication of Iran's and Syria's missile programmes, though, as noted in the previous pages, Iran and Syria still acquired substantial missile capabilities in the 2000s.

The Hague Code of Conduct

The creation of the Code

MTCR members discussed the idea of confidence-building measures to complement existing non-proliferation agreements after Pyongyang's 1998 Taepodong-1 launch. A growing awareness that supply-side controls were showing signs of strain in achieving their objectives, and a sense that implementing them would be enhanced by a widely accepted demand-side norm against missile possession, led to these discussions. Accordingly, MTCR signatories drafted elements of an International Code of Conduct for Missiles (ICOC) in 1999 and 2002. They invited non-member states to meetings to discuss the proposal in 2002. It was formally launched in November 2002, when it was renamed the Hague Code of Conduct Against Ballistic Missile Proliferation (Smith 2002/3).

The basic aim of the HCOC is to develop a common sensibility on ballistic missiles among its signatories (Smith 2006: 526–42). The code does not attempt to ban or proscribe ballistic missiles in part because they are widespread and some of the drafters themselves possess highly sophisticated capabilities. These missiles are widely used in conventional military strategy in addition to being linked to WMD and particularly nuclear weapons. Furthermore, HCOC covers ballistic but not cruise missiles. Nevertheless, the code requires member states to pursue "maximum possible restraint" in ballistic missile "development, testing, and deployment", and to adopt basic rules of transparency. "Restraint" is a vaguely defined term, but the transparency measures of the Hague Code are somewhat more explicit.

The two transparency- and confidence-building measures – aiming at regulating missile behaviour – required of states joining the HCOC are, first, a declaration of missile and space policy, and, second, pre-launch notification (PLN) of missile tests and space rocket launches. The former is still only vaguely defined in the Hague Code.[8] This is deliberate and probably necessary, in view of the sensitivity of such information, and since not all actors regard transparency on such matters as beneficial. For example, Israel referred to this aspect in 2002, when it pointed out that some states regard strategic ambiguity as an integral part of their deterrence strategy (*Haaretz*, 17 June 2002). Specifications on PLNs are more explicit. They call for "information on the generic class of the Ballistic Missile or Space Launch Missile, the planned launch notification window, the launch area and the planned direction", but still leave a good deal of the format and content of such reporting to the discretion of the submitting state (HCOC 2002).

The Hague Code of Conduct and the development of norms for missiles

The HCOC's basic aim of developing common norms for ballistic missiles has struggled to achieve fruition, both externally, and internally among its own membership. The code had 132 member states by 2011 (after 93 signed up when it was launched in 2002). Important missile-possessing signatories of the Hague Code include: four nuclear powers (the US, Russia, France, and the UK), states with space rockets like the Ukraine and South Korea, and Japan, the former missile aspirants Argentina and South Africa, Belarus, Canada (which operates sounding rockets), and Libya. Important states with missiles or space rockets outside the HCOC include Brazil, the Asian states China, Pakistan, and India, and the Middle Eastern/Gulf states Israel, Egypt, Iran, Syria, Saudi Arabia, and Yemen. In other words, by 2011, among missile-relevant states, only those which drafted the HCOC's norms had seen fit to subscribe to them. This was partly due to the code's origins in the MTCR, a regime many nonmembers regarded as discriminatory, and partly due to the fact that the drafting process excluded many of the actors that the draftees were hoping would join the HCOC.

While it is true that the draft code was discussed with prospective members at international meetings in 2002 prior to its opening for signature, participants were unclear about whether they were there simply to brainstorm, to negotiate, or to discuss. Egypt later commented to the UN First Committee that it "engaged in good faith in the efforts to consider the ICOC during two meetings held in Paris and Madrid in 2002, but somehow these efforts always fell short of the necessary requirements of a multilateral exercise" (PMEUN 2002). In internal terms, the story is similarly discouraging. The code's members did not always or adequately fulfil their obligations to report missile policies and launches. Also, in January 2008, Russia announced that it would no longer provide pre-launch notification to fellow signatories. It noted that "other participants, including the United States, are also not fully implementing their commitments" (quoted in Boese 2008b). Moscow's suspension was an unfortunate event for the HCOC, but arguably the greater problem was an almost tangible decline of interest in it among many of its members. Marius Grinius, the Canadian Ambassador to the UN Conference on Disarmament, noted to that body in February 2008 that the HCOC was "falling into disuse" (quoted in ibid.).

In short, by 2011, the code's signatories seemed to increasingly regard it as an optional extra rather than a significant element of global non-proliferation. Overall, while the European Union promoted the Hague Code in international diplomatic venues, and the regime held outreach activity among non-subscribing states, including a March 2008 seminar for Middle Eastern/Gulf states, major missile-possessing states still remained outside its framework.

The HCOC and the Middle East/Gulf

The HCOC failed to find a universally receptive audience in the Middle East as defined in this volume (see the chapter of Sven-Eric Fikenscher *et al.* on External Powers), where to date only Iraq and Jordan have subscribed to it. The factors contributing to this failure are similar to those which plagued the code elsewhere in the world: first, its provenance in the MTCR, which a number of Middle Eastern states regarded as a restriction on their ability to fulfil their national security obligations, and the related perceived lack of proper multilateral practice in its drafting; second, technical points such as its focus on ballistic missiles to the exclusion of cruise missile capabilities, and its perceived inadequate attention to the civilian peaceful uses of space; third, its failure to address all aspects of missile issues in a more balanced manner, aspects that some non-subscribing states sought to address through the UN, notably through the UN Panel of Governmental Experts on the Issue of Missiles in all its Aspects (see the Introduction by Bernd W. Kubbig).

Recognizing the first factor, HCOC members specifically sought both to position the code closer to the UN, and to have the UN recognize it, in order to align the treaty with global multilateral non-proliferation and disarmament discussions as well as to help to break ties with the MTCR. Thus, in the UN's 58th session in October 2003, several HCOC signatories mentioned the code in their statements to the First Committee. However, reflecting concerns that the HCOC was merely a device to respond to narrow concerns about certain ballistic missiles, Egypt stated that the credibility of the code "will depend largely on whether it is able to address the issues pertinent to the debate on missiles in a balanced and objective manner" (PMEUN 2003: 3). Iran did not mention the treaty in its statement, but (again reflecting the third aspect noted above) it put forward another resolution for a continuation of the UN Panel of Governmental Experts. The First Committee passed the resolution with 98 votes in favour, three against, and 59 abstentions. Israel, Micronesia, and the US voted against the measure, while the abstainers included the EU, Japan, and South Korea, all of whom cited the resolution's failure to mention the HCOC as one reason for their lack of support.

In the following years, resolutions regarding the Hague Code were tabled in the 59th, 60th, and 63rd sessions of the UN General Assembly. Still, Egypt, Iran, and other Middle Eastern states expressed concern about the code at these sessions. For example, at the United Nations First Committee sessions in October 2004, the HCOC Chair, Ambassador Luis Winter of Chile, submitted a draft resolution for consideration which asked the UN to recognize and welcome the HCOC, and to encourage its universalization. Ambassador Winter had expressed the hope that the draft could go through without a vote, but eventually one was taken. The resolution passed 137 to two with Egypt and Iran in opposition. This draft was subsequently adopted at the General Assembly, by a majority of 161 to two, again with Cairo and Tehran opposing. Egypt declined to support this HCOC Resolution and issued an explanation of its vote that mentioned a number of "lacunae" in the code:

> They relate to areas of peaceful uses, cooperation and assistance [with space technology]. They also relate more specifically to the lack of comprehensive scope ... [The Hague Code] only addresses the issue of proliferation of missiles but not their development, it addresses ballistic missiles but not cruise missiles ... and it ignores the most serious problem – that of the continued presence and development of nuclear weapons, of which ballistic missiles are only a means of delivery.
>
> (Government of Egypt 2004)

In this statement, which alludes to all three aspects previously mentioned, Cairo's reference to the development of missiles probably stems from a lingering suspicion that the HCOC's drafters, being MTCR states, harboured monopolistic intentions, and that they were more concerned with preventing states from acquiring ballistic missiles than with ensuring norms on their possession. Also, the reference to nuclear weapons is an indirect invocation of the Israeli nuclear weapon programme. Responding to the same resolution, Syria too condemned the Hague Code as "selective and discriminatory". Iran's opposition was a product of a refusal to accept its amendments (that would have somewhat addressed all missile issues) and was supported by Egypt, Kuwait, Qatar, and Syria. Unhappy at being presented with what it viewed as a 'take it or leave it' draft, the Islamic Republic argued that this "is unprecedented during the First Committee deliberations that a draft resolution is introduced and then co-sponsors reject to even consider and discuss improvements to the text" (Government of Iran 2004). In the following year (2005) a similar attempt was made at the UN to amend the UN HCOC Resolution's wording by seeking to address all aspects of missile issues. Egypt and Iran together with Indonesia and Malaysia submitted a package of amendments along the same lines as the previous year, but once again to no avail.

To summarize, three aspects influenced Middle Eastern/Gulf states to remain outside the HCOC, aspects that Egypt and Iran repeatedly emphasized in their statements at the UN. Still, a closer examination of recent views on the third aspect, the need to address all sides of the missile proliferation issue, suggests that there may be possible ways to break the impasse.

These views are found in discussions on the competitor initiative to the HCOC, which is submitted almost annually as the UN Resolution entitled "Missiles", that seeks a regular series of UN studies on "the issue of missiles in all its aspects". It is proposed by Iran and generally co-sponsored by Egypt. Typically, non-aligned states have favoured and Western countries and their allies have opposed or abstained from these resolutions. For example, the 2008 version was passed in the General Assembly with 120 in favour, 10 against, and 50 abstentions. In the Middle East, the co-sponsors Tehran and Cairo were supported in the vote by Iraq, Jordan, Kuwait, Oman, Qatar, Saudi Arabia, Syria, the UAE, and Yemen. Israel voted against. The US and the UK also rejected the resolution, but submitted an explanation of their decision that suggested some willingness to consider new initiatives. After rejecting the idea of another UN study as a waste of "scarce resources that could be better spent on more fruitful discussions", their joint statement said:

> We actively participate in many of the international efforts to curb the spread of missiles and related equipment and technology, and we plan to strengthen and reinforce this work. *These efforts have often proven highly effective when they were conducted on a regional basis and included the active participation of those States that were directly interested and affected.*
>
> (United States Delegation to the UN 2008, emphasis added)

This statement, which makes no direct mention of the HCOC, does suggest willingness to think about new, regionally based initiatives to missile proliferation rather than global ones. With the Obama administration's more cooperative approach to resolving proliferation problems, there may be some prospect of progress on this issue in the future, but in 2011 the signs were not promising. The following compromise might be feasible: Western states and Israel could accept UN resolutions seeking to focus on missiles in all their aspects, while Middle Eastern/Gulf states might, on a regional basis, accept existing missile non-proliferation instruments, as noted in the concluding section.

Conclusions

To what extent have the MTCR and supply-side strategies, as well as demand-side approaches, ultimately affected Middle Eastern missile programmes? Two regional states that bought missiles from external suppliers – Yemen and Saudi Arabia – have retained them but have not built missiles of their own. Other Middle Eastern/Gulf actors that sought to build missiles have, to varying extents, restrained their activities. Iraq was restricted to 150 km range missiles following the 1991 Gulf War, and its missile arsenal was eliminated after the Third Gulf War in 2003. Egypt and Syria have been restricted to Scud-Bs and -Cs, and have built or upgraded these missiles, partly with North Korean assistance. Iran and Israel have built single stage and two stage medium-range ballistic missiles and equivalent space rockets, and have the most advanced development programmes in the region.

To summarize, missile proliferation has been somewhat limited – it has been more restrained than it would have been in the absence of the MTCR, and the quality and quantity of Middle Eastern missile programmes have also been restricted.

Looking ahead, the MTCR's technology barriers may not thwart aspirants from building short- and medium-range missiles, since most of the technology is available to an actor with a modest industrial infrastructure. However, they may still restrain regional missile programmes in two ways. First, they can delay target states from building advanced intermediate- and intercontinental-range missiles, which require more sophisticated technologies (EastWest Institute 2009). Second, technology embargoes (and economic constraints) may considerably limit the volume of missile production and can hinder regional actors from developing sophisticated targeting, guidance, and manoeuvring systems. Ultimately, however, these constraining effects will diminish over time, and Middle Eastern states could build more and better medium-, intermediate-, and longer-range missiles. They could more rapidly advance their programmes if they acquired new – simple rather than sophisticated – technologies and better trained manpower (Karp 1996: 51–81), cooperated with and obtained technology from other regional powers, and if their governments allocated more resources to missile projects.

The likely effect of supply-side approaches will principally be to retard rather than curtail missile development, and while this result should not be underestimated, it may be a diminishing asset. If missile development among regional powers is not to advance along increasingly unrestrained and Hobbesian lines, the necessity for demand-side approaches will become correspondingly greater. Thus, as the effect of supply-side approaches slowly withers, we must ask ourselves what kind of 'missile politics' is likely to emerge as those controls retreat, and how it might be possible to encourage the emergence of a more cooperative approach. Possible cooperative approaches for tackling missile proliferation include the creation of flight test bans and space service regimes. Since flight testing is necessary in the development of new missiles, a test ban would hinder Middle Eastern/Gulf countries with short- and medium-range missiles from testing and developing more advanced longer-range missiles. And a space service regime would prevent regional countries from seeking space rockets that could also be used as missiles. Under such a regime, existing space-faring states (such as Russia, China, Europe, and the US) would launch satellites for other actors that renounce dual use rocket programmes. Thus, the MTCR and the HCOC can be supplemented by cooperative measures such as space service regimes and flight test bans in order to be more effective in the Middle East.

As of 2011, however, there is an almost complete absence of cooperative demand-side approaches in the regions of the world in which missiles most threaten security. Two potential futures can be envisaged here. In the first, and pessimistic, scenario, the continuing presence of the MTCR fosters a 'them-and-us' mentality in regional security complexes that is inimical to, and corrosive on, the emergence of demand-side approaches. This is likely to continue as long as some states are regarded as less affected than others, or not affected at all by the MTCR. The second, and more optimistic, scenario arises from the likely slow erosion of the regime's restraints on missile development. As regional missile programmes become shaped more by internal choice and by strategic and technological factors than by external export control ones, this might foster a recognition that such unrestrained proliferation is in nobody's interests. This, consequently, would promote the development of cooperative, rule-based approaches to missile behaviour and control.

Paradoxically, despite (and perhaps because of) the MTCR's and the HCOC's restrictions, the limited scope of their rules may be of interest to regional actors seeking to cooperate on developing rule-based approaches to missile behaviour and control. No other internationally prevalent rules on missile behaviour exist, and developing such rules will take time. Thus, if Middle Eastern states were to begin addressing missile issues, they could start by seeking regional adherence to the practices of the treaties. Despite their long-standing political opposition and objections to both agreements, the political resistance to them might be overcome by examining the regional costs and benefits of these instruments.

Under such an approach, Middle Eastern/Gulf states would not formally join the MTCR, but they could still agree to accept restraints on the transfer of missiles and relevant technology as mandated by UN Security Council 1540. Such restraints would be of security interest to all states in the region – each actor's security would be enhanced if they were all assured that other regional powers were not supplying missiles to their neighbours and adversaries. Simply put, Middle Eastern countries would be accepting the regime's general practices, without formally joining (or even formally declaring any adherence to) the MTCR under a regional missile technology control initiative. In parallel, Middle Eastern/Gulf states could pledge not to import missile technology from other actors, and external suppliers might pledge to respect such a regional agreement. Regarding the Hague Code, Middle Eastern states could agree to overcome their political objections to the code by recognizing its confidence-building and cooperation-promoting benefits. The regional political environment would be improved if all regional states successfully negotiated an initiative to join the HCOC (thus showing that regional cooperation on security issues such as missiles is possible). And such an agreement is feasible, in a military sense, because the code's transparency measures do not require revealing sensitive information about missiles but rather are non-sensitive and non-intrusive.

In conclusion, by subscribing to the practices of the Missile Technology Control Regime and the Hague Code of Conduct, neither of which would entail substantial security costs to themselves, Middle Eastern/Gulf states could begin developing habits of cooperation on missile issues. Such initial limited scope cooperation could pave the way for stronger longer-term cooperation to contain missile proliferation in the region.

Notes

1 This article deals with missiles covered by the MTCR and Hague Code, i.e. ballistic and cruise missiles of a certain range and payload. It does not address artillery rockets and anti-ship cruise missiles that generally have ranges under 100 km and generally carry light payloads, nor does it cover surface-to-air missiles.

2 MTCR members include Argentina, Australia, Austria, Belgium, Brazil, Bulgaria, Canada, the Czech Republic, Denmark, Finland, France, Germany, Greece, Hungary, Iceland, Ireland, Italy, Japan, Luxembourg, the Netherlands, New Zealand, Norway, Poland, Portugal, the Russian Federation, South Africa, South Korea, Spain, Sweden, Switzerland, Turkey, Ukraine, the United Kingdom, and the United States.

3 Additional assessments are found in Beck *et al.* (2003).

4 David Wright and Theodore A. Postol (2009) estimated that it could have a range of 7,000–8,000 km with a 1 ton payload.

5 Some sources note that Russian entities and engineers were the major source of medium-range missile technology to North Korea in the early 1990s, which was then transferred to Tehran. On this point, see especially assessments by Robert Schmucker as summarized by Theodore Postol (2009: 2, note 4).

6 Washington did not approve these transfers; it is unclear as to whether the supply of Arrow missiles would violate MTCR restrictions, see Ahlström (2004: 103–25).

7 Iranian and Syrian personnel were killed in a 23 July 2007 accident in Syria. This joint Syrian–Iranian team was reportedly attempting to place a chemical warhead on a Scud missile (*Jerusalem Post*, 18 September 2007).

8 The code calls on states to "make an annual declaration providing an outline of their Ballistic Missile Policies", that could include "relevant information on Ballistic Missile systems and land test launch sites", and "annual information on the number and generic class of ballistic missiles launched during the previous year" (HCOC 2002).

References

Ahlström, Christer (2004) 'Arrows for India? Technology Transfers for Ballistic Missile Defence and the Missile Technology Control Regime', *Journal of Conflict and Security Law*, 9(1): 103–25.

Beck, Michael D., Cupitt, Richard T., Gahlaut, Seema, and Jones, Scott A. (eds) (2003) *To Supply or to Deny: Comparing Nonproliferation Export Controls in Five Key Countries*, The Hague: Kluwer Law International.

Bermudez Jr, Joseph S. (1999) *A History of Ballistic Missile Development in the DPRK*, Monterey, CA: Center for Nonproliferation Studies, Monterey Institute of International Studies. Online, available at: http://cns.miis.edu/opapers/op2/op2.pdf (25 March 2011).

Boese, Wade (2008a) 'Interdiction Initiative Successes Assessed', *Arms Control Today*, 38(6): 33–4.

Boese, Wade (2008b) 'Russia Halts Missile Launch Notices', *Arms Control Today*, 38(2): 46.

CIA/United States Central Intelligence Agency (2008) *Unclassified Report to Congress on the Acquisition of Technology Relating to Weapons of Mass Destruction and Advanced Conventional Munitions, 1 January to 31 December 2006*, Langley, VA: CIA. Online, available at: www.fas.org/irp/threat/wmd-acq2006.pdf (25 March 2011).

EastWest Institute (2009) *Iran's Nuclear and Missile Potential: A Joint Threat Assessment by U.S. and Russian Technical Experts*, New York: EastWest Institute.

Etzioni, Amitai (2009) 'Tomorrow's Institution Today: The Promise of the Proliferation Security Initiative', *Foreign Affairs*, 88(3): 7–11.

Goodenough, Patrick (2009) 'Missile Collaboration between North Korea and Iran Goes Back Years', *Cybercast News Service*, 31 March. Online, available at: www.cnsnews.com/public/content/article.aspx?RsrcID=45885 (15 February 2011).

Gormley, Dennis M. (2008) *Missile Contagion: Cruise Missile Proliferation and the Threat to International Security*, Westport, CT: Praeger Security International.

Government of Egypt (2004) 'Explanation of Vote, L.50, Hague Code of Conduct Against Ballistic Missile Proliferation', Statement at the United Nations General Assembly, First Committee, New York: United Nations.

Government of Iran (2004) 'Explanation of the Vote on Draft Resolution A/C.1/59/L.50', Statement at the United Nations General Assembly, First Committee, New York: United Nations.

HCOC/Hague Code of Conduct Against Ballistic Missile Proliferation (2002) *International Code of Conduct Against Ballistic Missile Proliferation*, New York: United Nations. Online, available at: www.bmeia.gv.at/fileadmin/user_upload/bmeia/media/2-Aussenpolitik_Zentrale/114_hcoc.pdf (25 March 2011).

IISS/International Institute for Strategic Studies (2010) *Iran's Ballistic Missile Capabilities: A Net Assessment*, London: IISS.

Joseph, Jofi (2004) 'The Proliferation Security Initiative: Can Interdiction Stop Proliferation?' *Arms Control Today*, 34(5): 6–13.

Karp, Aaron (1996) *Ballistic Missile Proliferation: The Politics and Technics*, New York: Oxford University Press.

MissileThreat.com (2009) 'The Threat from Syria'. Online, available at: www.missilethreat.com/thethreat/pageID.252/default.asp (25 March 2011).

Mistry, Dinshaw (2003) *Containing Missile Proliferation: Strategic Technology, Security Regimes, and International Cooperation in Arms Control*, Seattle, WA: University of Washington Press.

MTCR/Missile Technology Control Regime (2008) 'Missile Technology Control Regime: Equipment, Software and Technology Annex', MTCR/TEM/2008/Annex/001. Online, available at: www.mtcr.info/english/MTCRTechnicalAnnexNov2008.pdf (25 March 2011).

Navias, Martin S. (1992) *Going Ballistic: The Build-up of Missiles in the Middle East*, London: Brassey's Ltd.

NTI/Nuclear Threat Initiative (2010) 'Iran Profile: Missile Overview'. Online, available at: www.nti.org/e_research/profiles/Iran/Missile/index.html (25 May 2011).

Parsi, Trita (2007) *Treacherous Alliance: The Secret Dealings of Israel, Iran, and the United States*, New Haven, CT: Yale University Press.

PMEUN/Permanent Mission of Egypt to the United Nations (2002) 'Statement by the Delegation of Egypt before the First Committee of the General Assembly: Disarmament and International Security', New York: United Nations. Online, available at: www.reachingcriticalwill.org/political/1com/1com02/speeches/2/egy100702.pdf (25 March 2011).

PMEUN/Permanent Mission of Egypt to the United Nations (2003) 'Statement by the Delegation of Egypt', General Debate of the First Committee of the 58th Session of the General Assembly, New York: United Nations. Online, available at: www.reachingcriticalwill.org/political/1com/1com03/genstate03/egypt.pdf (25 March 2011).

Pollack, Joshua (2010) 'Another North Korean Missile First'. Online, available at: http://pollack.armscontrolwonk.com/archive/3388/another-north-korean-missile-first (25 March 2011).

Postol, Theodore (2009) 'A Technical Assessment of Iran's Ballistic Missile Program' (no venue indicated, Typescript).

Report to the Security Council (2010) Report to the Security Council from the Panel of Experts Established Pursuant to Resolution 1874 (2009). Online, available at: www.armscontrolwonk.com/files/SCR1874.pdf (25 March 2011).

Shichor, Yitzhak (1998) 'Israel's Military Transfers to Taiwan and China', *Survival*, 40(1): 68–91.

Smith, Mark (2002/3) 'Stuck on the Launch Pad? The Ballistic Missile Code of Conduct Opens for Business', *Disarmament Diplomacy*, 68: 2–6.

Smith, Mark (2006) 'Pragmatic Micawberism? Norm Construction on Ballistic Missiles', *Contemporary Security Policy*, 27(3): 526–42.

United States Delegation to the United Nations (2008) 'Explanation of Vote on Draft Resolution A/C.1/63/L.27 on Missiles', United Nations First Committee, New York: United Nations. Online, available at: www.reachingcriticalwill.org/political/1com/1com08/EOV/USL27.pdf (25 March 2011).

UNSC/United Nations Security Council (2004) 'Resolution 1540', New York: United Nations.

UNSC/United Nations Security Council (2010) 'Report to the Security Council from the Panel of Experts Pursuant to Resolution 1874', New York: United Nations. Online, available at: www.fas.org/irp/eprint/scr1874.pdf (25 March 2011).

Wright, David and Postol, Theodore A. (2009) 'A Post-Launch Examination of the Unha-2', *Bulletin of the Atomic Scientists*, 29 June. Online, available at: www.thebulletin.org/web-edition/features/post-launch-examination-of-the-unha-2 (25 March 2011).

13 Caps and bans

Limiting, reducing, and prohibiting missiles and missile defence

Martin Senn, Jürgen Altmann, Bernd W. Kubbig, Jürgen Scheffran, Hans-Joachim Schmidt, and Oleg Shulga

Conceptualizing effective missile (defence) control

This article deals with the third milestone on the way towards the incremental realization of a regional Missile Free Zone (MFZ): the limitation, reduction, and eventual elimination of missiles and missile defence (MD). An adequate concept of effective control[1] has to cope with with the number and quality of the arsenals themselves. As a result, this milestone transcends both confidence- and security-building measures (CSBMs) as well as export controls and test bans, which were discussed in the two preceding articles. To be sure, the examples of the Near East Arms Coordinating Committee (1952), the Interim Agreements between Israel and Egypt (1974 and 1975), the Golan Heights Agreement (1974), and the working group on Arms Control and Regional Security (ACRS, 1992–95) show that confidence-building and arms control are not alien to at least parts of the Middle East (see the chapters of Emily B. Landau and Dalia Dassa Kaye as well as of Jürgen Scheffran *et al.*). More importantly, the Interim Agreements are ample proof that those confidence-building measures contributed to enhancing stability and to relaxing the specifically pronounced security dilemma in the region.

And yet, the idea that the reduction (not just the relocation) of existing arsenals may increase security, has only marginally entered the thinking of political and military elites in the region, not to mention the military postures of the respective countries. The overall Middle Eastern arms control experience has therefore rightly been characterized as being "rich in terms of initiatives, and largely barren with respect to results" (Steinberg 2005: 487). Countries in the Middle East/Gulf (and their external suppliers) have so far remained on the troublesome path of increasing their offensive weaponry in quest of greater security (Said 2002: 43). Moreover, the current build-up of missile defence capabilities by the United States and its allies has been uncritically portrayed as having a purely stabilizing effect on the regional arms dynamic.[2] However, because of their potentially destabilizing character – MD systems may for instance intensify the arms race – we take the ambivalent role of missile defence into account; this explains why we propose (selective) prohibitions in this area as well.

In this article, which explores possibilities for arms control in the Middle East, we use the achievements and deficits of trans-regional approaches developed in the East–West context as the conceptual background for envisioning successive phases of arms reductions in the Middle East/Gulf. In accordance with the emphasis on regional specifics, we also draw on the programmatic "Statement on Arms Control and Regional Security".[3] This document is the Magna Charta for any viable weapon reduction design in the Middle East; it was formulated in the context of ACRS and still represents the gold standard for contemporary and future discussions, such as for the proposed 2012 Conference on the establishment of a Middle East zone free of nuclear and all other weapons of mass destruction (Review Conference 2010: 30); this includes delivery systems which are explicitly mentioned in the mandate of the last Review Conference on the Nuclear Non-Proliferation Treaty with reference to the 1995 Resolution on the Middle East (ibid.). Our proposals for missile (defence) limitations and reductions are also informed by the analysis of military asymmetries in the region (see the chapter of Bernd W. Kubbig *et al.*).

While we do not claim to provide a blueprint for official negotiators, it is the aim of this article to propose an arms control process, which first stabilizes (Phase I), then reverses the regional arms dynamic (Phase II), and finally establishes a Missile Free Zone (Phase III). These phases consist of various building blocks such as CSBMs, limitations, reductions, and prohibitions; they should be embedded in the appropriate organizational setting and consist of measures that ensure the sustainability of the overall process. In addition, major external powers, above all the United States, will have to play a key role in this process due to their ambivalent role as friends of some regional states and foes of others, weapon suppliers, and – hopefully – as sponsors of a series of to-be-convened conferences which would be a core element of a regional peace strategy (Schimmelfennig 1991). The George H.W. Bush administration rightly assumed that "[o]nly a broad framework involving all the major supplier states can serve as a potential basis for conventional arms control" (Steinberg 1991: 65). In accordance with the contribution on military asymmetries in the Middle East, we shall primarily focus on states possessing missiles with ranges beyond 70 km that can be equipped with weapons of mass destruction (WMD) and are, at least in principle, adequately verifiable. Eventually, however, agreements on limitations, reductions, and (selective) prohibitions on missiles and missile defence systems between states would have to be supplemented by informal agreements with actors like Hamas or Hezbollah, if they posed a threat to the regional arms control process.

East–West arms control in theory and practice: providing adoptable and adaptable elements – building on regional experiences

The classical concept[4]

Thomas Schelling and Morton Halperin (1961: 2) have summarized the traditional objectives of arms control in their seminal definition as "all the forms of military cooperation between potential enemies in the interest of reducing the likelihood of war, its scope and violence if it occurs, and the political and economic costs of being prepared for it". Without a doubt, the prevention of war is the superordinate objective of arms control and therefore it takes precedence over damage and cost limitation as further primary goals. If, for example, a state wanted to limit the damage following a nuclear exchange by increasing the accuracy of its missiles and decreasing the yield of its warheads, the logic of arms control suggests these measures should not be implemented, because they could be seen as making the use of these weapons more likely, thereby undermining the objective of preventing war.

In order to achieve its primary goal, arms control has to meet the secondary objectives of establishing and maintaining strategic/arms race and crisis stability between actors. Strategic/arms race stability would foreclose the option of attaining or maintaining a military advantage which could be used for defeating or coercing other players. With effective arms control in place, a militarily inferior nation would not have to resort to force in order to neutralize the (perceived) military advantage of another country (or coalition of actors), nor would a militarily superior country feel the need to use its strength for political or military gain. In addition, strategic/arms race stability would also be achieved by decreasing the incentives and opportunities for actors to acquire new technologies, making arms dynamics more predictable. As far as crisis stability is concerned, arms control can shape forces in a way that reduces pressures to strike first during a crisis ('use or lose') or establish a range of CSBMs such as no-deployment zones, channels for communication or manoeuvre pre-notifications to prevent or limit the escalation of crises. Those CSBMs "aim not so much to diminish or restructure … forces as to introduce ways to make their use less likely" (Harvard Nuclear Study Group 1983: 210).

The instruments for meeting the above objectives include also limitations, freezes, reductions, and prohibitions, which can either be comprehensive or selective, i.e. affecting only a limited number of weapon systems, a specified region, or replacing destabilizing arms by weapon systems regarded as stabilizing.[5] Finally, verification is another indispensable instrument of arms control. However, the tools of limitations, freezes, and reductions do not automatically contribute to the objectives of arms control, because they may destabilize a deterrence relationship or increase the duration/damage of warfare. Therefore, in the final analysis, these instruments have to be judged by their contribution to stabilizing a deterrence relationship (Kemp 1994: 413; Schelling and Halperin 1961: 2) which "is considered the more stable, the less susceptible it is to political and technological events, information and mis-information, accidents, alarms, and mischief, that might upset it" (Schelling 1961: 723).

One can also distinguish between multi-, bi-, and unilateral forms of arms control. While bilateral and multilateral arms control obviously relies on "principles of conduct" (Ruggie 1992: 571), which are shared among two or more actors, unilateral measures, despite their frequent negative association with a withdrawal from bi- or multilateral forms of cooperation, can be constructive, not the least because they may initiate reciprocal responses by other actors.[6] These models of cooperation can coexist, overlap, or be pursued separately. They may come in the form of single initiatives or constitute a process lasting years and being pursued at adequate (i.e. neutral, inspiring, secret) venues with agreed-upon agendas and with multiple-vested bureaucratic and political interests.

The final conceptual dimension of arms control concerns conditions for (un)successful negotiation and implementation. The history of the East–West conflict demonstrates that favourable political conditions (especially a low degree of tension within a basically adversarial relationship) are undoubtedly the most central factor enabling the successful conclusion of arms control agreements. Moreover, the East–West experience shows that there is an asymmetrical relationship between political conditions and arms control; the fate of détente determines the fate of arms control, whereas arms control by itself can only contribute to initiating and facilitating détente. This implies two possible pitfalls for arms control: it is likely to fail if expectations are too high, and its chances of failing increase if it is too closely associated with other foreign policy areas ('linkage'), turning out to be a burden for the arms control process.

On a domestic level, the conclusion of agreements is facilitated by the existence of an arms control constituency – for instance an "epistemic community" as laid out by Peter Haas (1992) and Emanuel Adler (1992) – both inside and outside the government. Not only do the members of this constituency have expertise in the technical and political issues of arms control,[7] but they also pressure decision-makers to initiate or abide by arms control measures. Activism by grassroots organizations can also play an important role in shaping national positions as well as international negotiations on arms control, as was demonstrated on numerous occasions during and after the East–West conflict (Knopf 1998; Wittner 2009). Furthermore, the fate of arms control also depends on whether it is accepted by other domestic groups/institutions as in their own interests or in the national interest. Likewise, in order to be successful, an arms control agreement has to be seen by allied states as promoting or at least not undermining their interests. Beyond the realm of politics, the success of arms control also depends on how well actors can cope with (nearly inevitable) military asymmetries[8] and are able to constrain the technological momentum of arms development (Carter 1989: 19–20, 23–4). Finally, as mentioned above, adequate verification is another central condition for the success of arms control (see the article of Jürgen Scheffran *et al.*). The dimensions of the classical concept, which will be tailored to fit our regional proposal in a later section of this article, are summarized in Table 13.1.

The ambivalent reality of East–West arms control

Limitations and reductions

In the East–West context, the classical school of arms control with its dominant features, its achievements, essentials, and conditions for success/failure largely prevailed. One dominant feature of arms control during the Cold War was that restrictions on the quantitative expansion of strategic offensive arms were treated without regard to the qualitative 'modernization' of the arsenals. The Strategic Arms Limitation Talks (SALT) I Treaty was therefore criticized for having "worked as a catalyzer in shifting the weight of the arms race from quantity to quality" (Lodgaard 1977: 14). Both the US and the Soviet Union developed virtually every sophisticated weapon system that could feasibly be expected to maintain arms parity or achieve perceived (often short-lived) military advantages. New revolutionary technologies such as the introduction of Multiple Independently Targetable Re-entry Vehicles (MIRVs) not only led to a deterioration of strategic/arms race stability, but created immense problems for crisis stability as well.[9]

Moreover, limiting the quantitative arms expansion and only modestly restricting qualitative changes was incompatible with a comprehensive and unilateral arms freeze, which was designed in the early 1980s as an alternative approach to classical arms control. As a result, a weapons halt never became official policy. In view of the asymmetries between US and Soviet strategic weaponry in general, the "Freeze Now" approach increasingly focused on particularly destabilizing systems and thus came much closer to the classical school of arms control. Finally, the proponents of a nuclear halt in the U.S. Congress turned the original mandate of an immediate freeze into a call for an immediate start of bilateral negotiations on reductions (Kubbig 1988).

The second dominant feature of East–West bilateral treaties was the coexistence of arms control instruments and CSBMs. For instance, the two CSBM agreements (signed in May 1972) on the prevention of nuclear war and of accidents at sea strengthened crisis stability between the US and the Soviet Union. Within SALT I both sides accepted verification measures to

Table 13.1 Conceptual dimensions of East–West arms control

Primary objectives	*Secondary objectives*	*Instruments*	*Forms*	*Conditions of success*
Prevention of war Limitation of damage in case of war Limitation of costs	Strategic stability • No incentive for preventive use of force • Predictability of the arms dynamic Crisis stability • Survivability/invulnerability of forces • Communication mechanisms	Limitation Freeze Reduction Prohibition Substitution Verification measures	Unilateral Bilateral Multilateral	Favourable political/ideological conditions • Arms control constituency • Acceptability on domestic and international (alliance) level • Successfully coping with asymmetries • Constraints on momentum of arms technologies • Adequate verification

monitor treaty compliance by National Technical Means (primarily satellites) and by establishing the bilateral Standing Consultative Commission, mainly as a mechanism for resolving disputes. Later, within the Strategic Arms Reduction Treaties (START) framework, both parties could build on established arms verification mechanisms by reaching agreement on much more intrusive on-site inspections.

Finally, arms reduction proposals had entered the negotiations already during the Cold War, but according to the classical concept of arms control, reducing weapon arsenals was largely seen as jeopardizing East–West stability.[10] The reductions advanced during the series of START are summarized in Table 13.2.

Even under the New START signed in 2010, two decades after the end of the East–West conflict, the US and Russia will still have thousands of operationally deployed and inactive nuclear warheads. This is a telling example of how difficult real cuts in weapons arsenals are to achieve, i.e. comprehensive reductions without corresponding increases in other areas or hedges. Moreover, the extensive time lag between the easing of

Table 13.2 Reductions in the START series

	START I	*START II*	*START III*	*SORT*	*New START*
Status	Expired	Not entered into force	Not negotiated	Out of force	In force
Limits of deployed warheads (WHs)	6,000	3,000–3,500	2,000–2,500	1,700–2,200	1,550
Destruction of deactivated WHs	No	No	Yes	No	No
Limits of deployed delivery vehicles (DVs)	1,600	Prohibits MIRVs and heavy ICBMs	Yes	No	700[1]
Destruction of deactivated DVs	Yes	Yes	Yes	No	Yes[2]
Verification mechanisms	Yes	Yes	–	No	Yes
Expiry date	5 December 2009	–	–	31 December 2012	5 February 2021

Sources: ACA (2010), CNS (2009), Woolf (2010).

Notes
1 New START includes an aggregate limit of 800 deployed and non-deployed DVs.
2 Deactivated DVs may be converted or destroyed.

East–West tensions and the actual achievement of relatively (deep) arms reductions suggests that the acquisition of highly sophisticated, complex, and cutting edge weapon systems is a dynamic also driven by factors such as technological momentum and the influence of domestic interest groups, rather than being exclusively motivated by threat perceptions and mutual fear.

In the realm of conventional arms control, initial talks in Europe had no positive effect on crisis stability and war prevention. Whereas the Mutual and Balanced Force Reduction (MBFR) talks concluded in 1989 without results, negotiations on military confidence-building led to some agreed-upon measures by 1985. Nevertheless, the MBFR talks helped the parties understand the asymmetries of their military doctrines and arsenals and educate the Soviet leadership on Western approaches to arms control. Secretary General Mikhail Gorbachev eventually used these concepts to overcome the burden of Soviet quantitative conventional superiority in Europe. The second round of conventional arms control talks, concluding with the Treaty on Conventional Armed Forces in Europe (CFE) in 1990, proved to be a crucial step towards increasing crisis stability and preventing war. The CFE Treaty aimed at preventing large-scale and surprise attacks through numerical limitations and reductions in five weapon categories (tanks, artillery pieces, armoured combat vehicles, combat aircraft, and attack helicopters); regional sub-limits for land forces were also put in place to discourage destabilizing force concentrations.

The Soviet willingness to accept limits on its quantitative superior land forces was met by Western concessions on the number of its qualitatively superior air forces. The Eastern alliance members accepted on-site inspections and the Soviet Union agreed to a special sufficiency rule, which stipulated that no country should possess more than one-third of all restrained arms. Small artillery missiles were not restricted based on their range but on whether they exceeded 100mm calibre.[11] After the Dayton Peace Accords, the Sub-Regional Arms Control Agreement lowered the calibre of artillery pieces (including missiles) permitted in Croatia, Bosnia and Herzegovina, Montenegro, and Serbia to 75mm to cover the majority of artillery systems. In addition to arms limitations (CFE) conventional arms control included confidence-building (Vienna Documents of the Negotiations on Confidence- and Security-Building Measures, Treaty on Open Skies), enhancing the predictability of military activities and expand the areas to which transparency standards can be applied.

After the CFE reductions a new round of talks began to adapt the treaty to the changed security situation in Europe. Instead of adopting the principle of arms parity, which was used during the Cold War, the new CFE talks employed the concept of national and territorial ceilings, which seemed much better suited for regions and disputes beyond Europe since they place limits on both national forces and, under territorial ceilings, on stationary forces deployed by other countries. They also included additional restraints for temporary deployments (e.g. bi- or multilateral manoeuvres) in order to limit military flexibility. Finally, special political and transparency measures were added to enhance predictability and crisis stability for the case that temporary deployments are extended.[12]

Selected prohibitions

Unlike the manifold examples of CSBMs, limitations, reductions, as well as actual prohibitions and dismantlement of already existing weapons were rare events during the days of the East–West conflict. One of the few exceptions, the unilateral withdrawal of the multi-billion anti-ballistic missile (ABM) system in the United States, should be judged less by arms control and strategic standards than by economic and especially technological yardsticks. A coalition of ABM critics in the government, especially in the Senate, and a vocal group of scientists as their junior partners, considered this system to be technically problematic from the beginning. This was a politically powerful alliance in the wake of the Vietnam War and amidst strong public protests against the anti-ballistic missile system. The decision to strictly limit strategic MD stipulated by the ABM Treaty was made possible by the interest of both superpowers in preserving "*parity, mutual deterrence,* and *strategic stability*" (Garthoff 1978: 3). Finally, MIRV technology was presented as superior to anti-ballistic missile systems, a fact which may seem ironic from an efficient arms control point of view. In the end, there was a growing number of supporters for unrestricted ABM capabilities in addition to MIRV technology. Had they prevailed, the United States and Russia would probably have entered a costly and dangerous double arms race (Kubbig 2004: 289–322).

The Intermediate-Range Nuclear Forces Treaty (INF) of 1987 is a different case, since it led to the destruction of an entire class of the most modern land-based missiles with a range from 500–5,500 km. Its successful negotiation and implementation can be explained by its focus on specific symmetries (i.e. the same classes and about the same quantities of weapons), mutual anxiety regarding a possible destabilizing effect of these weapons, and the dialectical relationship between weapons and the broader political climate. While the signing of the INF Agreement would not have been possible without Mikhail Gorbachev's new policies of 'glasnost' and 'perestroika', it became part of the dynamics that paved the way towards the dissolution of the Soviet Union and the end of the East–West conflict. As far as the major features of East–West arms control are concerned, it can be noted in summary that a somewhat tamed arms growth with no restrictions on plans for weapon 'modernization' in coexistence with CSBMs was complemented by a policy of refusing to freeze or ban new weapons, and a dismantling of two already existing systems, with reductions coming in rather late in the process.

Achievements and conditions of success/failure

Overall, East–West arms control can be credited with major achievements. First, arms control treaties enhanced the predictability of the arms dynamic – both sides managed to work out counting rules and agree upon aggregates of parity consisting of various imbalances due to the asymmetrical composition of their force structures. Deplorable as it may be from a pure arms reduction point of view, even the higher levels were mainly agreed-upon cooperatively. Second, a large degree of transparency and accountability was established through verification and other communication mechanisms. Third, the arms control dynamic was to a high degree irreversible, since treaty-based agreements are more difficult to undermine and overcome than politically binding 'handshake' accords, which the George W. Bush administration favoured for some time. The arms control network proved to be robust enough that it could be renewed even after it was interrupted (as was the case at the beginning of the Reagan administration in the early 1980s).

As far as conditions for success/failure are concerned, the results are mixed. Arms control negotiations were not overnight affairs, but cumbersome, lengthy, and yet rewarding achievements. It was not clear at all that the initial efforts would develop into structured and institutionalized processes. Indeed, they required extraordinary patient effort. Taking into account the MBFR talks which preceeded it, the CFE Treaty took nearly 20 years to negotiate, the Intermediate-Range Nuclear Forces Treaty 12 years, and the Strategic Arms Reduction talks preceding the conclusion of START I required nine years (Chrzanowski 1994: 4–59). The limits and failures of arms control were in part due to the above-mentioned negative 'linkage' policy. SALT II, for instance, was never ratified by the U.S. Senate, and finally "buried in the sands of Ogaden" (Brzezinski 1983: 189). The agreement became a victim of the broader foreign policy conduct of the Soviet Union extending to Africa, which was considered unacceptable by the majority of the U.S. Senate. Arms control also had its inherent 'linkage' difficulties: START II was not implemented when the Russian government abandoned the agreement after the George W. Bush administration unilaterally withdrew from the ABM Treaty.

Building on conceptual elements and experiences in the ACRS context

The working group produced a "Statement on Arms Control and Regional Security" which was adopted by nearly all ACRS parties (with Egypt as the obstacle). The document reflects an innovative way of thinking about the issue of arms control which cannot be overrated. This platform endorses the limited yet vital role of regional security and arms control measures "to reduce tension or the risk of war".[13] It contains major elements of the classical arms control concept – crisis and strategic/arms race stability – specified in terms of "the risk of surprise attack" and the prevention of dynamics of (non-)conventional weaponry. This declaration, however, does not include the dominant feature of East–West arms control: the attempt to achieve stability by limiting the quantitative expansion of arms and only modestly requiring restrictions on qualitative changes. Instead – and this is utterly important – the programmatic "Statement" promotes the idea of reducing regional arms as a way of increasing security. It is linked to "the risk of surprise attack" and to "the stockpiles of conventional arms"; the objective is "to provide *enhanced* security at *lower* levels of armaments and militarization" (emphases added). It is also remarkable that the "Statement" does not address the issue of sequencing (CSBMs first, weapons later), since it simply enumerates both of them – this is entirely compatible with its emphasis on reducing the arsenals themselves.

Furthermore, the Israeli variant of the ACRS document explicitly mentions the term 'ballistic missiles' in connection with the goal of establishing a WMD Free Zone in the Middle East, while the United States and Egypt introduced the broader notion of delivery vehicles, implicitly including, above all, aircraft, an area in which Israel has been superior. Thus, the missile problématique is recognized as a challenge in the entire process towards achieving the ultimate objective of a comprehensive WMD Free Zone. In the "Statement", however, this zonal concept is implicitly connected to Israeli and Egyptian imperatives which remain politically explosive even today: the "Disarmament First!" and the "Regional Peace First!" claims are neither connected nor prioritized in this compromise "Statement". As a result, the platform does not outline specific conditions of success.

The actual outcome of the working group was limited to CSBMs, since it

> could not develop agreed practical steps to reduce concerns about military capabilities, strategic doctrines and political intentions. ACRS was only beginning to break down the walls of misunderstandings between the parties about each other's long-term security concerns and the reasons for the asymmetries between each state's force structure.
>
> (Yaffe 2001: 14)

This remains a challenge for any future limitation and reduction proposal in any setting.

A proposal for limiting, reducing, and prohibiting missiles in an adequate organizational setting

Combining the insights from East–West with regional results and experiences

The attempt to conceptualize a process of missile (defence) limitation and reduction towards an MFZ is informed by the Cold War experiences and by the regional approaches. These East–West concepts are relevant, because they contain the core elements of stability, conditions of success/failure (to be adapted to the Middle Eastern setting), and the appropriate organizational form for the envisaged development. In addition, a renewed and deepened arms control process in the Middle East would have to address a series of issues, ranging from current military asymmetries and their implications, conflict structures and alliances, domestic and external motives behind arms build-ups, to the complex and cumbersome counting and comparing of military inventories. Improving political relations is a vital condition for the successful process towards an MFZ. In this connection setting up normal and peaceful relations by taking up the Arab Peace Initiative would undoubtedly be an important step. A further and equally important condition for success is the adequate verification of the limitations, reductions, and prohibitions of WMD related missiles, at least those with a range of about 70 km (see the article of Jürgen Scheffran *et al.*). Moreover, the behaviour of external actors and their role as major suppliers of missiles, MD, and other types of armament have to be added to the classical approach (see Table 13.1) as a condition of success.

As in the context of assessing military asymmetries, the United States poses a major challenge: how does one integrate the highly mobile, regional capabilities of the 'sole superpower' into such a limitation and reduction endeavour? In addition to their role as major suppliers of weapons, extra-regional actors also have to be seen as indispensable sponsors of any multilateral arms control process and forum, at least initially. The insights gleaned from the East–West arms control experience are also important for the design of our own proposal, although we recognize the Middle East/Gulf poses different structural challenges. These differences are both an obstacle and facilitator of arms control in the region. A major difference and likewise an impediment is the principle of sovereignty. When Moscow and Washington signed SALT I in May 1972 virtually all borders were recognized. The principle of state sovereignty based on international law was crucial for the arms control process as the negotiating parties had agreed to forego the use of force, to alter borders only peacefully, and to refrain from violent interference in the domestic issues of other countries. The Middle East/Gulf has a different history. However, the fundamental principles of state building and sovereignty form part of the basic documents such as the Charter of the Arab League and the

Barcelona Declaration of 1995 which establishes the cooperation between Europe, North African, and Eastern Mediterranean states.

In fact, the ACRS "Statement" explicitly mentions all those "Fundamental Principles Governing Security Relations Among Regional Participants" like the parties' "commitment to the principles of the Charter of the United Nations" and their obligation to "refrain from the threat or use of force and from acts of terrorism and subversion", which are ultimately oriented towards a "just, lasting, and comprehensive peace settlement in the Middle East". A second difference and obstacle to effective regional arms control is the complexity of Middle Eastern/Gulf conflict dynamics (see the chapter of Sven-Eric Fikenscher *et al.* on Transparency). A third difference is the role of technological innovation as a driving force behind the arms dynamic and the corresponding premium placed on developing cutting edge and sophisticated weaponry. In the East–West context this process was driven in part by a strong and stable military-industrial-bureaucratic network. As outlined in the discussion of military asymmetries, the situation in the Middle East is vastly different. With the exception of Israel, Iran, and to some extent Saudi Arabia, the arms build-up in the Middle East/Gulf is largely numerical or quantitative rather than qualitative in nature, and often consists of an increase of fairly old and non-sophisticated weaponry. In general, a dynamic of action–reaction dominates domestic factors. Without calling into question the relevance of regional specifics and demand-side factors for effective arms control, the situation in the Middle East may not only offer some flexibility in starting such a process, but it certainly also puts a greater emphasis on the quality of armament supplied by the extra-regional powers and their policies.

Analogous to the East–West experience, a comprehensive freeze concept would most likely be a non-starter in the Middle East given the utterly asymmetrical situation. And yet, a specific ban of existing systems may be in the cards during regional negotiations if they are seen as especially destabilizing (INF example), technically inefficient, and cost-intensive (ABM site case). In addition, the CFE negotiations show that a trade-off between adversaries is possible by taking the asymmetrical military situation into account. The Arab states and Iran could accept limits on their quantitative superior land forces (as the Soviets did); in the Middle East context this would have to include missile capabilities, and Israel could agree to restrict its superior air forces (as the Western states did) and missile capabilities. This might also be a way of structuring and prioritizing arms control talks so as to cope with the negative ACRS experience Michael Yaffe (2001) has described. Finally, taking the ACRS "Statement" as a guide, one would start the regional arms control process with the reduction phase. But in view of the lack of experience in the region with lowering existing weapon arsenals, it would be prudent to follow the practice of arms control in the East–West context and begin with a stabilization phase at the beginning of the talks. The overall goal of this stabilization phase would be to transform the unrestrained regional arms dynamic into a managed process, based on cooperatively agreed-upon and stability-sensitive arms limitations. It is only after the successful stabilization of the regional arms dynamic and improvements in the political relations between the principal actors in the Middle East that a reduction/reversal process could develop and succeed.

Outlining three phases towards an MFZ

Phase I: stabilization

The first phase of arms control in the Middle East would pursue the twin goals of stabilizing the regional arms dynamic and of contributing to crisis stability. This *stabilization phase* could include efforts to cooperatively agree on fixed levels of missiles and/or other weaponry such as combat aircraft.[14] This is a way of offering priority lists of national and coalition concerns, allowing the most worrisome elements of the security dilemma to structure the dialogue, thereby avoiding the risk of repeating the failure of the ACRS talks. It has to be emphasized that achieving stability by agreed-upon levels would not only mean a freeze of certain kinds of missiles and conventional weaponry but would possibly also include an increase of weaponry in certain countries to offset asymmetries. In this context, it is also important to note that the probable inclusion of missile defence systems in these negotiations could mean accepting the possible acquisition of (additional) MD capabilities by Iran and Syria. In parallel, CSBMs in the form of de-alerting regional strike forces – and this would relate especially to crisis stability – could be introduced and implemented:[15]

- The launchers/aircraft would be kept separately from the strike weapons (missiles and bombs/air-to-surface missiles) such as to make their immediate use impossible. The launchers and combat aircraft could be used for personnel training purposes, but in an unarmed state.
- Tactical ground-based missile systems could be moved to areas from which they could not reach the neighbour(s). There would be an efficient inspection regime in place that would provide for verifiable baseline, reliable accountability, regular and challenge inspections, as well as overflights with reconnaissance aeroplanes in accordance with pre-agreed patterns. The Treaty on Open Skies could serve as a model for a regional regime regulating airborne verification (see the chapter of Jürgen Scheffran *et al.*).

In addition to agreed-upon limits of armaments, the stabilization phase would also have to include joint assessments of current capabilities, such as a regional arms catalogue, which could be based on the suggested Register of Conventional Arms in the Middle East (see the chapter of

Sven-Eric Fikenscher *et al.* on Transparency), to provide a common ground from which to start reductions (initially stabilizing measures such as a selective arms build-up may work as a hedge). As far as the ambivalent role of the external powers, especially the United States is concerned, one cannot exclude that in this phase the US could supply even more arms and intensify its military presence. However, the sale of arms would have to be within the agreed-upon limits and in parallel the above-mentioned redeployment process would have to be initiated, so that its warships could not fire missiles at zonal members such as Iran, which then might agree to relocate its land-based missile arsenal accordingly.

Phase II: first dimension – reversal and prohibition of offensive weaponry

The articulated threats to both variants of stability (as concretized in the ACRS "Statement") suggest taking them as additional building blocks in the *reduction/reversal phase* of our gradual approach:

Strategic/arms race stability – "lower levels of armaments and militarization". This dimension includes discussing the scope of the delivery systems to be incorporated in an agreement – ballistic and cruise missiles, rockets as well as air and missile defence systems, (armed) unmanned aerial vehicles (UAVs), and combat aircraft. In accordance with our approach outlined in the article on Military Asymmetries, we suggest focusing in a first phase on WMD related missiles with ranges extending 70 km.

In this phase the participants of a forum would discuss reducing the numbers of delivery vehicles and/or of ranges in a way compatible with stability and within a framework of reciprocity. The crucial criterion for both kinds of reductions (and a mix of them) would be to agree upon an incremental curtailment of the numbers and/or the range of weapons which can reach the adversary's territory or its main military, metropolitan, and industrial sites. In a subsequent phase the quantitative limitation/reduction of weapons with lower ranges down to 25 km as outlined in the UN Register of Conventional Arms (see the chapter of Sven-Eric Fikenscher *et al.* on Transparency) could be discussed. The aircraft issue will certainly appear on the list of concerns at the negotiating table and would have to be dealt with accordingly. This gradual approach does not exclude even further reductions in the numbers of missiles. The entire process would be even more complicated if actors such as Hamas and Hezbollah were taken into consideration.

As far as the external powers (again, above all, the United States) are concerned, their role as weapon suppliers has to be adapted accordingly, so that they support the goal of reducing the aforementioned arsenals. We can learn from past proposals for a more restrictive export policy, e.g. dollar ceilings for missile exports (CBO 1992: xiv) or limiting deliveries on the basis of their estimated impact on regional stability (Hunter 2010: 138). While dollar ceilings involve difficult pricing issues, the stability yardstick has the advantage of being tacitly observed especially by Washington and Moscow, as for instance the Russian reversal of its position on delivering the S-300 to Iran in response to US concerns illustrates.

Crisis stability – reducing "the risk of surprise attack". The topic of crisis stability includes analysing the specifically destabilizing effects of missiles, which – unlike aircraft – cannot be ordered back once they are launched and which may trigger preemptive firing ('use or lose') under pressure and time constraints in a crisis, since they are seen as vulnerable. The antidote to surprise attacks by accident or by design would – in Phase II – consist of a quid pro quo aimed at capping the number and range of missiles. A ban would be a more far-reaching measure and could be applied to the most destabilizing types of delivery vehicles. Surprise attacks by design are a matter of the respective national doctrines and so their offensive and preemptive character would also have to be discussed.

Phase II: second dimension – reduction/reversal and selective prohibition of defensive weaponry

Missile defence capabilities must be part of any overall incremental reduction strategy. Given their ambivalent and complicating role, one should be sceptical about the contributions of MD capabilities to both variants of stability:

Strategic/arms race stability – "lower levels of armaments and militarization". Missile defence systems have become an element of classical action–reaction arms dynamics, above all in the Gulf. The United Arab Emirates (UAE), Kuwait, and Saudi Arabia have opted for MD in response to the perceived Iranian missile threat. At the time of writing the consequences of these (planned) supplies are still unclear, but it is very likely that they will be ambivalent. Missile defence systems, as long as they work as intended, are to function as a non-proliferation measure by discouraging states – in this case the Islamic Republic – from acquiring missiles. However, it is probable that Tehran will in turn react to the efforts of the three Gulf states by enhancing its own missile force, in order to maintain its deterrence potential (Senn 2009: 61–3). The region would then be involved in a destabilizing and costly double arms race of missiles and missile defence capabilities. Perhaps at some point in time the countries stationing MD would revise their cost–benefit calculus, since these systems might be perceived as undermining their security or as being too expensive.

Adding to the general ambivalence of US exports of missile defence systems to Gulf countries is that they may be helpful in containing the demand for an indigenous nuclear weapons capability. This could be particularly important in the case of Saudi Arabia.[16] In the end, however, any stabilizing effect of missile defence (by contributing to the prevention of further nuclear and missile proliferation in the region) will mainly derive from US involvement and its military might. On the one hand, the exports symbolize the strengthening of ties between the US as a supplier and its allies as well as the corresponding determination to restrain Iranian regional ambitions. On the other, the deterring effect of MD per se is secondary, since its reliability remains limited; in addition, such systems are more costly than the acquisition of ballistic missiles. The potentially positive consequences of MD (in terms of a visible non-proliferation effect, either affecting the deploying state or the adversary) suggest allowing regional states to keep a reserve arsenal, unless the optimistic scenarios prove wrong and the offence–defence arms race intensifies. In this case negotiators should work out parallel reduction steps for missiles and MD systems.

Discussions of the costs and benefits of MD deliveries (from extra-regional actors) suggest that their impact on the overall balance can be perceived as jeopardizing stability. Thus, the stability yardstick would be a good starting point for a more transparent, comprehensive, and coordinated policy of restraint which may apply the criterion of limiting ranges more systematically and broadly (for instance by building on the US model of delivering weaponry to Saudi Arabia that cannot reach Israel). But undoubtedly, in the final analysis security and arms reduction arrangements between the US and Iran will be the most important single determinant for the format and process of MD reduction (see the chapters of Bernd W. Kubbig *et al.* and of Sven-Eric Fikenscher *et al.* on External Powers).

Crisis stability – reducing "the risk of surprise attack". The fact that MD is seen solely as part of the solution for the proliferation problématique is partially due to its image as being purely defensive. This claim, illustrated by the role of Patriot defence systems in preventing Israel from actively engaging in the 1991 Second Gulf War, depends for its validity on the scenario one envisions. A Syrian strike on Israel or an Iranian attack on Israel and its Arab neighbours in the Gulf would support claims about the purely defensive character of missile defence. However, the situation would change fundamentally if missile defence systems became part of a clearly offensive scenario in which, for example, the United States acting alone or in concert with Israel, struck nuclear and missile related facilities in Iran, which Tehran might counter by attacking US military sites in the Gulf states. This would imply that the expected deterrence effect of MD did not materialize in a crisis situation.

Furthermore, it is not unlikely that at least Iran saturates MD (by increasing the number of missiles, especially WMD tipped), overwhelms adversarial defence systems (e.g. with decoys, debris of missiles stages), or circumvents and neutralizes MD (with cruise missiles, UAVs, and light planes). Admittedly, ambivalence characterizes MD: from an Iranian perspective the Israeli Arrow systems may encourage the Jewish state to launch an air strike against its nuclear facilities, while from an Israeli perspective its missile defence capabilities may – in addition to its nuclear forces – raise the threshold for Iranian decision-makers. As the discourse within the Jewish state speaks against the thesis that MD systems might prevent Israel from initiating a strike against the Islamic Republic in a crisis situation, there is – from this point of view – obviously no reason for ignoring the missile defence dimension in the envisaged reduction process.

Phase III: comprehensive prohibitions – the MFZ

The third and final phase of effective arms control in the Middle East would have to ensure the stability and sustainability of the prohibitive regime by preventing countries from engaging in rule-breaking behaviour and by assuring and endorsing abiders.

(Post) strategic/arms race stability – prevent cheaters and assure abiders. First, rule-breaking behaviour and a resurgence of the regional arms dynamic could be prevented by tackling incentives for rearmament within the institutional framework of the MFZ. Following the example of the Organization for Security and Co-operation in Europe, a prospective regional organization could similarly establish comprehensive and issue-specific bodies (like the OSCE Permanent Council and the Conflict Prevention Centre) that would allow regional countries to address and manage emerging disputes at an early stage. In addition, potential cheaters would be deterred by a system of effective verification measures (see the article of Jürgen Scheffran *et al.*) and the prospect of rule enforcement by the great powers in their function as guardians of regional arms control. As the difficult history of passing sanctions against Iran in the UN Security Council shows, however, it is not always easy to enforce international rules. Nevertheless, it is reasonable to assume that any genuine threat to hard-won stability in the Middle East would spur the great powers into common action and induce them to exert political and economic pressure that would make cheating a costly endeavour.

Although missile defence has been credited by several of its proponents with the potential to stabilize nuclear or missile free regions[17] by deterring cheaters and assuring rule-abiding states, there are grounds to believe that it will probably not live up to these expectations. First of all, in order to be stabilizing, missile defence would have to be highly effective. However, given the technological challenges and the numerous options to saturate or overwhelm MD systems, their effectiveness will probably remain questionable for decades to come. As a consequence, regional states facing rule-breaking behaviour would probably resort to the tried and tested strategy of deterrence by offensive weaponry rather than entrusting their security to defensive systems with uncertain performance. As already mentioned, any stabilizing effect of missile defence would primarily result from the involvement and the military might of the United States rather than being a consequence of these systems per se. Furthermore, MD could also be a potentially destabilizing factor in an MFZ, since interceptors can be converted into missiles capable of carrying warheads. Even the possession of more or less unreliable MD systems could be a first stepping stone for breaking out of the zonal agreement.

Crisis stability – no crucial role for missile defence. As in the case of strategic/arms race stability, the institutional framework of an MFZ and great power intervention could maintain crisis stability. Provided that the above-mentioned requirements (redeployment and range reductions) in Phase I and II are met, missile defence would not be needed in Phase III to ensure crisis stability, assuming that a *rapid* redeployment of missiles in a situation of crisis is highly unlikely. The deployment of new and even mothballed missiles would have to undergo visible testing and military personnel would have to be trained in operating them, making missile rearmament a detectable long-term process. Put simply, an armed confrontation in a missile free environment would involve conventional means like ground or air forces against which missile defence would be useless.

At the end of the day, the stability of a Missile and WMD Free Zone would be assured by an appropriate institutional framework and the guardianship of the great powers to limit the benefits from cheating. Moreover, the existence of such a zone in the Middle East presupposes a high degree of cooperation and benign relations between actors in the region that would increase the costs of defection and render missile defence pointless. In this context, the following argument by Alexander Bollfrass (2010: 209) for the case of global nuclear disarmament is also applicable to the conflict region:

> Adopting the perspective of a potential cheater … reveals that the odds for success would be so slim, the path so long, and the rewards so meager that the safety and security of a disarmed world would overwhelm the risks of breakout in any rational actor's calculations.

Organizing and maintaining the process

Again, lessons learnt from the East–West and Middle East contexts will be important for establishing the adequate format among the participants of the suggested MFZ ranging from Egypt in the west, to Iran in the east, and Syria in the north, to Yemen in the south (see the chapter of Sven-Eric Fikenscher *et al.* on External Powers). As with the military asymmetries, it is the conflict structures, the sub-regional specifics, the bilateral adversarial context, and the role of the vital external actors (especially the United States), that will be decisive for finding the appropriate structure. Without ruling out an 'assembly' of all the parties to the zone, two sub-regional settings come to mind:

- *Levant Forum*: Israel, Egypt, Syria, Lebanon, and Jordan;
- *Gulf Forum*: Saudi Arabia, the UAE, Kuwait, Qatar, Oman, Yemen, Iran, and Iraq.[18]

The participants in these two forums will have to find ways to realign and coordinate themselves because of existing overlapping conflicts and alliances (the coalition between Damascus and Tehran transcends both forums). Within these two settings the capping of missile ranges, for instance, can be properly addressed: mutually limiting the range of medium-range Iranian and Israeli missiles would not take the Syrian–Israeli conflict into account, as long as their short-range types are not included in a framework. This example illustrates also the importance of the zonal approach.

The role of external actors cannot be reduced to that of weapon suppliers. As the ACRS talks have shown, it is imperative that they also sponsor and help steer the entire process. Providing incentives and occasionally gentle pressure is required to initiate and maintain it. There is one more lesson to be learnt from the ACRS talks: without the participation of all the relevant parties it is difficult both to address the full scope of security concerns and to conduct comprehensive negotiations (see the article of Emily B. Landau and Dalia Dassa Kaye). In the 1990s, Damascus and Beirut did not accept an invitation to the talks pending a breakthrough in their bilateral relations with Israel, while Tehran and Baghdad were not invited. This did not prevent the parties from moving ahead: today, the challenge is to get Israel, Iran, Syria, and Lebanon involved. Based on the lengthy East–West arms control processes, the extra-regional players could counsel and encourage regional participants: in this respect, the imperative to "'have patience' is perhaps the most important lesson offered by veterans of East–West arms control negotiations" (Chrzanowski 1994: 2).

The ACRS experience showed that even under extremely favourable political circumstances the United States was successful in starting, mediating, and keeping the process alive only for a couple of years – at a time when Washington was powerfully involved in Middle Eastern politics. This suggests the process requires additional input from the outset, including the establishment of an informal 'Arab Quartet' consisting, for example, of Egypt, Jordan, Saudi Arabia, and the UAE (Meyer 2007: 17). It could develop the necessary expertise, infrastructure, and political will to facilitate the work of the external players. Maybe the Secretariat of the Arab League can play a constructive role as well. While the conference process has to be maintained primarily by the regional actors themselves, the external powers could help provide political and economic incentives, e.g. international recognition (within the UN) and additional funding, respectively.

The ACRS working group provided an important means that could foster or even institutionalize the (missile) reduction process and strengthen the security cooperation of regional actors through the establishment of Regional Security Centres (RSCs). A good starting point for revitalizing this idea would be the draft mandate from the 1990s covering goals, structure, operations, immediate objectives, and division of labour between the three centres. In September 1995 all the parties with the exception of Egypt had agreed to establish such an institution in Amman with associated centres in Tunis and Doha. Of course the participants finally ought to include those states that were not part of the ACRS process. Should the conference activities resume, the RSCs could be ready for immediate use, having already received minimal funds – a figure less than $200,000 has been mentioned – and equipment for operation from several extra-regional sponsors (Kaye 2001: 92, 265, note 65).

Establishing RSCs would be a good focal point not only for Track II initiatives, but for interested non-nuclear actors such as Austria, Finland, Germany, Norway, Qatar, Sweden, and Switzerland. Arms control specialists both from inside and outside government as well as from international organizations could build on and strengthen regional expertise on arms control (Yaffe 2001: 15) and provide assistance during negotiations. RSCs could also be the meeting point of constructive interaction between both tracks.[19] The hurdles for participation in a formal series of conferences could be decreased by installing informal preparatory processes, which the parties may use for extensive yet preliminary consultations, as the ACRS "Statement" suggested.

In sum, the major task is to create the political will for action and to keep up the momentum. To go beyond CSBMs in the Middle East/Gulf by addressing the arsenals themselves would be a quantum leap forward in its own right. To be sure, the conditions facilitating arms control in the East–West context such as the Cuban Missile Crisis cannot be found in the Middle East and the hurdles for successful arms control are high. Yet, the governments in the region may come to the conclusion that the non-cooperative way of dealing with the security dilemma by unilateral arms build-ups is "self-defeating" (Feldman and Toukan 1997: 73) and too costly.

Conclusions

The aim of this article has been to explore possibilities for effective regional arms control in the Middle East/Gulf by drawing on the experiences and theoretical background of the East–West conflict, and, to a lesser extent, of the Middle East itself. The two elements of the classical arms control concept – strategic/arms race and crisis stability – were adopted and adapted to the Middle East context as were the tools of effective arms control, the central conditions of success/failure, as well as an appropriate organizational format. The insights into the overall asymmetrical military setting, driven by a dynamic of action and reaction (to be structured according to the existing conflict formations and alliances) were also important. Improving the overall political climate in the region would of course be a central positive condition for making substantial progress in the gradual process of missile disarmament.

Without trying to provide a detailed agenda for official negotiators, the authors have presented an outline of the process in three phases, each focusing on strategic/arms race stability and crisis stability as a way of structuring a possible series of talks. An agenda that starts with a list of mutual concerns may be a productive way of coping with the most pressing issues, thereby avoiding the ACRS experience of getting lost in the jungle of imbalances. At some point, a shared understanding of the various arsenals in the region will have to be worked out. This process might be based on a specific element of the first milestone, namely, transparency (however limited and focused).

The path towards an MFZ will not be a linear one. East–West arms control suggests an initial stabilization phase may even witness a jointly agreed-upon increase of certain kinds of weaponry. However, this step would be unnecessary if one took the overly optimistic starting point of reducing the existing arsenals as contained in the ACRS "Statement" for granted. It is hoped that such a cooperative framework might at some point turn into a second phase of reduction/reversal and prohibition of both offensive and defensive weaponry. We are aware that the goal of reaching and maintaining an MFZ as the third and final phase is extremely ambitious.

Creating the political will for beginning and maintaining this phased process will be a major challenge. In addition, external players will be crucial, in large part due to their role as weapon suppliers. As the composition of the national military inventories, and above all the realm of missile defence, have shown, imported weaponry is increasingly becoming relevant for the regional arsenals. This points to the importance of the extra-regional powers in steering the flow of weapons and the regional arms build-ups, and to a lesser extent to the relevance of indigenous factors (with the exception of Israel and to some extent Iran). If the suppliers decide to cooperatively pursue a policy of restraint, which is also a central element of the second milestone within our incremental approach, the reduction could be facilitated in a way compatible with security.

Containing the flow of weaponry could proceed with a new series of conferences as a regional peace strategy whose objective is an MFZ which might build on the latest proposals for a WMD Free Zone that explicitly includes delivery vehicles. This process admittedly requires sustained effort, but as the overcoming of the East–West conflict has demonstrated, it is worth embarking upon. We suggest a triggering element in this process should be a new thinking about security. In order to reduce the security dilemma, we must recognize that cooperative thinking about security includes the insight that fewer weapons can lead to more security. In a nutshell, making this point is what our entire academic effort is all about.

Notes

1 In contrast to the classical definition of arms control the term 'effective' is more restrictive for arms race dynamics in that it is gradually geared towards disarmament; this process would include a reduction or reversal phase in which current arsenals are lowered, and the qualitative arms race is restricted by (selective) bans, i.e. prohibitions of certain weapon systems.
2 For an overview and discussion of recent developments related to missile defence, see Senn (2009).
3 The following quotations are also taken from this "Statement" of the Arms Control and Regional Security Working Group (ACRS 1997: 103–7).
4 Parts of this section are based on Kubbig (1985, 1988).
5 The Harvard Nuclear Study Group (1983: 209–10) used the term 'force restructuring' to refer to the combination of limitations, reductions, and substitutions.
6 Admittedly, the inclusion of unilateral forms stretches the concept of arms control as defined by Schelling and Halperin (1961: 2). Nevertheless, unilateral forms, which initiate reciprocal behaviour and/or constructively contribute to the goals of arms control, can be subsumed under the definition.
7 For instance, ACRS was troubled by a lack of expertise on arms control:

> At the outset of ACRS, no regional party had a government office devoted strictly to arms control. Generally, the bureaucracies' orientation to regional security was left to the province of the military and diplomats assigned to international organizations. … But by 1995, a regional cadre of experts in security and arms control was beginning to emerge and some parties had set up offices to focus on arms control. This cadre of officials has now virtually disappeared in most regional states.
>
> (Yaffe 2001: 15)

8 As Kemp (1994: 411) notes in this context,

> cooperative security will not be possible in the Middle East until key parties come to believe what the western Europeans have accepted: that a state may be secure within a system that permits vast asymmetries in power among its members.

9 The US, which introduced MIRVs first, was later haunted by the Soviet MIRVed SS-18.
10 Because of its emphasis on stability and its different basic assumptions, the second arms control school never became dominant. Within the strategic frame of "Mutually Assured Destruction" it put a strong emphasis on actually controlling (in the sense of restricting) arms dynamics and reducing the arsenals with the final goal of disarmament. It understood nuclear deterrence only in terms of preventing a first strike, and, hence, defined stability as having sufficient countervalue reserves.
11 The decision to use the calibre rather than the range as a basis for verification was due to the possibility of modifying the range of artillery missiles by changes in fuel and payload.
12 The Adapted CFE Treaty was signed in Istanbul in 1999 and later ratified by Russia, Ukraine, Belarus, and Kazakhstan. The other 26 state parties have not ratified the treaty due to differences about the fulfilment of certain Russian Istanbul Commitments. As a consequence, the Adapted CFE Treaty has not entered into force and Russia suspended the CFE Treaty in late 2007. Since April 2010, all treaty parties together with six new NATO members, which are not yet parties to the CFE accord, have expressed interest in the revival and modernization of conventional arms control in Europe.

13 The following quotations are also taken from this "Statement" of the Arms Control and Regional Security Working Group (ACRS 1997: 103–7).
14 It is beyond the scope of this article to address regional arsenals of WMD and their delivery systems other than missiles (aircraft for instance) in greater detail. These issues will have to be dealt with by the delegations during their negotiations on the building blocks of arms control in the Middle East/Gulf (on the problem of WMD, see the article of Gawdat Bahgat *et al.*).
15 We admit that as long as rocket arsenals of organizations such as Hamas and Hezbollah are not taken into account, Israel would hardly accept such de-alerting measures.
16 As convincingly argued by Al-Marashi (2010), domestic factors will probably play an increasingly important role in future Saudi Arabian nuclear policy.
17 The potentially stabilizing role of MD has been emphasized during the debate about global nuclear disarmament, see, for example, Cortright and Väyrynen (2010: 150–4).
18 See the chapter of Bernd W. Kubbig *et al.* on Military Asymmetries, see also Eriksen and Holøien (2010: 285–7) as well as Holøien (2006: 24–8).
19 The Asian countries have developed a very sophisticated set of interlocking Track One and Track Two dialogues (and discussions which are known as 'Track one and a half') which allow both for formal negotiations on issues that are ripe and less formal venues for officials (acting in their private capacities) to discuss and explore issues their governments are not yet prepared to address officially. These methods also allow states that do not recognize each other to sit together.

(Jones 2003: 149)

References

ACA/Arms Control Association (2010) 'U.S.-Russian Nuclear Arms Control Agreements at a Glance', *Arms Control Association Fact Sheet.* Online, available at: www.armscontrol.org/system/files/USRussiaGlance_March26_2010_0.pdf (25 March 2011).

ACRS/Arms Control and Regional Security Working Group (1997) 'Statement on Arms Control and Regional Security', in Shai Feldman and Abdullah Toukan (eds) *Bridging the Gap: A Future Security Architecture for the Middle East,* Lanham, MD: Rowman & Littefield, 103–7.

Adler, Emanuel (1992) 'The Emergence of Cooperation: National Epistemic Communities and the International Evolution of the Idea of Nuclear Arms Control', *International Organization,* 46(1): 101–45.

Al-Marashi, Ibrahim (2010) 'Saudi Petro-Nukes? Riyadh's Nuclear Intentions and Regime Survival Strategies', in William C. Potter and Gaukhar Mukhatzhanova (eds) *Forecasting Nuclear Proliferation in the 21st Century: A Comparative Perspective,* Vol. 2, Stanford, CA: Stanford University Press, 76–99.

Bollfrass, Alexander K. (2010) 'Breaking out of Zero: Would Cheating Be Worth the Risk?' in Barry M. Blechmann and Alexander K. Bollfrass (eds) *Elements of a Nuclear Disarmament Treaty,* Washington, DC: Henry L. Stimson Center, 209–54.

Brzezinski, Zbigniew (1983) *Power and Principle: Memoirs of the National Security Adviser 1977–1981,* New York: Farrar, Straus and Giroux.

Carter, April (1989) *Success and Failures in Arms Control Negotiations,* Oxford: Oxford University Press.

CBO/Congressional Budget Office (1992) *Limiting Conventional Arms Exports to the Middle East*, Washington, DC: CBO. Online, available at: www.cbo.gov/ftpdocs/105xx/doc10579/1992_09_limiting.pdf (25 March 2011).

Chrzanowski, Paul L. (1994) 'Summary Report', Workshop on Arms Control and Security in the Middle East II, San Diego, CA: Institute on Global Conflict and Cooperation. Online, available at: http://igcc.ucsd.edu/pdf/policypapers/pp.07.pdf (25 March 2011).

CNS/James Martin Center for Nonproliferation Studies (2009) *Inventory of International Nonproliferation Organizations and Regimes*, Monterey, CA: CNS.

Cortright, David and Väyrynen, Raimo (2010) *Towards Nuclear Zero*, London: Routledge.

Eriksen, Sara Kristine and Holøien, Linda Mari (2010) 'From Proliferation to Peace: Establishing a WMD-Free Zone in the Middle East', *Nonproliferation Review*, 17(2): 281–99.

Feldman, Shai and Toukan, Abdullah (1997) 'Bridging the Gap: Resolving the Security Dilemma in the Middle East', in Shai Feldman and Abdullah Toukan (eds) *Bridging the Gap: A Future Security Architecture for the Middle East*, Lanham, MD: Rowman & Littefield, 73–96.

Garthoff, Raymond L. (1978) 'SALT I: An Evaluation', *World Politics*, 31(1): 1–25.

Haas, Peter (1992) 'Introduction: Epistemic Communities and International Policy Co-ordination', *International Organization*, 46(1): 1–35.

Harvard Nuclear Study Group (1983) *Living with Nuclear Weapons*, Toronto: Bantam Books.

Holøien, Linda Mari (2006) *Weapons of Mass Destruction Free Zones in the Middle East*, Kjeller: Forsvarets Forskningsinstitutt. Online, available at: http://rapporter.ffi.no/rapporter/2006/02488.pdf (25 March 2011).

Hunter, Robert E. (2010) *Building Security in the Persian Gulf*, Santa Monica, CA: RAND Corporation.

Jones, Peter (2003) 'Negotiating Regional Security and Arms Control in the Middle East: The ACRS Experience and Beyond', *Strategic Studies*, 26(3): 137–54.

Kaye, Dalia Dassa (2001) *Beyond the Handshake: Multilateral Cooperation in the Arab-Israeli Peace Process 1991–1996*, New York: Columbia University Press.

Kemp, Geoffrey (1994) 'Cooperative Security in the Middle East' in Janne E. Nolan (ed.) *Global Engagement: Cooperation and Security in the 21st Century*, Washington, DC: Brookings Institution, 391–418.

Knopf, Jeffrey W. (1998) *Domestic Society and International Cooperation: The Impact of Protest on US Arms Control Policy*, Cambridge: Cambridge University Press.

Kubbig, Bernd W. (1985) '(Re-)Defining and Refining the Criteria for Nuclear Arms Control', *Bulletin of Peace Proposals*, 16(3): 199–223.

Kubbig, Bernd W. (1988) *Amerikanische Rüstungskontrollpolitik: Die innergesellschaftlichen Kräfteverhältnisse in der ersten Amtszeit Reagans (1981–1985)*, Frankfurt am Main: Campus.

Kubbig, Bernd W. (2004) *Wissen als Machtfaktor im Kalten Krieg: Naturwissenschaftler und die Raketenabwehr der USA*, Frankfurt am Main: Campus.

Lodgaard, Sverre (1977) 'The Functions of SALT', *Journal of Peace Research*, 14(1): 1–22.

Meyer, Berthold (2007) 'Sub-Regional or Multilateral? New Approaches to Conflict Management in the Middle Eastern Region under Scrutiny', Paper presented at the Request of Friedrich-Naumann-Stiftung Foundation for Freedom.

Online, available at: www.la.fnst-freiheit.org/uploads/896/Meyer-Nahost-engl.pdf (15 February 2011).

Review Conference of the Parties to the Treaty on the Non-Proliferation of Nuclear Weapons (2010) 'Final Document', Vol. 1. Online, available at: www.un.org/ga/search/view_doc.asp?symbol=NPT/CONF.2010/50%20%28VOL.I%29 (25 March 2011).

Ruggie, John Gerard (1992) 'Multilateralism: The Anatomy of an Institution', *International Organization*, 46(3): 561–98.

Said, Mohamed Kadry (2002) 'Layered Defense, Layered Attack: The Missile Race in the Middle East', *Nonproliferation Review*, 9(2): 37–46.

Schelling, Thomas C. (1961) 'The Future of Arms Control', *Operations Research*, 9(5): 722–31.

Schelling, Thomas C. and Halperin, Morton H. (1961) *Strategy and Arms Control*, New York: Twentieth Century Fund.

Schimmelfennig, Frank (1991) *Konferenzdiplomatie als regionale Friedensstrategie: Läßt sich das KSZE-Modell auf den Vorderen Orient übertragen?* Hamburg: Institut für Friedensforschung und Sicherheitspolitik.

Senn, Martin (2009) 'The Arms-Dynamic Pacemaker: Ballistic-Missile Defense in the Middle East', *Middle East Policy*, 16(4): 55–67.

Steinberg, Gerald M. (1991) 'Toward Real Arms Control in the Middle East', *Issues in Science and Technology*, 7(4): 63–9.

Steinberg, Gerald M. (2005) 'Realism, Politics and Culture in Middle East Arms Control Negotiations', *International Negotiation*, 10(3): 487–512.

Wittner, Lawrence S. (2009) *Confronting the Bomb: A Short History of the World Nuclear Disarmament Movement*, Stanford, CA: Stanford University Press.

Woolf, Amy F. (2010) *The New START Treaty: Central Limits and Key Provisions*, Washington, DC: Congressional Research Service.

Yaffe, Michael D. (2001) 'Promoting Arms Control and Regional Security in the Middle East', *Disarmament Forum*, 2: 9–25.

14 Conclusions: the stage reconsidered

Findings, concluding remarks, and research perspectives

Bernd W. Kubbig

The research question posed by this book has been: under what circumstances are gradual steps towards a Missile Free Zone (technically) feasible and politically acceptable? This question has served to collectively guide our probing of the conditions for and the conceptuality behind the implementation of a gradual arms reduction process towards an MFZ. The approach to addressing the security dilemma in the Middle East/Gulf we have adopted is an alternative to the traditional (and in our view counterproductive) one featuring unilateralism, self-help, and zero-sum thinking. We by contrast aim at reducing the security dilemma and replacing the dominant mind-set by the outlined Cooperative Security Concept (CSC, see my Introduction). In view of the results achieved, both in terms of empirical findings and analysis, we are now in a position to reconsider the Middle East 'missile stage'. This will form the basis for drawing conclusions and for outlining perspectives for further research.

Against this backdrop, we assume that the incremental arms reduction approach as the core element of the CSC is academically, normatively, and politically superior to the traditional one. Two of the region's leading analysts support this assumption by acknowledging the latter as "self-defeating" (Feldman and Toukan 1997: 73). But given the lamentable situation in the Middle East, an exploration of the merits of alternatives and the conditions under which they could be implemented is surely warranted. As stated in my Introduction the three major pillars of the authors' overall approach in this volume are: first, the recognition of missiles as a research area in its own right without excluding weapons of mass destruction (WMD), especially nuclear ones; second, the regionalization of the missile problématique without ignoring the important role played by the major external powers; third, the identification of the conditions for success and failure for designing the gradual reduction process which strikes a balance between implementing modest confidence- and security-building measures (CSBMs) and realizing the vision of a Missile Free Zone (the MFZ is defined in this volume as ranging from Egypt in the west to Iran in the east and from Syria in the north to Yemen in the south).

In my view, the (groups of) contributors to this book, irrespective of their national origins, academic or political orientations have endorsed and substantiated the centrality of these three pillars. In what follows, I will present the distinctive contribution of this common project by summarizing this study's major results in accord with the overall structure of the volume. This amounts to making good on the claim that this book fills a gap in the research on the subject by focusing on missiles, that it provides new insights by putting the missile problématique in the context of largely regional conflict structures, and that it outlines in a transparent way the conditions, difficulties, and chances of initiating such a gradual but comprehensive approach to missile reduction and disarmament.

Ten major findings

First finding

Missiles and rockets matter. They are deadly instruments of war, killing many people in the region. Because they are fundamentally destabilizing, (un)guided rockets, ballistic and cruise missiles exacerbate the already pronounced security dilemma, and in principle so do missile defence (MD) systems. If left unchecked, the situation could spiral out of control. Missiles equipped with nuclear, biological, and chemical warheads would be even more devastating (see Fourth Finding). This result not only corroborates the academic claim that missiles should be a research area in their own right but also the politico-practical effort to look for ways of reducing and finally eliminating them.

In their contribution, Dennis M. Gormley, a worldwide authority in the missile realm and his colleagues Colin P. Clarke and Jürgen Altmann, have vividly presented the ubiquity of missiles and the pervasiveness of their use in the Middle East/Gulf region: nearly 91 per cent of the 5,025 ballistic and cruise missiles fired in combat since the end of the Second World War have been launched in this part of the world. They were used in five different conflicts in the Middle East, including the Egyptian and Syrian missile attacks on Israel in the 1973 Yom Kippur War, the 1985–88 ‘War of the Cities’ between Iraq and Iran, the 1991 Second Gulf War, the Yemen Civil War of 1994, and the 2003 US-led invasion of Iraq. The most notable were the ‘War of the Cities’ and the two Gulf Wars of 1991 and 2003. In the Iraq–Iran War, both sides used ballistic missiles sparsely until 1985, when Baghdad began to attack Tehran with its new longer-range weapon, the al-Hussein. This campaign culminated in a period of six weeks during 1988 in which about 190 Iraqi missiles killed some 2,000 Iranians. Tehran's responsive attacks on Baghdad involved about 90 Scud missiles. In addition, the use of rockets by groups like Hamas and Hezbollah (the Party of God) is not merely a nuisance to be ignored. From 2001 until 2008, some 8,000 rockets and mortar bombs were launched at Israeli territory. Altogether 16 Israelis were killed. But aside from the casualties, the rockets have had a devastating effect on the sense of security of a quarter of a million residents of the western Negev. As Margret Johannsen, Ghassan Khatib, and Anat Kurz have emphasized, most of Hamas' weapons, namely the Qassams, which are generally inaccurate and mostly fall harmlessly into uninhabited areas, are manufactured locally and have a range of up to 10 km; by contrast, the Katyusha-type Grad rockets are imported (see the Table compiled by Michael Haas and Bernd W. Kubbig in the Appendix).

Hezbollah's missile arsenal has an entirely different destructive capability from that of Hamas. This conclusion is one that can easily be drawn from Judith Palmer Harik and Walid Abu-Dalbouh's contribution on Hezbollah. Some experts maintain that since July 2006 the organization has not only managed to recover but to strengthen its military arm, amassing rockets and missiles with greater ranges. Missile saturation using highly mobile cadres and hard-to-locate stationary launchers will probably continue to be the backbone of any Hezbollah strategy although this certainly does not exclude other military options. The Party of God's effective use of missiles and modernized guerrilla warfare tactics have not been lost on Syria, inasmuch as the restructuring and redeployment of its military has been developing along similarly unconventional lines. Additionally, Hezbollah's cadres have probably been equipped with Syrian weapons as well as more advanced guided missiles (see the Table compiled by Michael Haas and Bernd W. Kubbig in the Appendix).

Two trends could make this bad situation even worse. The first concerns the spread of intra-regional missile capabilities (so-called horizontal proliferation). To be sure, as Dinshaw Mistry and Mark Smith point out, Middle Eastern states have been minor suppliers of missile related technology within (and in some cases outside) the region. And yet these transfers illustrate the potential of the countries to become more substantial sources of missile proliferation as they develop more and better missile capabilities (so-called vertical proliferation). In fact Gormley and Clarke have identified "a truly global contagion" of new missile procurement by regional states. They also issue an unequivocal warning regarding the linking of these delivery vehicles to preemptive or preventive strike doctrines. Military strategists have always favoured offensive missiles as preemptive as well as preventive instruments because of their capacity to shock the opponent with little warning and because this strategic preference shows no signs of abating. The Middle East is certainly no exception to global trends in this regard. Preemptive strikes have been an option for the Israeli military since the Six Day War of 1967. Jürgen Altmann, an expert on unmanned aerial vehicles (UAVs), concurs with his fellow contributors about the destabilizing effects of UAVs: they will tend to exacerbate preemption fears and pressures.

Turning to missile defence systems, the United States and its allies have uncritically portrayed them as having a purely stabilizing effect on regional security. But their role is ambivalent at best. In fact, MD has become part of an intensified arms race in the region, probably aggravating its specific security dilemma. This applies especially to the classical action–reaction arms dynamics between the Gulf states and Iran. Saudi Arabia, the United Arab Emirates (UAE), and Kuwait have been importing MD in response to the perceived Iranian missile threat. Since it is likely that Tehran will react to these efforts in kind, the region will be involved in a destabilizing and costly arms race between offensive and offensive–defensive weapons. Therefore MD must be part of any overall incremental reduction strategy.

Perhaps at some point the countries on the receiving end of missile defence systems will revise their cost–benefit calculations, since these developments challenge MD supporters' claims that these systems discourage states from acquiring missiles. Adding to the general ambivalence of US exports of missile defence systems to Gulf countries is the fact that it may discourage the demand for an indigenous nuclear weapons capability. This could be particularly important in the case of Saudi Arabia. Like other exports to the region, missile defence systems symbolize the strengthening of ties between the US as a supplier and its allies, as well as the corresponding determination to restrain Iranian regional ambitions.

Time will tell whether Tehran will be deterred from interfering in the domestic politics of its Arab neighbours and whether it will restrain its inflammatory rhetoric towards Israel and the United States. In the end, however, any stabilizing effect of MD by contributing to the prevention of further missile and WMD proliferation in the region, will on the one hand derive from the US involvement – and on the other hand from the deterrence provided by its military strength rather than from the qualities of MD per se, since missile defence is still in many respects insufficiently reliable and is more costly than the acquisition of ballistic missiles. Most experts in this volume contest whether Israeli missile defence programmes are a viable response to the rocket threats posed by Hamas and Hezbollah. Adaptability and innovation are the hallmarks of these actors.

For Gormley and Clarke missile defence systems are important for psychological reasons, especially because of the political obligation of a country to protect its citizenry from the rocket threat. This creates enormous pressure on politicians to deploy defence systems, regardless of their costs and levels of effectiveness. Other experts have criticized the Israeli Iron Dome which is designed to detect incoming short-range rockets and to destroy them mid-air. They doubt whether these systems will eliminate the threat to Israel posed by Qassam rockets and comparable weapons. In addition to the technological challenges, these critics point to the financial costs compared to the expense of producing those rockets. Indeed, accelerating expenditures could make Israel lose the arms race, given the fact that an interceptor can cost between $30,000 and $100,000, whereas a Qassam rocket can be made for somewhere between $200 and $400.

Second finding

Despite their inherently destabilizing features, missiles and rockets do not exist in a vacuum. They become a threat once countries factor them into the overall foreign policy, reflecting structures of conflict and alliance. These constellations constitute the core conditions of success for any arms control/reduction process.

One group of authors, primarily from the Middle East/Gulf region – Bernd W. Kubbig, Mohamed Noman Galal, Michael Haas, Murhaf Jouejati, Sabahat Khan, Ahmed Saif, and Mahmood Sariolghalam – has proposed a nuanced definition of a threat. They define two conditions for the presence of a menace: if, first, a state is perceived as destabilizing by other countries/regimes by interfering in their domestic politics or even threatening them with intervention; second, if a country is taken to be impeding a regional order which takes the security needs of all states in the Middle East into consideration. The emphasis on perceptions of external threats includes not only weapon capabilities, but also military strategies/doctrines, whether predominantely offensive or defensive, as well as political intentions – a fact that explains why both the basic foreign policy orientation of a country and its actual behaviour are at issue. Regarding missiles as part of the conflict constellations and alliances, the contributors identify Israel and Iran as the major reason why Arab countries are procuring missiles and missile defence systems: they have a long-standing fear of Israel as a de facto nuclear weapon state and, more recently (with the exception of Syria), concerns about the Islamic Republic as a possibly emerging one.

The latter aspect holds especially true for the Arab Gulf states, while Iran has its own concerns about Israel and even greater fears about the United States in its role as a regional power. But as the contributors point out, even among the Arab states threat perceptions vary regarding Tehran. Four members of the Gulf Cooperation Council (GCC) – Saudi Arabia, the UAE, Kuwait, and Bahrain – are mainly fearful because with its nuclear potential the Islamic Republic might feel less vulnerable and therefore be even more willing to project its power and influence abroad; in the case of Bahrain the uprisings in 2011 contributed to this perception. The other two GCC states – Qatar and Oman – have managed to maintain relatively good relations simultaneously with Iran and the United States. Cairo shares the Arab world's assessment that to date Israel is the only existing nuclear threat in the entire region. A nuclear Iran would pose an additional challenge to Egypt's status and role. Whereas for Syria, Iran's pragmatic ally, Israel constitutes the greatest menace. On the regional level the constellations of conflicts and alliances illustrate the limits to what can be gained within the region by engaging the Islamic Republic as well as Syria. This conclusion is a challenge to our advocacy of a CSC. Our emphasis on the importance of conflict and alliance structures touches upon a leitmotiv of this volume: the adequate assessment of the limited (albeit still relevant) role of arms control in this asymmetrical structure.

Third finding

The conflict and alliance structures in the region explain the overall military asymmetries and account for the specific arms dynamics. They also provide the incentives for countries to procure missiles/weapons. Conversely, all governments possessing these arms will accept a drastic reduction process towards an MFZ only if they conclude that it does not jeopardize their own countries' security.

According to Kubbig *et al.* missile procurement, aimed at increasing the security of a state, can have the unintended effect of undermining the security of its neighbours, who respond in kind by arms build-ups, leading to an interactive dynamic of escalation, which reinforces the specifically pronounced security dilemma in the conflict region. This situation is complicated by military asymmetries which justify self-serving explanations of why one's own missiles are 'good' and the neighbours' missiles are 'bad' (see on this my Introduction). Iran's interest in acquiring missiles, in addition to enhancing its regional aspirations is, for instance, a response to the American presence in the region. Syria's procurement activities are motivated by Israel's military strength, but also by the US-dominated environment (within the Egyptian arsenal, missiles do not play an important role). The military imports of the five small Gulf states are mainly a reaction to Tehran's nuclear and missile efforts. The situation is more complex in the case of Saudi Arabia, because its own hegemonic aspirations as well as its unprecedented and ongoing arms build-up are perceived as a threat by Tehran.

The authors have considered the extent to which these asymmetries are a matter of perception (see the Second Finding) that are often expressed in lists of concerns which parties discuss at a later stage at the negotiation table. Here one has to factor in the role of the United States as the partner of all GCC members into the overall regional picture. While taken by itself it makes sense to juxtapose Saudi Arabia's combat aircraft and missile defence with Iran's missiles at the regional level, this ignores that the US arsenal shifts the balance in favour of Riyadh since the United States is by all standards militarily superior to the Islamic Republic of Iran. Meanwhile, Saudi Arabia, but also the UAE, are starting to acquire the military-industrial structures necessary to produce missiles domestically (this is the case to a much greater extent in Iran and the reason why the domestic power constellations are briefly considered). At present, indigenous factors in the Arab states examined by the authors do not constitute a major obstacle towards an arms reduction process with the MFZ as the ultimate objective. Yet the ongoing regional arms races (concerning offensive/offensive and offensive/defensive build-ups) have intensified, and they exacerbate the specifically pronounced security dilemma. This is a trend that clearly does not lead in the direction of an MFZ.

No Arab state will agree to embark on the path towards reduction, unless Israel's capabilities are adequately taken into account, even if Iran were to replace the Hebrew state as the paramount threat. Therefore,

Israel's position especially with regard to its nuclear arsenal will remain relevant for whether Arab states are willing to consider an alternative security concept. As Israeli scholar Uri Bar-Joseph has noted, there is room for reducing the role of nuclear weapons vis-à-vis Arab states. A sober analysis of the mixed success of Israel's nuclear deterrence policies shows that there are security-compatible alternatives. To be sure, the Jewish state has been able to build up an impressive nuclear arsenal, but it played only a minor role in deterring Arab countries from starting a conventional war against Israel and in convincing them to ultimately accept its right to exist. Instead, these objectives were achieved by conventional superiority.

As Bar-Joseph makes clear, Israel's situation looks different with respect to a potentially nuclear Iran, since the Islamic Republic's nuclear activities constitute the most severe challenge to the regional nuclear monopoly of the Jewish state. Given the extreme anti-Israeli rhetoric coming out of Tehran, its nuclear activities have become a strategic nightmare for the 'one bomb state' which could be destroyed by a singular nuclear strike. How attractive would an MFZ be to Israel against this backdrop? In Bar-Joseph's view a regional accord might be acceptable for the Jewish state, but for Iran, which considers its strategic arsenal a means by which to deter the US, such a zone would hardly be security compatible. In identifying the core of any Israeli policy vis-à-vis a nuclear Iran, Bar-Joseph underscores the crucial element of nuclear deterrence. Since a first strike by Tehran would probably be with ballistic missiles, the Jewish state can be expected to focus on an anti-missile strategy. Despite Israel' s advantages – Tehran would in any case have a far smaller and less sophisticated nuclear arsenal, less developed means of delivery, and few active defence capabilities – its status as a 'one bomb state' might still leave it in a weak position.

And yet, the author sees the most effective means to reduce the burden on Israel's deterrence to be significant progress in the Arab–Israeli peace process. An accord with Damascus would weaken the cooperation between Syria and Iran and hinder the latter's ability to provide military support for Hezbollah. And a settlement with the Palestinians would neutralize a key element of Tehran's justification for its call to destruct the Jewish state since even the Iranian fundamentalist regime would find it difficult to be more committed to Palestine than the Palestinians themselves. Turning to the political sphere, Bar-Joseph (like several other authors on different issues in this volume) highlights how bilateral cooperative arrangements can contribute to solving seemingly narrow security problems.

Fourth finding

The reduction and elimination of missiles cannot be achieved without parallel measures towards a zone free of WMD. WMD cannot only be delivered by missiles, but as part of a future arms control agenda would allow for a trade-off between the supporters of missile and nuclear disarmament. Because of the perennial endorsement of a WMD Free Zone (WMDFZ), which so far has remained confined to the level of concepts and rhetoric, a more moderate step-by-step process similar to our gradual path in the missile area is suggested.

The authors of the corresponding chapter – Gawdat Bahgat, Avner Cohen, Sven-Eric Fikenscher, Giorgio Franceschini, Bernd W. Kubbig, and Patricia Lewis – have examined the posture and positions of Egypt, Israel, and Iran (the most important regional states in this regard), and their attitudes regarding the following building blocks of the gradual process towards a WMDFZ: advancing the ratification process of the Comprehensive Test Ban Treaty (CTBT); broadening the support of the Additional Protocol; negotiating a regional Fissile Material Cut-off Treaty (FMCT); and reaching (regional) fuel cycle arrangements as well as accepting the Biological and Toxin Weapons Convention (BTWC) and the Chemical Weapons Convention (CWC). As the scholars make clear in their country profiles, the obstacles for making progress in all these areas are to a considerable extent due to the historically grown policies, especially in the nuclear field which is the major focus of this article.

With respect to Egypt, instead of seeking to match Israel in this area, it has sought to build a strong conventional military power and to develop (undeclared) chemical weapons capabilities. Additionally, Cairo has employed diplomatic means to pressure Israel to join the Nuclear Non-Proliferation Treaty (NPT), to give up its nuclear weapons, and to place its facilities under full scope safeguards of the International Atomic Energy Agency (IAEA). The confrontation over Iran's nuclear programme since the early 2000s has further fuelled Egyptian efforts to address the proliferation issue and prompted Cairo to press for a WMDFZ within the framework of the NPT Review Conference in 2010. In order to avoid double standards, Egypt has favoured a comprehensive approach to nuclear disarmament that includes both Iran and Israel as part of a WMDFZ, as Gawdat Bahgat writes. Cairo has so far refused to ratify the Additional Protocol, the CTBT, and the BTWC; it also opposes the CWC and a FMCT as well as fuel bank agreements, since all the proposals fail to address Israel's nuclear superiority.

As to Israel, Avner Cohen and Patricia Lewis emphasize that this country – in contrast to its Arab neighbour – has always insisted that the negotiation of an NWFZ can only be initiated *after* peace, i.e. after full and mutual political recognition of Israel by all the states of the region. From an Israeli perspective the NWFZ proposal is a diplomatic instrument which eases the pressure on the country to sign the NPT, without the need to make any concrete commitments. Israel thus far seems firmly committed to maintaining its undeclared regional nuclear monopoly, regarding it as an indispensable and fundamental feature of its national security, and relying on its policy of nuclear opacity as the means of keeping discussion of these matters off limits.

As far as the individual building blocks are concerned, Israel, which as a non-member to the NPT, is not obliged to adopt full scope safeguards including the Additional Protocol. It is a signatory to the CTBT, but has maintained an ambivalent attitude towards the FMCT. Its concerns seem to be both political (commitments towards premature nuclear disarmament) and technical (difficulty to maintain its policy of opacity because of the treaty provisions for transparency and verification). Regional fuel cycle arrangements have not been officially endorsed by Israel, which has signed but not ratified the CWC (it is not clear whether the country still maintains an operational chemical weapons capability). Although it is widely believed that the country does not have an offensive B-weapon programme, it has yet to sign and ratify the BTWC.

Iran's WMD policy has been examined by Sven-Eric Fikenscher, Giorgio Franceschini, and Bernd W. Kubbig between domestic infighting and external pressure. The interests and driving forces behind its missiles programmes (see Third Finding) apply even more so to the nuclear area, i.e. the multitude of domestic factors (among them the wish to demonstrate the capacity to master a so-called cutting edge technology and the associated national prestige) and regional considerations (especially hegemonic ambitions, but also the fear of the US military presence). The dominant feature of Iran's nuclear activities is international doubts regarding a clandestine nuclear programme for achieving a nuclear weapons capability; they have always persisted and turned into strong suspicions since the disclosure in 2002 of an undeclared underground enrichment facility and of a heavy water plant. The international community has responded to Tehran's behaviour with resolutions of the UN Security Council which prohibit enrichment, reprocessing, and heavy water related activities in Iran.

The reformist government of Mohammad Khatami suspended the enrichment activities, and it agreed to apply the – signed (December 2003) but not ratified – Additional Protocol. However, the ultraconservative President Mahmoud Ahmadinejad, elected in June 2005, began enriching uranium again – supported by the central figure, Supreme Leader Ali Khamenei. To justify its nuclear activities, Iran has insisted on the "inalienable right" (Art. IV of the NPT) of all treaty members to use nuclear energy for peaceful purposes and has stressed that it has no intention whatsoever of acquiring a nuclear weapons capability. But in practice Tehran has failed to report a number of sensitive nuclear activities to the IAEA, and has not been able to provide sufficient guarantees that its nuclear programme is for exclusively peaceful purposes.

As far as Iran's attitude towards the building blocks is concerned, Tehran has signed but not ratified the CTBT for reasons which are largely similar to Cairo's. Sven-Eric Fikenscher, Giorgio Franceschini, and Bernd W. Kubbig stress that the actual application or even ratification of the Additional Protocol will depend on the development of a mutually satisfactory modus vivendi especially with the United States. The Iranian government is not interested in exclusively enriching its fuel abroad, but it may be in principle open for regional fuel cycle centres if they do not impinge on its national enrichment capabilities. The position on a regional FMCT remains unclear and so does its explicit attitude towards a WMDFZ. The future of talks with regional countries on such a zone will depend to a large extent on the power constellations within the Islamic Republic. Any government is likely to retain a partial nuclear capability, i.e. especially some enrichment capability – the zero enrichment option demanded by the UN Security Council is not on the table for Tehran.

In accordance with the overall design of the book the authors conclude that the building blocks, consisting of both confidence-building and arms controls measures, would be formidable steps on the long path to the elimination of all WMD in the Middle East – and despite all difficulties both a practical and effective way of reducing the security dilemma. If initiated and conducted successfully, the Middle East Conference on the establishment of a WMDFZ, including its delivery vehicles, which the NPT Review gathering envisaged in May 2010, could make a difference in this respect.

Fifth finding

While focusing on the regional level, it is imperative to address the politics of major external powers if a missile reduction process towards an MFZ is to proceed. Although the US, Russia, and China have by and large exercised restraint in exporting missiles, the American military presence complicates the situation considerably. A security arrangement between the United States and Iran is a central condition for successful missile reduction.

As the cumbersome and lengthy processes of establishing treaty-based NWFZs in Latin America/the Caribbean and the South Pacific (Treaties of Tlatelolco and Rarotonga, respectively) have shown, external powers can influence the success or failure of zonal arms control, depending on the resolution of issues such as territories under external influence, foreign military bases, as well as transit rights and testing activities. In the Middle East, the United States, by virtue of its extensive and concentrated military presence, is of far greater relevance than other external actors, such as Russia and China. In their chapter Sven-Eric Fikenscher, Sameh Aboul-Enein, Gülden Ayman, Anton Khlopkov, and Hui Zhang argue that the US could be a spoiler of missile-focused disarmament efforts not only by virtue of its military facilities, especially in the Gulf states, but also because of its employed missile defence systems to protect its most important bases and allies. This in turn fuels Iran's threat perceptions since its deterrence strategy mainly relies on missile capabilities. Washington's decision to co-finance Israeli MD systems might reinforce the way Arab countries perceive the threat posed by their counterpart. The same holds true for the Iranian view on the US–Israeli alliance.

The authors argue that the challenge consists of convincing the United States to negotiate an agreement that leads to a substantial withdrawal of its armed forces from the region and Afghanistan and to give up its regional missile (defence) exports. A mutually satisfactory security agreement between Washington and Tehran would have to be worked out. But the basic rationale for multilateral MD cooperation continues to exist as long as Tehran retains its missile ambitions. The US would probably insist on severely limiting Iran's missile programmes in parallel with, if not in advance of, such measures. Both Washington and Tehran could in principle agree to limit the Iranian influence in both countries, while significantly reducing American forces.

While the success of an MFZ would also depend on the prevention of future missile transfers from Russia and China, both actors have shown increasing restraint in exporting missiles since the 1990s. This has expressed itself in the Russian membership to the Missile Technology Control Regime (MTCR), and while Beijing has yet to become a formal member, it has pledged to adhere to the regime's guidelines. Unlike the US and Russia, China is not a member of the Hague Code of Conduct Against Ballistic Missile Proliferation (HCOC) which was opened for signature in 2002 and calls upon states to make annual declarations about their missile and space rocket policies and to give advance notice of missile/space rocket testing. The authors emphasize that since the indigenous capability for missile production in the countries involved is fairly limited, greater restraint could shape regional politics even more. Such a course would contribute to vitalizing the export control regimes and thereby strengthening norm-building in the virtually norm free missile area (see also tenth finding, the second milestone).

In the final analysis a coordinated policy among the three major extra-regional players is necessary. As the five authors point out, the US–Chinese dimension is complicated by a trans-regional linkage – one of Beijing's motives for potential missile exports to the Middle East is its concern about the United States' arms deliveries, especially the export of MD systems, to Taiwan and Japan. If Washington were to restrict these weapon deals, it would facilitate China's support for an MFZ. In addition, a coordinated policy among all nuclear weapon states is needed, since the signatories to an MFZ would need negative security assurances including guarantees not to use or threaten to use the weapons banned in the zonal states. However, one challenge might be that Washington and Moscow are obviously unwilling to limit their deliveries of anti-air missiles, which would have a negative impact on the success of any MFZ agreement including such systems.

Sixth finding

Adequate verification during the reduction process and in the context of ultimately creating an MFZ in the Middle East is possible in principle if technical monitoring devices are supplemented by cooperative human inspections.

The contributors – Jürgen Scheffran, Bharath Gopalaswamy, Dennis M. Gormley, Bernd W. Kubbig, Uzi Rubin, and Hartwig Spitzer – who are mainly technical experts have emphasized in their article that the progressive realization of an MFZ as well as its sustainability will depend upon a system of effective verification measures. Verification usually consists of the gathering or fact-finding of data (monitoring) and the interpretation (evaluation) of the information collected. Both elements are indispensable for providing the protagonists with the necessary quantum of confidence/trust to assume that the other parties are willing to comply with the agreed-upon reduction steps and the MFZ itself. The verification challenges in this context will be enormous and cannot be established overnight. In the early stages of the process, a flight test ban would prove to be a useful tool to halt or at least to slow down the development of delivery vehicles, since flight testing is essential to building confidence that the system under development will work as intended.

Such a ban would be relatively easy to verify, as missile systems have to be tested in the open. Additional measures for strengthening crisis stability include de-alerting weapon systems by separating missiles from warheads and adequately verifying them through on-site inspections. The contributors have made clear that as far as the destruction of the missile related production, testing, repair/maintenance, and related storage facilities are concerned, verification is more difficult. Restraints on research and development are hard to verify, particularly due to their overlap with legitimate space projects. Once a binding MFZ agreement has been implemented, the task of verification would be to prevent violations and breakouts, whether through indigenous capabilities or imports. All this requires the permanent verification of critical sites, facilities, and cross-border activities.

If an MFZ agreement should allow for the production, testing, and deployment of missiles up to a certain range (at least 70 km, as the experts assume in agreement with the authors of other relevant contributions in this regard), the verification measures would have to ensure that no legally permitted missiles or components/technologies (including imports) are diverted to the manufacture and testing of prohibited delivery vehicles. The verification demands increase with a reduction in the permitted range threshold, since delivery systems at low ranges (like the unguided short-range artillery rockets or small UAVs) are hard to detect and track. Control of these systems would be left to later stages in the process, once progress had been achieved on the larger systems. In an environment of trust that is conducive to cooperative security, monitoring of the smaller systems is not excluded since functioning on-site inspections can check any suspected location for the existence and the infrastructure needed to maintain them. Since a perfect level of confidence cannot be achieved in practice, standards of adequacy or effectiveness are necessary (see the Seventh Finding), as well as cooperative measures; among them established counting rules regarding weapons and delivery vehicles, but also on-site inspections which complement unilateral measures, i.e. National Technical Means – basically satellites – and human sources (e.g. spies and defectors).

Here, the contributors conclude that while aerial and satellite surveillance can play a supportive role, ground-based surveillance will have to be the backbone of a sufficiently credible and effective monitoring system. Optical ground-based sensors for remotely controlled monitoring of distant sites are usually derived from the home security, anti-burglary industry, and they are cheap and easy to obtain. Yet monitoring technologies, advanced as they might be, cannot be expected to replace human inspections and mutually agreed-upon data exchanges. It can be safely assumed that on-site inspections could provide the most trustworthy tool of verifying the step-by-step process towards an MFZ and of verifying the compliance with a zone related agreement itself. In addition, the authors conclude that the combination of technology and human activities, especially ground-based monitoring and inspections, could provide a solution to the enormous verification demands – as long as this is embedded in a cooperative setting and adequate verification standards are applied that do not demand the impossible, i.e. absolute or perfect security against all imaginable worst-case scenarios.

Seventh finding

Adequate verification as an indispensable condition for the establishment of an MFZ is at the centre of the relationship between confidence-building, arms control, and the overall political situation. In the final analysis, verification measures will only succeed in a politically favourable environment. This touches on the issue of their domestic political acceptability as a major condition for facilitating arms reductions.

For Scheffran *et al.* the relationship between effectiveness and intrusiveness highlights the crucial verification challenge: the more effective the verification measures, the more intrusive they are, and the more they depend upon the trust and goodwill of the country in question. Space-based surveillance technologies are the least intrusive mechanisms, but they are limited to objects in open view. Ground-based technologies, especially when combined with inspections, can observe and detect more, but they are highly intrusive, and therefore depend on prior trust and must be confidence-generating. At the heart of the entire surveillance problem is the limited role of verification which "cannot act as surrogate for trust" but needs in turn "some initial trust" (Krass 1985: 750). Indeed, paradoxically but not surprisingly, confidence is required to generate confidence. The implications are twofold. First, if history is any guide here, the long way of arms reductions towards disarmament gets its momentum in the wake of peace, not vice versa. Second, this insight nevertheless allows a limited but vital role for verification: increasing the readiness of the adversaries and competing domestic actors to accept mutual on-site inspections, might in turn facilitate and increase interest in an arms reduction process compatible with security. The question of how to start the verification related process of building confidence and generating interest is a matter that concerns extra-regional actors, independent experts, and governments alike (see the tenth finding, third milestone).

Eighth finding

By focusing on missiles and rockets, not only state actors but also so-called hybrid organizations such as Hamas and Hezbollah have become part of the analysis (see the First Finding), and all the elements we have described so far must also be applied to them. Working out agreements on the non-use of missiles with these hybrid actors could be helpful incremental steps in this regard. An engagement strategy is suggested for Hamas and an Israeli–Syrian agreement for Hezbollah is proposed.

As for Hamas, the three authors – Margret Johannsen, Ghassan Khatib, and Anat Kurz – realize that all the strategies that have been employed to curtail Hamas' use of rockets have so far failed: neither various Israeli missile defence programmes nor offensive measures such as boycotts or the use of force have seriously hampered its military build-up and diminished the potential for another escalation of rocket fire. Therefore, these German, Palestinian, and Israeli experts suggest engaging Hamas as an alternative to punitive approaches. In their view, this is the most promising way of reaching a 'no use' agreement or even the more ambitious goal of disarmament. Having analysed Hamas' acquisition and use of rockets in great detail as an intractable part of the overall interwoven Israeli–Palestinian dynamics, the authors also make clear how complex and cumbersome the way to providing incentives to Hamas to become less militant and more politically accountable is. This transformation, the three authors have argued, would be aided by further encouraging Palestinian national reconciliation and unity, for instance by Hamas' signalling endorsement of the two-state solution.

Given the sorry state of the peace process, a formal agreement to end the conflict and remove the rocket threat (for instance by collecting and destroying them or by putting them under the control of a legitimate authority) appears to be an unlikely possibility. A more pragmatic policy would be to take incremental steps to reduce the violence in Israeli–Palestinian relations, thereby making meaningful negotiations possible. A pledge by Hamas not to use its rockets would be a substantial confidence-building measure and facilitate negotiations on the final issues. No matter what the terms of an eventual ceasefire look like, the commitment not to use the weapons at one's disposal would be solely a declaration of intent on the condition of reciprocity. Facing what it sees as Western and Israeli hostility, Hamas will remain armed during negotiations. However, strengthening the pragmatic forces within the organization cannot be achieved by isolation and boycotts. As long as it is threatened with annihilation, Hamas will not forsake weapons it sees as an 'equalizer'. Therefore, within the framework of a truce, gradual demilitarization, which would include curbing rearmament and modernization of Hamas' rocket arsenal, may be the more promising path to follow. One way of implementing this option would be to absorb Hamas militants and their weaponry into a national security framework as part of a comprehensive reform of the Palestinian security sector.

Margret Johannsen, Ghassan Khatib, and Anat Kurz have underscored that such a process could be facilitated by enhancing economic development in Gaza and by capping Hamas' support from its regional allies Syria and Iran. Finally – and here the dense interaction of the state and substate levels comes into play – an easing of regional tensions would help mitigate the danger of escalating violence. In particular, the relaxing of American–Iranian tensions might weaken Tehran's, and consequently Damascus', motivation to support Hamas, and settling the Israeli–Palestinian and the Israeli–Syrian conflicts could dry up Damascus' and Tehran's support for Hamas' military posture. Tangible progress towards a resolution of the Arab–Israeli conflict would also make it increasingly difficult for the Islamic Republic to use the organization as a bargaining chip against the United States and Israel.

In the case of Hezbollah, the prospects for disarmament, although a necessary measure for achieving a comprehensive MFZ, are dim. Nevertheless, Judith Palmer Harik and Walid Abu-Dalbouh suggest that changing local, regional, and international circumstances and/or a new approach to the Hezbollah–Israel conflict could create conditions under which participation in an MFZ might be viewed more favourably. Focusing on the Party of God, the authors explore potential changes in regional politics – especially Hezbollah's ties to Syria and Iran – or attitudes that could induce Hezbollah to reverse its present adamant rejection of any initiative requiring its disarmament. The scenarios they consider are the use of missile defence and military force, the possibility of Lebanon regaining its sovereignty, a missile 'no use' agreement, and withdrawal of the support of Hezbollah's allies Iran and Syria – in the latter case this implies successful 'land for peace' negotiations.

Politically, the Party of God is the product of an alliance between Syrian President Hafiz al-Asad and Iran's Ayatollah Ruhollah Khomeini struck in 1980. This coalition provided each partner with an unparalleled opportunity to influence the Middle East peace/war equation from the sidelines. Yet it would be a mistake according to the authors to assume that the organization is completely dependent on its benefactors. This fact, as well as the level of commitment of all three players to the fight against Israel and the United States, is of great importance for understanding the conditions under which Hezbollah might eventually agree to limit or even destroy its missile arsenals. Judith Palmer Harik and Walid Abu-Dalbouh have emphasized that as a militant Lebanese resistance force, the Party of God gained its credentials in its struggle to control the border area. Although Israel withdrew from Lebanon in May 2000, its troops continue to retain outposts in the Shebaa Farms area. Their ongoing presence is the major *raison d'être* of Hezbollah's resistance campaign.

As to the first scenario of the Party of God's military restraint and disarmament – intercepting incoming missiles – the authors have stated that Israel is preparing a multi-layered missile defence umbrella, featuring the Iron Dome short- and medium-range missile defence system to meet a possible future threat from Hezbollah rocket fire. But the many technological problems of defending cities, towns, and villages from Katyusha short- and medium-range rocket diversification and saturation have yet to be resolved. Arms limitation by force (the second scenario) is not a convincing option either – Israel's use of massive conventional force to roll back the Party of God failed in 2006, as it had in 1993 and 1996. Intensive Israeli aerial bombardment did not succeed in destroying Hezbollah's command and control structure, while at the same time it fanned popular support for the resistance. Restoring Lebanon's sovereignty (the third scenario) is a major goal of US foreign policy in the region and to this effect, efforts have been made to strengthen the Lebanese Army with upwards of $500 million in recent years.

Perhaps a more daunting obstacle to using the army to disarm the Party of God and recover Lebanese sovereignty is that the military will not fight against its fellow citizens and may balk at any action to curb Hezbollah, given the fact that 65 per cent of its troops are Shiites. For the Party of God to adopt a 'no use' agreement (the fourth scenario) some requirements would have to be met – among them that Syria would have to regard such an accord as a measure worth trying and that Israel would have to recognize that Hezbollah is its real antagonist. Further it is likely that such a 'no use' accord would have to apply to all types of missiles. The key to the Party of God's acquiescence to such an agreement would most likely be Israel's willingness to cease its espionage including aerial surveillance and activities by Mossad on Lebanese soil. If a 'no use' accord could be concluded that met the requirements, it might provide a reduction of tensions while diplomacy is allowed to take its course.

Under what conditions might Iran and Syria reduce or withdraw their support for Hezbollah (the fifth scenario)? The answer given by the two authors is similar to the response of the three Hamas experts. Since Hezbollah's ability to engage the more powerful Israeli force with relative impunity is an important asset to Tehran (especially in case of an Israeli attack on Iranian nuclear facilities), a new American–Iranian modus vivendi could at some point motivate Tehran to reconsider its alliance with the Party of God. But the most likely element of this scenario regards the 'land for peace' option between Syria and Israel, i.e. a Syrian decision to suspend its support for Hezbollah, once Damascus has struck a deal with Israel on the Golan Heights; both sides have already conducted a number of (in)direct negotiating rounds. This final scenario also presents a leitmotiv of this study: the creation of new, concrete political circumstances which can have the most positive impact on increasing the chances to eventually disarm.

Ninth finding

Although arms control averse, the Middle East has conceptual and practical experiences in the CSBM area. Provided that one avoids past mistakes, intra-regional learning remains important, and so does inter-regional learning from the East–West context – if it is tailored to the specific circumstances in the Middle East/Gulf without abandoning the normative achievements associated with East–West arms control.

One important point of reference is the Arms Control and Regional Security (ACRS) process, as Emily B. Landau and Dalia Dassa Kaye have emphasized. ACRS became one of the five working groups that made up the multilateral track of the Madrid peace process. Fourteen regional participants – Israel and Arab states from the Maghreb to the Gulf (including Palestinians, beginning in 1993) – met for the first time from 1992–95. Orchestrated especially by the United States, but also by Russia, the participants worked out important confidence- and security-building measures (even though they were not implemented) – among them agreements about information exchange and the establishment of three Regional Security Centres in Jordan (main facility), Qatar, and Tunisia.

One of the negative results of the ACRS talks was the fact that major states were absent – Iran as well as Iraq were not invited, and Syria as well as Lebanon opted to remain outside the framework. According to the two authors this limited regional participation made it more difficult to address the security concerns of those countries that did take part. Also, while created as a device to improve the regional climate for bilateral peace-making, very little attention was devoted to the content of the discussions themselves; the multilateral talks were always subordinated to the more high-profile bilateral peace tracks. Asymmetries of power (and perceptions of power disparities) as well as classic security dilemmas proved to be fundamental barriers to the ACRS process.

Regional arms races fed into the security dilemma, whereby measures taken by one state to bolster its security have inevitably been viewed by others as threatening their own. Emily B. Landau and Dalia Dassa Kaye have emphasized that in addition regional political dynamics and rivalries

– in particular Egypt's self-understanding as the leading power in the entire Middle East – played a major role in the deterioration of the ACRS talks. Egypt and Israel stood for different emphases on the two elements of the working group – arms control and security, respectively. This led to an unfruitful dispute on how to deal with the issue of WMD, especially nuclear ones. Cairo wanted to combine a WMDFZ with the demand that Israel join the NPT and adhere to inspections of its nuclear facilities by the IAEA. Israel, by contrast, refused to give up its preference for CSBMs and to turn to the sensitive nuclear issue. Instead it advocated moving incrementally at a non-specified stage in the future to reduce conventional forces and ban long-range missiles.

In addition to the seminar-like ACRS format, Middle Eastern states have had practical and generally positive experiences in implementing CSBMs, as Sven-Eric Fikenscher, Patrick Flamm, Lina Golob, Bernd W. Kubbig, and Christian Weidlich have concluded. Three accords that were struck after the October War in 1973 show that future CSBM related talks do not necessarily need to start from scratch: two accords between Israel and Egypt in 1974 and 1975 – the Separation of Forces Agreement (Sinai I) and the Interim Agreement (Sinai II) – and an accord between Israel and Syria (1974) on the Golan Heights can serve as models for how to proceed. The Sinai agreements for instance included a demilitarized buffer zone and militarily thinned-out areas as well as a complex verification system implemented by two extra-regional third parties, the United Nations and the US, that allowed the conflict parties to participate in most of its monitoring activities (see the article of Jürgen Scheffran *et al.*). The Sinai accords were a political success, since they paved the way for the Israeli–Egyptian Peace Treaty (1979). But they also contributed to the proposed goal of confidence-building, as any (minor) reported violations could be settled through the established consultative mechanisms.

Learning from the East–West context is another leitmotiv in this volume. While the Conference on Co-operation and Security in Europe has in practice served as the role model for ACRS, the arms control/reduction experiences and achievements of the US, Europe, and the Soviet Union/Russia as well as the security conceptions and rationales behind them are in principle useful for future Middle East initiatives. But their application will be limited, and must be tailored to the specific conflict and power constellations in the Middle East, as Sven-Eric Fikenscher *et al.* have argued. Nevertheless, the tension between the positive achievements and models from the East–West context – fewer weapons can in principle provide more security – and the Middle East aversion to arms control is a fruitful one, provided that the conflict region adopts and adapts European experiences such as balancing interests/priorities of the parties involved, the willingness to take risks, and patience throughout the negotiations. In fact, as several articles have shown, various facets of cooperation, however rudimentary they might be, have already played an important role – as was the case with the multilateral working group on Arms Control and Regional Security, and in the bilateral Sinai and Golan Heights accords, all of which involved extra-regional partners. It goes without saying that the conflict region can draw important lessons from past pitfalls and build on its own positive experiences with CSBMs, for instance with respect to the envisaged 2012 Middle East Conference.

Tenth finding

As the three milestones below demonstrate, when we turn to the weapons themselves, the missile/arms reduction process towards an MFZ has to be gradual and stability-oriented. Modest confidence-building measures can go hand in hand with tackling the quantities and 'qualities' of regional weapon arsenals. An adequate organizational setting has to be found in order to make the process and an MFZ in the broader context of a WMDFZ sustainable.

First milestone: the role of transparency within a register of conventional arms in the Middle East

The disclosure of existing missile capabilities in the Middle East/Gulf would be a crucial CSBM for any gradual arms control/reduction process towards an MFZ. Against the backdrop of the positive European experience in the East–West context Sven-Eric Fikenscher *et al.* note that transparency facilitates mutual confidence/trust and therefore is a necessary condition for such a missile reduction process. The authors regard the UN Register of Conventional Arms (UNROCA) as a promising starting point for considering openness with regard to missiles/launchers since it covers these weapons and has, in principle, been accepted by almost all countries in the Middle East.

But the UNROCA can only serve as a constructive CSBM if it is reformed and expanded. One of its major weaknesses is its focus on arms/missile exports (thus putting countries which are dependent on foreign deliveries on the spot) while ignoring national weapon stockpiles and indigenous production (thereby favouring Israel and to a certain extent Iran). While it is necessary that regional actors provide information on missiles, such a demand will only be implemented if the states agree on an overall framework that takes military asymmetries into account. In particular, those countries which have developed missile capabilities in response to Israel's air superiority will insist on such a framework. In addition, Israel's nuclear arsenal will have to be factored into the arms control equation in order to prevent transparency measures from failing.

Second milestone: the MTCR and the Hague Code of Conduct as the context for enhanced norm-building

In their chapter on the second milestone, Dinshaw Mistry and Mark Smith note, first, that the MTCR has contributed to reducing the transfer of missile technology to the Middle East from Russia and China, previously the region's major suppliers, while North Korea continues to provide missiles for regional arms programmes. Second, external suppliers have aided some Middle Eastern states (e.g. Egypt and Syria) with short-range Scud class missiles and they have assisted the development of Iran's medium-range missiles. Third, multilateral attempts beyond the MTCR, i.e. the HCOC, have not been very effective in the region. To be sure, missile proliferation has been constrained somewhat – the increase is not as great as it would have been in the absence of the MTCR, and the quality and quantity of Middle Eastern programmes in this area have also been restricted. But no missile-relevant state has subscribed to the Hague Code, and the MTCR is mostly seen as a cartel whose members operate using a double standard by preventing the procurement of technologies which they themselves possess.

Therefore, Dinshaw Mistry and Mark Smith have argued, that in the future the MTCR's technology barriers may not prevent aspirants from building short- and medium-range missiles, since most of the technology is available to states with a modest industrial infrastructure. Thus, export control approaches are likely to have a retarding impact on missile development, but probably will not stop it. While the MTCR's achievement should not be underestimated, it may be a diminishing asset. In addition, necessary norm-building endeavours will have to take the strong sentiments in the region against both the MTCR and HCOC into account. The weak HCOC regime neither bans ballistic missiles nor does it cover cruise missiles, but it nevertheless requires member states to pursue "maximum possible restraint" in ballistic missile "development, testing, and deployment". As the two authors point out, Middle Eastern countries have nevertheless criticized the HCOC above all because of its roots in the MTCR, its perceived lack of attention to peaceful, civilian uses of space, and its failure to address missile issues in a more balanced fashion, preferably within the UN framework, such as the UN Panel of Governmental Experts (see my Introduction).

Against this backdrop the authors have suggested a more cooperative approach for tackling regional missile proliferation, i.e. to supplement the MTCR and the Hague Code by measures such as space service regimes and flight test bans. Since flight testing is necessary in the development of new missiles, a test ban would make vertical proliferation more difficult, i.e. it would hinder countries in the region with short- and medium-range missiles from testing and developing more advanced long-range missiles. Dinshaw Mistry and Mark Smith base their alternate approach on the assumption that it is precisely the limited scope of both regimes that may be of interest to regional actors. If Middle Eastern states were to begin addressing missile issues, the political resistance to both regimes might be overcome by examining the regional costs and benefits of these instruments. The two authors suggest that Middle Eastern countries would be able to accept MTCR's general practices, without formally joining (or even formally declaring any adherence to) the regime under a regional missile technology control initiative.

In parallel, external suppliers could pledge to respect a regional missile technology control initiative and Middle Eastern countries could commit to not importing such technology from other actors. Such commitments and restraints would be of interest to all states in the region and in fact reduce the security dilemma – each actor's security would be enhanced if they were all assured that other regional powers were not supplying missiles to their neighbours and adversaries. Regarding HCOC, Middle Eastern states could agree to overcome their political objections to this regime by recognizing its confidence-building and cooperation-promoting benefits. The political environment would be improved if all regional states successfully negotiated an initiative to join the Hague Code thus showing that cooperation on security issues such as missiles is possible.

Third milestone: a proposal for limiting, reducing, and prohibiting missiles and missile defence

The attempt by six arms control experts – Martin Senn, Jürgen Altmann, Bernd W. Kubbig, Jürgen Scheffran, Hans-Joachim Schmidt, and Oleg Shulga – to conceptualize a process of missile limitation and reduction towards an MFZ is informed by Cold War experiences and by regional approaches. These East–West concepts are relevant because they include the core elements of stability – strategic/arms race and crisis stability – conditions for success and failure (to be adapted to the Middle Eastern setting), and the appropriate organizational form, for instance, the planned Middle East Conference on a zone free of WMD and delivery vehicles (DVs). Major differences between both regional contexts are recognized. A renewed and deepened arms control process in the Middle East would have to address a series of issues, ranging from current military asymmetries and their implications, conflict structures and alliances, domestic and external motives behind arms build-ups, to the complex and cumbersome counting and comparing of military inventories.

In addition, major external powers, above all the United States, play a key role in this process due to their ambivalent role as weapons suppliers, friends of some regional states and foes of others, and as sponsors of a series of conferences which would be a core element of a regional peace strategy. In accordance with the contribution on military asymmetries in the Middle East, the authors have primarily focused on states possessing missiles with ranges beyond 70 km that can be equipped with WMD. Eventually, however, agreements on limitations, reductions, and prohibitions of missiles between states would have to be supplemented by informal agreements with actors like Hamas or Hezbollah, if they should pose a threat to the regional arms control process. Like the contributors to other relevant articles in this regard the authors of this chapter emphasize that improving the overall political climate in the region would of course be a central condition for making substantial progress in the gradual process of missile disarmament.

Addressing missile defence must be part of any overall incremental reduction strategy. Countries with MD capabilities should be sceptical about the claim that this technology contributes either to arms race or crisis stability (see First Finding). The fact that currently MD is largely seen as part of the solution for the proliferation problématique is to some extent due to its image as being purely defensive. But, in fact, whether MD is defensive or offensive depends on the scenario in which it is embedded. The view that missile defence is indeed purely defensive is illustrated by the role of Patriot defence systems in preventing Israel from actively engaging in the 1991 Second Gulf War. Moreover, a Syrian strike against Israel or an Iranian attack on Israel and its Arab neighbours in the Gulf would also support claims about the purely defensive character of missile defence. However, the situation would change dramatically if missile defence systems became part of a clearly offensive scenario such as an American or joint US–Israeli attack on nuclear and missile related facilities in Iran.

Without trying to provide a detailed agenda for official negotiators, the authors have presented an outline of the missile reduction process in three phases, each focusing on arms race/strategic and crisis stability as a way of structuring a possible series of talks. An agenda that starts with a list of mutual concerns may be a productive way of coping with the most pressing issues, thereby avoiding the ACRS experience of getting lost in the jungle of imbalances. Creating the political will for beginning and maintaining this phased process will be a major challenge. Constructive external players will be vital for this purpose.

Five conclusions and research perspectives

The major findings of this volume suggest a number of conclusions and research perspectives. They regard issues of substance, the transfer potential and the crucial regional determinants, the building on positive regional developments, the improvement of the Cooperative Security Concept, and the need to find an adequate institutional forum such as the Middle East Conference on a zone free of WMD and DVs. In the following I list five conclusions and research perspectives.

First conclusion and research perspective

Since missiles matter in the Middle East, they need to become the subject of academic research and a serious part of the discourse about WMD to a greater extent and in a more organized way than they have been in the past.

Missiles, rockets, and UAVs constitute a viable research area in their own right. The latter category of weapons – in which considerable advancements have been made and can also be expected in the foreseeable future – is a technological novelty deserving more systematic efforts, especially because of their destabilizing effects. In addition, one can expand the scope of inquiry as some missiles can serve as delivery vehicles for WMD warheads. We hope the conceptual analysis presented in this volume stimulates further research and discussion. This aspect will be vital for the envisaged Middle East Conference on a zone free of WMD and DVs. With regard to the realm of politics, the comparative advantages regarding missiles should not be neglected: since they are often considered a peripheral issue in the highly emotional discussion of WMD, one could attempt to use missiles as a tactical tool to soften those debates without ignoring other important issues like the nuclear question. The strategic posture of the United States, which is characterized by a growing emphasis on conventional weapons, including long-range missiles, is a case in point. While this may have a destabilizing effect, it could also facilitate the thinking about reducing the strategic role of nuclear arsenals.

Second conclusion and research perspective

It is possible to learn from the experience of the East–West conflict as well as from previous success in the region but the transfer potential from such examples is limited and therefore any borrowing from other contexts needs to be tailored to current regional conditions and needs. In the Middle East/Gulf conflict constellations, alliance structures, and arms dynamics are the most important factors characterizing the security dilemma. Quick fix arms control/reduction proposals that ignore these specifics are bound to fail.

Proposals from outside the region can easily be dismissed as self-serving and discriminatory if they deny Middle Eastern countries those weapons extra-regional countries themselves have in their inventory. The same holds true if they propose banning a category of missiles (for instance the Intermediate-Range Nuclear Forces Treaty prohibits missiles between 500 and 5,500 km) which two major external powers – the US and Russia – no longer possess, but which play an important role in the military postures of some Middle Eastern countries.

Third conclusion and research perspective

Circumstances can change for the better or worse, as International Relations theories recognize and as a result it would be important to identify appropriate 'actors of change' at the state (Track I) and non-governmental (Track II) levels. The analysis of possible obstacles or opportunities is equally relevant.

As the upheavals in the Arab world, above all in Egypt, have shown, domestic politics are likely to play an increasingly greater role in arms/missile procurement and, conversely, in reduction processes. Previous studies have focused on the effects of domestic politics on arms control in the East–West context (e.g. Kubbig 1983, 1988) and in other regions including the Middle East (Potter and Mukhatzhanova 2010a, 2010b; Solingen 2007). Therefore much more systematic work needs to be done on using actor-centred approaches for analysing domestic power constellations among fractionated elites and other social groups for determining how their foreign policy orientations contribute to or impede arms control/reductions. The attempt to build on positive regional developments which require further systematic research and shift the focus to potential 'actors of change' could include the analysis of the 'culture of mediation' by actors in the region which is gradually developing, whether in Egypt, Qatar, or Turkey. Also, the potential impact of the path-breaking Arab Peace Initiative (API) is a possible area of inquiry, as are the domestic factors that have so far impeded a discussion about the API in Israel.

Fourth conclusion and research perspective

This volume is not about missile head counts but a new way of thinking about decreasing the security dilemma through a cooperative approach that has an incremental reduction strategy as its analytical core. It is precisely this approach that needs to be pursued more thoroughly and systematically.

There are a number of issues that need to be explored in greater depth. First, expanding the missile focus to include not only WMD, but also the potential for cooperation both in the missile defence area, which is often exaggerated, and in the realm of civilian space programmes. Second, there needs to be an effort to embed arms control/reductions and long-term objectives such as a Missile or a WMD Free Zone, in more comprehensive regional security arrangements. However, such analyses will have to avoid the elusiveness and the duplication, which has often characterized studies on regional order.

Third, to date the missile area has largely been a 'norm free zone'; enhancing norm-building is therefore a necessary step for improving the CSC. Norm-building could be encouraged by further restricting an already tight missile export policy from major extra-regional powers – a task weakened enormously by the US missile defence deliveries. The best place to start with rudimentary norm-building in the Middle East is by embarking on an interest-based approach which will show the cooperative approach to security to be less costly (financially and politically) than the traditional one. Fourth, actors could focus on seeking common ground by identifying common adversaries such as al Qaida. Fifth, and finally, norm-building could be enhanced by establishing social cooperative practices, processes, and networking, preferably in a structured and institutionalized way (Kubbig in cooperation with Fikenscher *et al.* 2010).

Fifth conclusion and research perspective

Since conferences have proved to be a useful regional peace strategy, Track I and II actors need to find an appropriate multilateral organizational setting to discuss and negotiate a path towards an MFZ or a WMDFZ, both of which form the core of a Cooperative Security Concept. In doing so, it will be important to learn from pitfalls and successes of the past, including the working group on Arms Control and Regional Security and the Sinai accords as major CSBMs.

For Track II research the challenge is to create innovative ideas for tackling at least the following challenges. First, with regard to the institutional framework it needs to be decided whether, for instance, the UN would be the appropriate setting. Second, which organization(s) would convene and be responsible for the conference(s)? Third, concerning the process, should only a single conference be planned, or is a series of structured meetings more effective? The latter could provide a process for setting clear goals, time frames, and criteria for success. Fourth, with regard to procedural and substantive issues: what would be the scope of the discussions and which issues could be negotiated later? Fifth, how can the conference process be maintained and the setting institutionalized? Sixth, it is necessary to develop at least a modest and concrete Plan B – for instance taking up the efforts of the mid 1990s to build Regional Security Centres. Again, such an initiative (which could be implemented in parallel with a conference process) could be optimized by learning from the experiences with arms control related centres in other regions (such as Latin America and Europe), but also by assessing the success of Middle Eastern ecological institutions and analysing their transfer potential to the arms control area.

To sum up: future efforts which build on the academic research of this volume and its authors would develop the idea of the Cooperative Security Concept and the core analytical focus on a gradual path towards a Missile/WMD Free Zone in ways that are more sustainable, e.g. by embedding the CSC in a more formal and maybe even institutionalized conference process. This seems to be the most promising way of initiating a shift away from the predominant mind-set, which is a necessary precondition for decreasing the Middle East security dilemma.

References

Feldman, Shai and Toukan, Abdullah (1997) *Bridging the Gap: A Future Security Architecture for the Middle East*, Lanham, MD: Rowman & Littlefield.

Krass, Allan S. (1985) 'The Politics of Verification', *World Policy Journal*, 2(4): 731–52.

Kubbig, Bernd W. (1983) *Gleichgewicht oder Überlegenheit: Amerikanische Rüstungskontrollpolitik und das Scheitern von SALT II*, Frankfurt am Main: Campus.

Kubbig, Bernd W. (1988) *Amerikanische Rüstungskontrollpolitik: Die innergesellschaftlichen Kräfteverhältnisse in der ersten Amtszeit Reagans (1981–1985)*, Frankfurt am Main: Campus.

Kubbig, Bernd W. in cooperation with Fikenscher, Sven-Eric, Flamm, Patrick, Golob, Lina, Knussmann, Marieke, Mildner, Annika, Nieke, Sebastian, and Weismann, Hanne (2010) *ATLAS: Track II Research Activities on the Middle East/Gulf Region, in the Middle East/Gulf, Europe, and the USA*. Online, available at: http://hsfk.de/fileadmin/content-ABM/ATLAS-gesamt.pdf (25 March 2011).

Potter, William C. and Mukhatzhanova, Gaukhar (eds) (2010a) *Forecasting Nuclear Proliferation in the 21st Century: The Role of Theory*, Vol. 1, Stanford, CA: Stanford University Press.

Potter, William C. and Mukhatzhanova, Gaukhar (eds) (2010b) *Forecasting Nuclear Proliferation in the 21st Century: A Comparative Perspective*, Vol. 2, Stanford, CA: Stanford University Press.

Solingen, Etel (2007) *Nuclear Logics: Contrasting Paths in East Asia and the Middle East*, Princeton, NJ: Princeton University Press.

Appendix

The arsenals of actors relevant to a missile free zone in the Middle East/Gulf

Michael Haas and Bernd W. Kubbig (Compilers)

Unless otherwise stated, all data is taken from IISS (2010a), which is our main source because it analyses the most recent inventory, from SIPRI (2011) and from INSS (2011). All links referred to in the tables were last accessed 27 November 2010.

The data does not include the following: enablers other than tanker aircraft; ballistic missiles, cruise missiles, and rockets of less than 70 km range; MANPADS unless fixed or vehicle-mounted; AA guns not part of systems that also include SAMs (SPAAGMs).

Missiles: numbers refer to number of missiles unless otherwise mentioned.
SAMs: numbers refer to launchers unless otherwise mentioned.
For Russian and Chinese systems: name = NATO reporting name; indigenous designations are given in parentheses where advisable.
Otherwise, supplier/indigenous designations are used.

The airpower, missile, and air and missile defence inventories of selected states

Table A.1 Bahrain

Name	*Type*	*Number in possession*	*Delivery/introduction*	*Comments*
F-16C/D *Fighting Falcon*	Multirole fighter	21 (17/4)	1990–2000	Block 40. Single- (C) and two-seat (D) versions.
F-5E/F *Tiger II*	Multirole fighter	12	1985–87	Single- (E) and two-seat (F) versions. Limited ground-attack capability. Obsolete.
MGM-140A ATACMS	SRBM	9 launchers, 30 missiles	2002	Tactical SSM with ≥165 km range and 560 kg cluster warhead for the M270 MLRS.[1]
MGM-168 ATACMS	SRBM	30	2010 (ordered)	T2K (TACMS 2000) version with 300 km range and 213 kg or 247 kg unitary HE warhead.[2]
RGM-84C *Harpoon*	ASCM	8	1997/98	Anti-ship missile with 120–30 km range and 222 kg warhead.[3]
MM-40 *Exocet*	ASCM	50	1984/85	Anti-ship missile with 70 km range and 165 kg warhead.[4]
MIM-23B *I-HAWK*	Medium-range SAM	8	1997	Ageing medium-range SAM. Range: max. 40 km, altitude: low to 20 km.[5]
R440 *Crotale*	Short-range SAM	7	1995	Capable, but ageing French-made AD system. Supplier uncertain.

Notes
1 www.missilethreat.com/missilesoftheworld/id.73/missile_detail.asp; www.lockheedmartin.com/data/assets/12818.pdf.
2 www.missilethreat.com/missilesoftheworld/id.76/missile_detail.asp; www.dsca.osd.mil/PressReleases/36-b/2010/Bahrain_10-57.pdf; http://defense-update.com/wp/20101111_atacms_gulf.html.
3 www.missilethreat.com/cruise/id.134/cruise_detail.asp.
4 www.missilethreat.com/cruise/id.10/cruise_detail.asp.
5 www.fas.org/spp/starwars/program/hawk.htm.

Table A.2 Egypt

Name	*Type*	*Number in possession*	*Delivery/introduction*	*Comments*
F-16A/B *Fighting Falcon*	Multirole fighter	38 (26/12)	1982–85	Basic, single- (A) and two-seat (B) versions.
F-16C/D *Fighting Falcon*	Multirole fighter	119 (113/6)	1991–2002	Plus 24 Block 50/52 on order as of 2009, including 4 F-16D, status of deal unknown.
Mirage-2000 EM/BM	Multirole fighter	21 (18/3)	1986–88	Modern, 4th generation fighter aircraft. Single- (EM) and two-seat (BM) versions.
Mirage-5 SDE/SDR/E2	Ground-attack aircraft	75 (53/6/16)	1973–80	Obsolete attack version of Mirage III. 16 modernized to E2. 6 combat-capable reconnaissance version.
F-4E *Phantom II*	Fighter-bomber	29	1979–93	Obsolete.
F-7A	Interceptor	74	1962–86	Chinese copy of MiG-21F *Fishbed* C for export. Obsolete.
MiG-21R *Fishbed H*	Reconnaissance aircraft	14	1960s/70s	Combat-capable, but obsolete.
J-6/F-6	Fighter aircraft	44	1979–83	Chinese-built MiG-19 *Farmer B*. Obsolete.
Alpha Jet	Light attack aircraft	12	1982–85	Licensed by Egypt.
SS-1C Scud-B/Hwasong-5	SRBM	9–24 (launchers), 100 (missiles, est.)	1960s–90s	North Korean version of inaccurate and obsolete material w/300–20 km range and 1,000 kg payload (usually unitary HE warhead).
Scud-C/Hwasong-6	SRBM	?	?	Range increased to 500 km, smaller payload of 700–70 kg.[1] Status unclear, but has received components from DPRK during 1990s.[2]
Project T	SRBM	≤90 missiles (est.)	1990s	Indigenous derivative of Scud-B with 450 km range and 1,000 kg payload.[3]

continued

Table A.2 Continued

Name	*Type*	*Number in possession*	*Delivery/introduction*	*Comments*
Frog-7	Long-range artillery rocket	≥9 launchers	1960s	Nuclear-, chemical-capable artillery rocket with 70 km range. Payload: 450–550 kg (incl. unitary HE, mines, or cluster).[4] Probably retired.
Sakr-80	Long-range artillery rocket	>24 launchers, ≤several hundred missiles (est.)	1980s	Partially French designed, indigenous replacement for Frog-7. Range: 80 km, payload: 200 kg, cargo warhead.[5]
SS-C-2B *Samlet*	LACM/ASCM	≤48	1963/64	Obsolete, early Soviet cruise missile with 90 km range and 600 kg HE warhead.[6] Status uncertain.
AGM/RGM/UGM-84C *Harpoon*	ASCM	≤200 (est.)	1983–2003	Air-, surface-, and submarine-launched versions. Including 25 RMG-84L (Block II), but without land-attack capability.[7]
SS-N-2C/SS-C-3 *Styx*	ASCM	≤70 (est.)	2006/07	Obsolete, with 80 km range and 513 kg HE warhead.
SY-1/CSS-N-1 *Scrubbrush*	ASCM	≤15 (est.)	1984/85	Obsolete copy of Soviet P-15/SS-N-2 *Styx* with 80 km range and 454 kg HE warhead.[8]
HY-1/CSS-N-2 *Safflower*	ASCM	≤20 (est.)	1984	Improved, but obsolete SY-1 with 80 km range and 454 kg HE warhead.[9]
AS-39 *Exocet*	ASCM	≤35 (est.)	1982/83	Air-launched version with 70 km range and 165 kg payload.[10]
Otomat Mk 1/2	ASCM	≤120 (est.)	1975–82	Italian-made missiles with 60 km (Mk 1) and 180 km (Mk 2) range and 210 kg HE warhead.[11]

AS-5 *Kelt*	ASCM	20–30 in store?	1970–72	Air-launched, nuclear-capable cruise missile with 180 km range and 1,000 kg payload.[12] Probably retired together with Tu-16 bombers.
MIM-23B *I-HAWK*	Medium-range SAM	12 batteries, ≥78 launchers	1982–2001	Some modernized ex-US.
SA-2 *Guideline*	Medium-range SAM	?	1969–72	Obsolete.
SA-3 *Pechora/Goa*	Medium-range SAM	≥212	1971/72, 2002–06 (*Pechora*-M2)	30 modernized by Russia to *Pechora*-M2 standard. Rest obsolete.
SA-6 *Gainful*	Self-propelled, medium-range SAM	≥56	1971–73	Ageing mobile SAM.
SA-9 *Gaskin*	Self-propelled, short-range SAM	20	?	2 × twin *Strela* mounted on BRDM-2 vehicle.
R440 *Crotale*	Short-range SAM	14 batteries, ≥24	1977–85	
M-48 *Chaparral*	Short-range SAM	38 batteries, ≥50	?	
Amoun/Skyguard/RIM-7F *Sea Sparrow*	Short-range SAM/AAA system	≥72	1984–87	Swiss *Skyguard* radar with 2 × twin 35 mm *Oerlikon* AAA and US *Sea Sparrow* SAM.

Notes

1 www.missilethreat.com/missilesoftheworld/id.103/missile_detail.asp; Bermudez Jr (1999).
2 Bermudez Jr (1999: 19); Cordesman (2006: 196–8).
3 Cordesman (2006: 196).
4 www.fas.org/man/dod-101/sys/missile/row/frog-7.htm.
5 Cordesman (2002: 549, 2006: 197).
6 www.missilethreat.com/cruise/id.105/cruise_detail.asp.
7 www.boeing.com/defense-space/missiles/harpoon/docs/HarpoonBlockIIBackgrounder.pdf.
8 www.fas.org/man/dod-101/sys/missile/row/hy-1.htm.
9 www.fas.org/man/dod-101/sys/missile/row/hy-1.htm.
10 www.missilethreat.com/cruise/id.5/cruise_detail.asp.
11 www.missilethreat.com/cruise/id.37/cruise_detail.asp.
12 www.missilethreat.com/cruise/id.78/cruise_detail.asp.

Table A.3 Iran

Name	*Type*	*Number in possession*	*Delivery/introduction*	*Comments*
F-14A *Tomcat*	Air superiority fighter	44	1976–78	79 originally delivered. Status uncertain, about 20 operational. Updated by Iran.
MiG-29A/U/UB *Fulcrum*	Air superiority/ multirole fighter	35	1989–91	
Mirage F1E	Multirole fighter	10	1991	Formerly Iraqi Air Force, captured when flown to Iran in 1991.
F5B *Freedom Fighter*	Multirole fighter	20	1965–72	Including some of indigenously built
F-5E/F *Tiger II*	Fighter	60	1973–77	*Azarakhsh/ Saegheh* variants.
Su-24MK *Fencer*	Ground-attack aircraft	30	1990/91	Majority ex-Iraqi.
Su-25A/T/UBK *Frogfoot*	Ground-attack aircraft	13	1991–2006	6 delivered by Russia in 2000s, rest ex-Iraqi.
F-7M *Airguard*	Interceptor	24	1993–96	Improved, Chinese-built MiG-21.
F-4D/E *Phantom II*	Interceptor/fighter-bomber	≥71	1968–79	Obsolete. Including ≥6 RF-4E combat-capable reconnaissance version.
P-3MP *Orion*	ASW/patrol aircraft	5	1975	
R'ad	UCAV	?	2010–	Announced February 2010.[1] Status uncertain.
Karar	UCAV	?	2010–	Announced as long-range bomber drone with 1,000 km range and 250 kg payload.[2] Status uncertain.
Boeing 707	Tanker aircraft	3	1975–78	
Boeing 747	Tanker aircraft	≤5	1975–77	

CSS-8	SRBM	≤200	1990–94	Obsolete, Chinese-made SSM-version of SA-2. Range: ≤150 km.[3] Iranian designation: Tondar-69.
Fateh-110	SRBM	≥50 (est.)	2002–	Improved Zelzal with guidance system, greater accuracy. 200+ km range, 500 kg warhead.[4]
Shahab-1/Hwasong-5	SRBM	12–18 launchers, 300 missiles (est.)	1987–	Initially delivered by DPRK and partly assembled in Iran, indigenous production capability since late 1990s.[5]
Shahab-2/Hwasong-6	SRBM	12–18 launchers, 300 missiles (est).	1991–	
Shahab-3/Nodong	MRBM	6+ launchers, ≤several hundred missiles (est.)	2003–	Indigenously produced copy of North Korean Nodong with 800–1,000 km range and 750–1,100 kg payload.[6]
Ghadr-1	2004–	Improved Shahab-3 with 1,500–600 km range and 750 kg payload.[7]		
Sajjil/Sajjil-2	MRBM		2012 (est.)	Indigenous two-stage, solid-fuel design with 2,200 km range and 750 kg payload. Successfully tested four times in 2008/09.[8]
Musudan	SLBM/IRBM	≤19	2006–	North Korean derivative of obsolete Soviet R-27/SS-N-6 *Serb* SLBM. Range: 2,500–3,200 km. Payload: 1,000–200 kg. Delivered to Iran according to diplomatic cable obtained through WikiLeaks.[9] Status uncertain.
Fajr-5	Long-range artillery rocket	≥100 (est.)	1990s	Highly inaccurate, 75 km range, 90 kg HE warhead.[10]

continued

Table A.3 Continued

Name	*Type*	*Number in possession*	*Delivery/introduction*	*Comments*
Nazeat-6	Long-range artillery rocket	≥100 (est.)	Mid 1990s	100 km and 140–50 km range, 150 kg and 250 kg HE warhead, respectively.[11]
Nazeat-10				
Zezal-1	Long-range artillery rocket	≤several thousand (est.)	1998–	Spin-motor-stabilized, but still inaccurate. Ranges: 125 km and 200 km, respectively. 600 kg warhead.[12]
Kh-55/AS-15 *Kent*	LACM	6	2001	Advanced, nuclear long-range cruise missile illegally obtained from Ukraine without warheads. Range: 2,500 km, payload: 410 kg.[13] Status uncertain, but probably not deployed.
CSS-N-8 *Saccade*	ASCM	≤320	1994–	C-802 export version of Chinese cruise missile with 120 km range and 165 kg HE warhead.[14] Produced in Iran as Noor.
CSS-C-3 *Seersucker*	ASCM	≤150	1987–98	Obsolete anti-ship missile with 100 km range and 513 kg HE warhead.[15] Including CAS-1 air-launched version.
MIM-23B *I-HAWK*	Medium-range SAM	≥150 launchers, ≤2000 missiles	1974–86	Post-1979 deliveries part of Iran-Contra deal.
SA-2 *Guideline*	Medium-range SAM	45 launchers, ≤350 missiles	1985–91	Chinese HQ-2 version.
SA-5 *Gammon*	Long-range SAM	10 launchers	1991	Obsolete.
SA-15 *Gauntlet* (Tor-M1)	Self-propelled, short-range SAM	29 launchers, 750 missiles	2006/07	Very capable short-range AD system.

SA-22 *Greyhound* (Pantsir-S1)	Short-to-medium-range SPAAGM	10	2008/09	Obtained from Syria.
FM-80/R440 *Crotale*	Short-range SAM	≤250	1999–2004	Chinese copy of French *Crotale*, a.k.a. HQ-7 (domestic version).[16] Probably produced in Iran as Shahab Thaqeb.
Rapier	Short-range SAM	30 launchers, ≤2,000 missiles	1972–75	
Tigercat	Short-range SAM	15 launchers, ≤400 missiles	1969/70	

Notes

1 http://en.rian.ru/mlitary_news/20100208/157809895.html.
2 www.defensenews.com/story.php?i=4753618&c=MID&s=AIR; www.bbc.co.uk/news/world-middle-east-11052023,
3 www.missilethreat.com/missilesoftheworld/id.30/missile_detail.asp,
4 IISS (2010b: 51–3).
5 IISS (2010b: 14–16).
6 IISS (2010b: 19–20).
7 IISS (2010b: 24–5).
8 IISS (2010b: 55–63).
9 www.missilethreat.com/missilesoftheworld/id.185/missile_detail.asp; http://pollack.armscontrolwonk.com/archive/3351/north-korea-debuts-an-irbm#more-656; www.nytimes.com/2010/11/29/world/middleeast/29missiles.html?_r=1&partner=rssnyt&emc=rss.
10 IISS (201b0: 45).
11 IISS (2010b: 48).
12 IISS (2010b: 49–50).
13 www.missilethreat.com/cruise/id.81/cruise_detail.asp.
14 www.missilethreat.com/cruise/id.62/cruise_detail.asp.
15 www.fas.org/man/dod-101/sys/missile/row/c-201.htm.
16 www.sinodefence.com/army/surfacetoairmissile/hongqi7.asp.

Table A.4 Israel

Name	*Type*	*Number in possession*	*Delivery/introduction*	*Comments*
F-15A/B *Baz*	Air superiority fighter	34 (27/7)	1991/92	Basic, single- and two-seat air-to-air versions, ex-American.
F-15C/D *Akef*	Air superiority fighter	28 (17/11)	1981–92	Improved, single- and two-seat air-to-air versions.
F-15I *Ra'am*	Multirole (strike) fighter	25	1998/99	Based on F-15E *Strike Eagle*. Combat radius: 2,200 km.[1] Can carry all current US air-to-air and air-to-ground munitions, including the BLU-109 and GBU-28 hard target penetrators (up to 11,000 kg overall).[2]
F-16A/B *Netz*	Multirole fighter	106 (90/16)	1980–94	
F-16C/D *Barak*	Multirole fighter	101 (52/49)	1986–94	
F-16I *Sufa*	Multirole fighter	102	2004–09	Based on F-16D Block 52+. Combat radius: max. 2,100 km. Can carry most US air-to-air and air-to-ground munitions of up to 2,000 lbs including the BLU-109 hard target penetrator (7,800 kg overall).[3]
F-35A *Joint Strike Fighter*	Multirole fighter	20 ordered, option for 55 more	delivery 2015–17[4]	Fifth-generation, stealthy multirole fighter. Combat radius: 1,080 km on internal fuel. Can carry up to 8,150 kg of air-to-air and air-to-ground weapons.[5]
Hermes 450	UCAV	1 squadron	2003–	Possibly adapted as UCAV according to media reports.[6]
Harpy	UCAV	≥100 (est.)	?	Loitering SEAD drone-cruise missile hybrid with 500 km range and integrated HE warhead.[7]

Harop/Harp-2	UCAV	≥20 (est.)	2009–	Larger version of *Harpy* for use against wider range of targets. Range: 1,000 km.[8]
KC-707	Tanker aircraft	4	2001 (?)	Boeing 707 modified after delivery.
KC-130H *Hercules*	Tanker aircraft	5	1976/77 (?)	Turboprop tanker.
Jericho-1	SRBM	estimates vary widely	Approx. 1973	Obsolete, probably retired. 500 km range, 450 kg payload.[9]
Jericho-2	IRBM	50–90 (in underground silos and caves)	1990–	1,500 km range, 1,000 kg payload
Jericho-3	IRBM/ICBM	?	?	Tested 2008, current status unknown, possibly deployed. 4,800–6,500 km range, 1,000–300 kg payload, single warhead or 2–3 MIRV possible.[11]
MGM-52 *Lance*	SRBM	7	?	Tactical, nuclear-capable SRBM with 70–120 km range and different warhead options.
Popeye/AGM-142 *Have Nap*	LACM	≥100 (est.)	1989–	Air-launched cruise missile with 80 km range. Can be carried by F-15.[12]
Popeye-2/*Have Lite*	LACM	≥50 (est.)	1995–	Smaller, lighter version of original Popeye. Can be carried by F-16. Range: 75 km.[13]
Popeye Turbo	LACM	?	2000 (?)	Popeye cruise missile with extended range of 300–1,500 km. Payload: ≤865 kg.[14] Can be fired from 650 mm torpedo tubes on German-built *Dolphin*-class submarines.[15]

continued

Table A.4 Continued

Name	*Type*	*Number in possession*	*Delivery/introduction*	*Comments*
Delilah	LACM/ASCM	≥30 (est.)	1995–	Small, very precise, loitering cruise missile with ≥250 km range and 30 kg HE warhead.[16]
Gabriel I	ASCM		Early 1970s	Family of missiles with rocket or turbojet motors. Number of all Gabriel type missiles is estimated to be in the 100s. Up to 200 km range.[17] Deployed on various Israeli Navy surface vessels.
Gabriel II	ASCM		1970s	
Gabriel III	ASCM		1982	
Gabriel IV	ASCM		Early 1990s	
Gabriel V/Advanced Naval Attack Missile	ASCM		Under development	
RGM/UGM-84C *Harpoon*	ASCM	≥200 (est.)	1981–2001	Some UGM-84 (submarine-launched) version.
MIM-23B *I-HAWK*	Medium-range SAM	17 batteries	1976–84	
MIM-104 *Patriot* Pac-2	Medium-to-long-range SAM/MD system	6 batteries	1990–93, 2007–10 (GEM+)	Improved (GEM+/GEM-T) Pac-2. Israeli designation: Yahalom.[18]
MIM-104 *Patriot* Pac-3	Short-to-medium-range MD system			Tested by Israel October 2006. Future acquisition likely.[19]
Arrow-2	MD system	3 batteries	1998	Third battery under construction as of October 2010.[20]
Arrow-3	MD system		2014 (est.)	Joint US–Israeli project. Under development.[21]

David's Sling/Magic Wand	Medium-range anti-rocket/MD system		2014 (est.)	Joint US–Israeli project. Capable of intercepting long-range rockets and slow cruise missiles. Range: 40–300 km.[22]
Iron Dome	Short-range anti-rocket system	2 batteries (to be deployed in late 2010)	2010	US provides full funding.[23] Capable of intercepting short-range rockets and 155 mm artillery shells. Range: 3–70 km.[24]

Notes

1 www.af.mil/information/factsheets/factsheet.asp?id=102.
2 www.aerospaceweb.org/aircraft/bomber/f15e/.
3 www.aerospaceweb.org/aircraft/fighter/f16/.
4 www.defensenews.com/story.php?i=4866067&c=AME&s=TOP.
5 www.lockheedmartin.com/products/f35/f-35A-ctol-variant.html.
6 www.timesonline.co.uk/tol/news/world/africa/article5993093.ece; www.worldtribune.com/worldtribune/07/front2454229.238888889.html.
7 http://defense-update.com/directory/harpy.htm.
8 http://defense-update.com/products/h/harop.html.
9 www.missilethreat.com/missilesoftheworld/id.56/missile_detail.asp.
10 www.missilethreat.com/missilesoftheworld/id.57/missile_detail.asp; www.nti.org/e_research/profiles/Israel/Missile/facilities.html#zachariah; www.nti.org/e_research/profiles/Israel/Missile/index.html; www.globalsecurity.org/wmd/world/israel/sedot_mikha.htm.
11 www.missilethreat.com/missilesoftheworld/id.58/missile_detail.asp.
12 www.missilethreat.com/cruise/id.25/cruise_detail.asp.
13 www.missilethreat.com/cruise/id.30/cruise_detail.asp.
14 Gormley (2008: 179).
15 www.globalsecurity.org/wmd/world/israel/popeye-t.htm; www.fas.org/nuke/guide/israel/missile/popeye-t.htm.
16 www.janes.com/articles/Janes-Air-Launched-Weapons/Delilah-AL-and-Delilah-HL-Light-Defender-Israel.html.
17 www.fas.org/man/dod-101/sys/missile/row/gabriel.htm; http://defense-update.com/newscast/1007/news/021007_gabriel.htm; Carus (1992: 35).
18 www.worldtribune.com/worldtribune/WTARC/2010/me_israel0405_05_12.asp.
19 www.ynetnews.com/articles/0,7340,L-3484655,00.html.
20 www.jpost.com/Israel/Article.aspx?id=192350.
21 www.defensenews.com/story.php?i=4548013&c=FEA&s=SPE; www.fas.org/sgp/crs/mideast/RL33222.pdf.
22 www.fas.org/sgp/crs/mideast/RL33222.pdf.
23 www.fas.org/sgp/crs/mideast/RL33222.pdf.
24 www.rafael.co.il/marketing/SIP_STORAGE/FILES/6/946.pdf.

Table A.5 Kuwait

Name	*Type*	*Number in possession*	*Delivery/introduction*	*Comments*
F/A-18C/D *Hornet*	Multirole fighter	39 (31/8)	1991–93	
Hawk Mk-64	Trainer/light attack aircraft	19	1985–87	
AGM/RGM-84 *Harpoon*	ASCM	60	1997–2003	
AM-39/MM-40 *Exocet*	ASCM	95	1980–90	
MIM-23B *I-HAWK*	Medium-range SAM	24 launchers	1977/78	*I-HAWK* Phase III.
MIM-104 *Patriot* Pac-2	Medium-to-long-range SAM/MD system	40 launchers	1995/96, 2007 (requested)	Upgrading existing systems to GEM-T. 290 missiles on order.[1]
MIM-104 *Patriot* Pac-3	Short-to-medium-range MD system		2007 (requested)	Upgrading existing systems to Pac-3. 80 missiles on order.[2]
Skyguard/Amoun	Short-range SAM/AAA system	12 systems	1989–2001	Egyptian *Amoun version.*
Aspide	Short-range SAM/AAA system			Italian SAM used with *Skyguard* instead of *Sea Sparrow.* Being upgraded to *Spada* 2000.

Notes
1 www.dsca.osd.mil/PressReleases/36-b/2007/Kuwait_08–23.pdf; www.dsca.osd.mil/PressReleases/36-b/2007/Kuwait_08-23.pdf.
2 www.dsca.osd.mil/PressReleases/36-b/2007/Kuwait_08–23.pdf.

Table A.6 Oman

Name	*Type*	*Number in possession*	*Delivery/introduction*	*Comments*
F-16C/D *Fighting Falcon*	Multirole fighter	12 (8/4)	2005–08	Block 50.
Jaguar OS/OB	Ground-attack aircraft	24 (20/4)	1977–98	Single- (OS) and two-seat (OB) versions.
Hawk Mk 103/203	Light multirole combat aircraft	16 (4/12)	1993–95	
AGM-84D *Harpoon*	ASCM/LACM	≤20	2006	Range: 220 km.[1]
AM-39/MM-40 *Exocet*	ASCM	≤160	1982–2006	Including 50 MM-40 Block 3 with 180 km range.[2]
Rapier	Short-range SAM	≤600	1976/77	
Rapier-2	Short-range SAM	≤800	1995	Propably included rebuilding of earlier *Rapiers* to this standard.
Mistral-2	Short-range SAM	≤230	2001–03	Vehicle-mounted version.

Notes

1 www.designation-systems.net/dusrm/m-84.html.

2 www.missilethreat.com/cruise/id.10/cruise_detail.asp.

Table A.7 Qatar

Name	*Type*	*Number in possession*	*Delivery/introduction*	*Comments*
Mirage 2000–5	Multirole fighter	12	1997–99	Single-seat (EDA) and two-seat (DDA, combat-capable trainer) versions.
Alpha Jet	Trainer/light attack aircraft	6	1980/81	
Apache	LACM	≤50	1999–2003	Stealthy, air-launched anti-runway munition with 140 km range for Mirage 2000.
AM-39/MM-40 *Exocet*	ASCM	≤225	1982–97	Air- and ship-launched versions.
Roland II	Self-propelled, short-range SAM	9	1986–88	Some on AMX-30 chassis, some in shelters.
Mistral	Short-range SAM	24	1992–96	Naval and portable versions.

Table A.8 Saudi Arabia

Name	*Type*	*Number in possession*	*Delivery/introduction*	*Comments*
F-15C/D *Eagle*	Air superiority fighter	84 (66/18)	1981–92	
Tornado ADV	Interceptor	15	1989/90	
F-15S *Strike Eagle*	Multirole (strike) fighter	70	1995–99	Almost identical with US F-15E *Strike Eagle*. To be upgraded to F-15SA configuration.[1]
F-15SA	Multirole (strike) fighter	84	2010 (ordered, delivery expected 2014–)	Improved F-15E, similiar to South Korean F-15K version.
EF *Typhoon*	Multirole fighter	8 (72 ordered)	2009–11	Tranche 2.
Tornado IDS	Multirole (strike) fighter	85	1986–98	Including 10 reconnaissance version.
F-5F *Tiger II*	Multirole fighter	22	1974–86	Including some F-5B *Freedom Fighter* and RF-5E *Tigereye* (combat-capable reconnaissance version).
A330 MRTT	Tanker/transport aircraft	9 (ordered)	2011 (expected)	State-of-the art tanker aircraft.
KC-130H *Hercules*	Tanker aircraft	8	1977–80	
KE-3A	Tanker aircraft	7	1986/87	
CSS-2/DF-3	IRBM	≥10 launchers, 40–60 missiles	1987/88	Obsolete, liquid-fuelled Chinese IRBM with 2650–800 km range and 2150 kg conventional HE warhead.[2]
Storm Shadow/SCALP EG	LACM		?	Modern cruise missile with 300 km range and 500 kg warhead.[3]

continued

Table A.8 Continued

Name	*Type*	*Number in possession*	*Delivery/introduction*	*Comments*
AGM-84L *Harpoon*	ASCM/LACM	400	2010 (ordered)	Air-launched Block II version. For F-15SA.
RGM-84C *Harpoon*	ASCM	≤160	1980–83	
AM-39/MM-40 *Exocet*	ASCM	≤70	1990–2005	Mostly MM-40 Block 2.
Otomat Mk-2	ASCM	≤225	1985–88	
Sea Eagle	ASCM	≤100	1993	Range: ≥110 km
MIM-23B *I-HAWK*	Medium-range SAM system	16 batteries, 128 launchers	1978–85	
MIM-104 *Patriot* Pac-2	Medium-to-long-range SAM/MD system	16 batteries, 96 launchers	1993–97	
MIM-104 *Patriot* Pac-3	Short-to-medium-range MD system			Future acquisition probable.
Shahine	Self-propelled, short-range SAM	17 batteries, 73 launchers	1980–90	Developed from *Crotale* for Saudi Arabia. Mounted on AMX-30 chassis.
R440 *Crotale*	Short-range SAM	40 launchers	1980–91	
Shahine/ Crotale	Short-range SAM	73 static defence units with 68 launchers	1980–91	
Mistral	Short-range SAM	≤500 (missiles?)	?	Naval version of MANPADS.

Notes
1 www.dsca.osd.mil/PressReleases/36-b/2010/Saudi_Arabia_10–43.pdf.
2 www.missilethreat.com/missilesoftheworld/id.21/missile_detail.asp.
3 Gormley (2008: 179); www.globalsecurity.org/military/world/europe/casom.htm; www.mbda-systems.com/mbda/site/ref/scripts/siteFO_contenu.php?lang=EN&noeu_id=120&page_id=115.

Table A.9 Syria

Name	*Type*	*Number in possession*	*Delivery/introduction*	*Comments*
MiG-31E *Foxhound*	High-speed interceptor/ reconnaissance			2007 deal including up to 8 MiG-31s mentioned in some reports. Officially denied October 2010.[1]
MiG-29A *Fulcrum*	Air superiority/ multirole fighter	≥40	1987/88	Some two-seat MiG-29UB version.
MiG-29M/M2/SMT	Multirole fighter	?	?	Modernized MiG-29. Status uncertain.[2]
MiG-25PD/R/RB *Foxbat*	Interceptor/ reconnaissance/light bomber	40	1979–85	Some interceptor, some reconnaissance with bombing capability. Includes 2 MiG-25U non-combat trainer/reconnaissance version.
MiG-23BN/MLD/UM *Flogger*	Interceptor/fighter-bomber	146	1979–85	Including some combat-capable trainers.
MiG-21H/J/U *Fishbed*	Interceptor/ fighter-bomber/ reconnaissance	219	1967–83	Different versions, including reconnaissance and trainers (*Mongol A*).
Su-24 *Fencer D*	Ground-attack aircraft	20	1990	
Su-22 *Fitter D*	Fighter-bomber	50	1979–85	Export version of Su-17M2.
L-39 *Albatros*	Light attack aircraft/ trainer	40	1980–86	
Scud-D	SRBM		2000s	North Korean Hwasong-7, range increased to 700 km. Tested by Syria in 2005 and produced at a rate of 15–30/year.[3]

continued

Table A.9 Continued

Name	*Type*	*Number in possession*	*Delivery/introduction*	*Comments*
SS-21 *Scarab*	SRBM	≥18 launchers, ≥100 missiles	1983–88	Capable SRBM with 70–120 km range.[4]
M-600	SRBM	100s of missiles (est.)	2000s	Improved Fateh-110 with 300 km range and 500 kg warhead. Large number allegedly transferred to Hezbollah.[5]
Frog-7	Long-range artillery rocket	18 launchers	1973–78	Possibly retired.
WS-1	Long-range artillery rocket	?	?	302 mm artillery rocket with 90–100 km range. Also transferred to Hezbollah.[6]
SS-C-1B *Sepal*	ASCM/coastal defence	35 (est.)	1987	Obsolete, nuclear-capable cruise missile with 450 km range.[7]
SS-N-2/SS-C-3 *Styx*	ASCM/coastal defence	250 (est.)	1963–86	Obsolete. Range: 80 km.[8]
CSS-N-8 *Saccade*	ASCM	25	2009	
SS-N-26 *Strobile* (P-800 *Yakhont*)	ASCM	72	2010 (ordered)	Very modern, supersonic ASCM with 120–300 km range.[8]
SA-2 *Guideline*	Medium-range SAM	320	1967–69	
SA-3 *Goa*	Medium-range SAM	148	1972–74	
SA-5 *Gammon*	Long-range SAM	44	1982–85	In static emplacements.
SA-6 *Gainful*	Self-propelled, medium-range SAM	195	1973–81	

SA-8 *Gecko*	Self-propelled, short-range SAM	14	1979–81	
SA-9 *Gaskin*	Self-propelled, short-range SAM	20	1975–85	
SA-11 *Gadfly*	Self-propelled, medium-range SAM	20	1986	
SA-13 *Gopher*	Self-propelled, short-range SAM	30	1985/86	
SA-17 *Grizzly*	Self-propelled, medium-range SAM	?	2007?	Modern and highly capable SAM. Status unknown.
SA-22 *Greyhound* (Pantsir-S1)	Short-to-medium range SPAAGM	36–50	2008/09	

Notes

http://en.rian.ru/mlitary_news/20101027/161102905.html; www.defensenews.com/story.php?i=4983525&c=AIR&s=TOP.

2 www.defenseindustrydaily.com/syria-buying-mig31s-mig35s-for-1-billion-03391/.

3 www.missilethreat.com/missilesoftheworld/id.104/missile_detail.asp.

4 www.missilethreat.com/missilesoftheworld/id.140/missile_detail.asp; www.missilethreat.com/missilesoftheworld/id.141/missile_detail.asp.

5 www.aviationweek.com/aw/generic/story_generic.jsp?channel=defense&id=news/awst/2010/05/17/AW_05_17_2010_p28-226325.xml&headline=Israel%20Sees%20Increased%20Hezbollah%20Capability; www.aviationweek.com/aw/generic/story_generic.jsp?channel=awst&id=news/awst/2010/08/09/AW_08_09_2010_p44-243730.xml&headline=null&next=0; www.jpost.com/MiddleEast/Article.aspx?id=174792.

6 Rubin (2007: 5–6).

7 www.fas.org/nuke/guide/russia/theater/ss-n-3.htm.

8 www.missilethreat.com/cruise/id.106/cruise_detail.asp.

9 http://defense-update.com/wp/20100920_yakhont_in_syria.html.

Table A.10 United Arab Emirates

Name	*Type*	*Number in possession*	*Delivery/introduction*	*Comments*
F-16E/F *Desert Falcon*	Multirole fighter	80 (55/25)	2004–08	Most modern (Block 60) F-16 variant available. Single- (E) and two-seat (F) versions.
Mirage 2000–9	Multirole fighter	62	2003–07	Most modern Mirage 2000 variant, including EAD (single-seat), DAD (two-seat) and RAD (single-seat, combat-capable reconnaissance) versions. 32 rebuilt to this standard.
Hawk Mk-102	Light attack aircraft	13	1993/94	
A330 MRTT	Tanker/transport aircraft	3 (+3 possible)	2011 (expected)	
SS-1 Scud-B	SRBM	6 launchers, up to 20 missiles	?	
MGM-168 ATACMS	SRBM	100	2010 (ordered)	T2K (TACMS 2000) unitary warhead version with 300 km range.[1]
M30/31 GMLRS	Long-range artillery rocket	20 launchers, 1,560 rockets	2006 (requested)	Guided artillery rocket with ≥70 km range for HIMARS MLRS. Status uncertain.
Black Shaheen	LACM	?	2003–	Variant of *Storm Shadow*/SCALP EG, range restricted to 280 km.
AGM/RGM-84 *Harpoon*	ASCM	36	1998–2005	Including Block II version.

AM-39/MM-40 *Exocet*	ASCM	≤300	1980–2006	Including 150 MM-40 Block 3 with 180 km range.
THAAD	MD system	3 batteries, 9 launchers, 147 missiles (planned)	2008 (requested)	Congress was notified of possible sale in September 2008.[2] Acquisition in the near future probable.
MIM-104 *Patriot* Pac-3	Short-to-medium-range MD system	9 batteries, 37 launchers, 504 missiles (planned)	2007 (requested), 2015 (delivery)	Congress was notified of the possible sale in December 2007. Includes 216 Pac-2/GEM-T missiles.[3] Will replace MIM-23B.
MIM-23B *I-HAWK*	Medium-range SAM	6 battalions	1986/87	
R440 *Crotale*	Short-range SAMs	3 battalions	1977/78	
Rapier	Short-range SAMs	3 battalions	1976–78	
Mistral	Short-range SAMs	3 battalions		Years of delivery unknown.
SA-22 *Greyhound* (Pantsir-S1)	Short-to-medium range SPAAGM	8 (50 ordered)	2009	

Notes
1 www.dsca.osd.mil/PressReleases/36-b/2010/UAE_10–58.pdf; http://defense-update.com/wp/20101111_atacms_gulf.html.
2 www.dsca.osd.mil/PressReleases/36-b/2008/UAE_08–19.pdf.
3 www.dsca.osd.mil/PressReleases/36-b/2007/UAE_08–17.pdf.

Table A.11 Yemen

Name	*Type*	*Number in possession*	*Delivery/introduction*	*Comments*
MiG-29SMT *Fulcrum*	Multirole fighter	18	2002–05	Upgraded MiG-29, incl. 2 combat-capable trainers (UBT version).
Su-20 *Fitter/Su-22 Fitter D*	Fighter-bomber	30	1995–2007	Different versions.
F5E *Tiger II*	Fighter aircraft	12	?	Incl. 2 combat-capable trainers (F5B *Freedom Fighter*).
MiG-21N *Fishbed*	Interceptor/ fighter-bomber	19	2006/07	Incl. 4 combat-capable trainers (*Mongol A*). Status uncertain.
Scud-C	SRBM	45 (?)	2001/02	Possible sale of North Korean Hwasong-6.
SS-21 *Scarab*	SRBM	10 launchers	?	
Frog-7	Long-range artillery rocket	12	?	
SS-N-2C *Styx*	ASCM	16	1990/91	
SA-2 *Guideline*	Medium-range SAM	10	?	
SA-3 *Goa*	Medium-range SAM	15	?	
SA-6 *Gainful*	Self-propelled, medium-range SAM	?	?	
SA-9 *Gaskin*	Self-propelled, short-range SAM	120 (missiles?)	?	
SA-13 *Gopher*	Self-propelled, short-range SAM	?	?	

The rocket arsenals of selected non-state actors

Hamas

No exact, independently verifiable figures exist for either Hamas' total inventory or specific systems. Its arsenal was severely decimated in Israel's 2008/09 "Operation Cast Lead", but has since been rebuilt.[1] There is reason to believe that Hamas' arsenal now exceeds the group's pre-"Operation Cast Lead" inventory by a significant margin.[2] At the time of the 2008/09 War, the total rocket holdings of Gaza-based militant groups were estimated at 3,000 by Israeli officials.[3] All figures given here are rough estimates, based on existing data and the compilers' informed judgements of what is plausible. All systems operated by Hamas are highly inaccurate and of limited military value. Nonetheless, these systems are (from an Israeli perspective) strategically significant.

Numbers refer to rockets (as opposed to launchers).
Short-range rockets: ≤20 km range.
Medium-range rockets: 20–50 km range.
Long-range rockets: ≥50 km range.

Table A.12 Hamas

System	*Type*	*Estimated inventory*	*Comments*
Qassam-1	Short-range rocket	Overall inventory of Qassams is estimated to include as many as 2,500 rockets.	Self-made rocket. Approx. 60 mm in diameter, range: 3–4.5 km, warhead: 0.5 kg. Introduced in 2001.[1] Probably superseded by other types.
Qassam-2	Short-range rocket		Self-made rocket. Approx. 150 mm in diameter, range: 8–9.5 km, warhead: 5–9 kg. Introduced in 2002.[2]
Qassam-3	Short-range rocket		Self-made rocket. Approx. 170 mm in diameter, range: 10–12 km, warhead: 10–20 kg. Introduced in 2005.[3]
Type 63/ Haseb	Short-range artillery rocket	Several hundred	Chinese-designed 107 mm rocket with 8.5 km range and 8.3 kg HE warhead.[4] Part of several seized arms shipments bound for Gaza.[5]
BM-21 *Grad* Arash	Short-range artillery rocket	≤ several hundred	122 mm *Katyusha*-type MLRS with 20 km range. Rocket probably derived from the Soviet 9M22U with 18 kg fragmentation warhead.[6] Introduced in Gaza 2008.[7]
175 mm rocket	Medium-range artillery rocket	≤100	First used by Palestinian Islamic Jihad in 2008, possibly Iranian design. Range: 18–26 km, warhead: 10 kg.[8] Unclear if part of Hamas arsenal as well.

continued

Table A.12 Continued

System	*Type*	*Estimated inventory*	*Comments*
WS-1E	Medium-range artillery rocket	≤several hundred	Chinese-designed 122 mm rocket with 40 km range and 18–22 kg fragmentation warhead. Introduced in Gaza 2008.[9]
Unknown Iranian type	Long-range artillery rocket	≤several dozens	Rocket of Iranian origin and unknown type with 60 km range, possibly a Fajr-3 or Fajr-5 variant. Tested in 2009.[10]

Notes

1 www.aerospaceweb.org/question/weapons/q0279.shtml; www.globalsecurity.org/military/world/para/hamas-qassam.htm.
2 www.globalsecurity.org/military/world/para/hamas-qassam.htm; www.aerospaceweb.org/question/weapons/q0279.shtml.
3 www.globalsecurity.org/military/world/para/hamas-qassam.htm; www.aerospaceweb.org/question/weapons/q0279.shtml.
4 www.globalsecurity.org/military/world/iran/mrl-iran-specs.htm; www.fas.org/man/dod-101/sys/land/row/type-63-r.htm; www.sinodefence.com/army/mrl/type63.asp.
5 www.upi.com/Top_News/Special/2010/11/15/Evidence-Iran-is-behind-seized-arms-cargo/UPI-30221289854759/; www.mfa.gov.il/MFA/Government/Communiques/2002/Seizing%20of%20the%20Palestinian%20weapons%20ship%20Karine%20A%20-.
6 www.fas.org/man/dod-101/sys/land/row/bm-21.htm.
7 *Jane's Defence Weekly* (30 April 2008: 18); www.globalsecurity.org/military/world/para/hamas-qassam.htm.
8 *Jane's Defence Weekly* (12 March 2008: 5, 30 April 2008: 18); Moshe (2008: 10).
9 www.globalsecurity.org/military/world/para/hamas-qassam.htm; www.sinodefence.com/army/mrl/weishi.asp.
10 *Jane's Defence Weekly* (3 March 2010: 25); www.jpost.com/Home/Article.aspx?id=159461; http://uk.reuters.com/article/2009/11/03/idUKL3622445; www.haaretz.com/print-edition/news/analysis-israel-preparing-public-for-a-new-war-in-gaza-1.4839.

Hezbollah

Hezbollah's total rocket inventory is estimated at 40,000–55,000 by US and Israeli sources.[4] Most of these are short-range artillery rockets of the Grad type, but, according to the Israeli government, the number of long-range rockets exceeds 10,000.[5] No exact, independently verifiable figures exist for either the total inventory or specific systems. All figures given here are rough estimates, based on existing data and the compilers' informed judgements of what is plausible.

All types, except for the Fateh-110/M-600, are highly inaccurate and of limited military value. However, as part of a massive arsenal able to threaten a very large fraction of Israel's territory, they are of great strategic significance. Most, if not all, rockets used in the 2006 War carried anti-personnel warheads.[6]

Table A.13 Hezbollah

System	*Type*	*Estimated inventory*	*Comments*
Type 63/ Haseb	Short-range artillery rocket	Thousands	Chinese-designed 107 mm rocket with 8.5 km range and 8.3 kg HE warhead.[1]
BM-21 *Grad*/Arash	Short-range artillery rocket	Thousands	122 mm *Katyusha*-type MLRS with 20 km range. Rocket probably derived from the Soviet 9M22U with 18 kg fragmentation warhead.[2]
Improved *Grad*	Medium-range artillery rocket		*Grad* variants with up to 50 km range.[3]
Shahine I	Short-range artillery rocket	≤100	Iranian 333 mm rocket with 13 km range and 190 kg warhead.[4]
Shahine II	Short-range artillery rocket		Iranian 333 mm rocket with 20 km range and 190 kg warhead.[5]
BM-27 *Uragan*	Medium-range artillery rocket	Several thousand	220 mm improved, *Katyusha*-type MLRS with 35 km range. Rocket in use possibly a version of Soviet 9M27K3 with 90 kg anti-personnel warhead.[6]
Fajr-3	Medium-range artillery rocket	Several thousand	Iranian 230 mm rocket with 45 km and 45 kg warhead. Possibly called *Ra'ad* by Hezbollah.[7]
WS-1	Long-range artillery rocket	≤several hundred	302 mm rocket with 90–100 km range, probably derived from Chinese WS-1.[8] Apparently highly unreliable.[9]
Fajr-5	Long-range artillery rocket	Several hundred	Iranian 333 mm rocket with 75 km range and 90 kg HE warhead.[10] Possibly called *Khybar/Haibar* by Hezbollah.[11]
Zelzal-2	Long-range artillery rocket	≤several hundred	Iranian 610 mm rocket. Range: 200 km, warhead: 600 kg. Spin-motor-stabilized, but still inaccurate.[12]
Fateh-110/M-600	SRBM	40–50 (low est.), several hundred (high est.)	Improved Zelzal with guidance system, higher accuracy. Range: 250 km, warhead: 500 kg. 500 m estimated CEP.[13]
Scud-D	SRBM	≤10	North Korean Hwasong-7, range increased to 700 km. Delivered to Hezbollah according to US and Israeli sources.[14] Other sources indicate Scud-B version.[15] Status uncertain.[16]

continued

Table A.13 Continued

System	*Type*	*Estimated inventory*	*Comments*
C-701	ASCM	≤20	Chinese anti-ship missile with 15–20 km range and 29 kg warhead. Iranian *Kosar* version.[17] Possibly used against INS *Hanit* during 2006 Lebanon War.[18]
CSS-N-8 *Saccade*	ASCM	≤20	C-802 export version of Chinese cruise missile with 120 km range and 165 kg HE warhead. Produced in Iran as *Noor*. Used against INS *Hanit* in 2006 Lebanon War.[19]

Notes

1 *Jane's Defence Weekly* (16 July 2008: 30); www.globalsecurity.org/military/world/iran/mrl-iran-specs.htm; www.fas.org/man/dod-101/sys/land/row/type-63-r.htm; www.sinodefence.com/army/mrl/type63.asp.
2 *Jane's Defence Weekly* (16 July 2008: 30); www.fas.org/man/dod-101/sys/land/row/bm-21.htm; www.aerospaceweb.org/question/weapons/q0279.shtml.
3 Rubin (2006, 2007: 5); *Jane's Defence Weekly* (28 October 2009: 5).
4 www.globalsecurity.org/military/world/iran/mrl-iran-specs.htm; www.aerospaceweb.org/question/weapons/q0279.shtml.
5 www.globalsecurity.org/military/world/iran/mrl-iran-specs.htm; www.aerospaceweb.org/question/weapons/q0279.shtml.
6 www.fas.org/man/dod-101/sys/land/row/9p140.htm.
7 *Jane's Defence Weekly* (16 July 2008: 30); www.globalsecurity.org/military/world/iran/mrl-iran-specs.htm.
8 Rubin (2007: 6); www.sinodefence.com/army/mrl/weishi.asp.
9 Rubin (2007: 11).
10 *Jane's Defence Weekly* (16 July 2008: 30); IISS (2010b: 45); www.globalsecurity.org/military/world/iran/mrl-iran-specs.htm.
11 www.globalsecurity.org/military/world/para/hizballah-rockets.htm.
12 *Jane's Defence Weekly* (7 May 2008: 22); McGregor (2006: 2); IISS (2010b: 49–50).
13 *Jane's Defence Weekly* (28 October 2009: 5); *Jane's Defence Weekly* (26 May 2010: 5); www.nytimes.com/2010/12/07/world/07wikileaks-weapons.html?pagewanted=1&_r=1&ref=hezbollah; www.aviationweek.com/aw/generic/story_generic.jsp?channel=defense&id=news/awst/2010/05/17/AW_05_17_2010_p28-226325.xml&headline=Israel%20Sees%20Increased%20Hezbollah%20Capability&next=10; IISS (2010b: 51–3).
14 *Jane's Defence Weekly* (21 April 2010: 5); www.guardian.co.uk/world/us-embassy-cables-documents/249979; www.nytimes.com/2010/12/07/world/07wikileaks-weapons.html?pagewanted=1&_r=1&ref=hezbollah; www.nytimes.com/2010/04/15/world/middleeast/15israel.html; www.missilethreat.com/missilesoftheworld/id.104/missile_detail.asp.
15 IISS (2010c).
16 *Jane's Defence Weekly* (20 October 2010: 27).
17 www.globalsecurity.org/military/world/iran/kosar.htm.
18 www.strategycenter.net/research/pubID.117/pub_detail.asp.
19 www.missilethreat.com/cruise/id.62/cruise_detail.asp; Rubin (2007: 6).

Notes

1 *Jane's Defence Weekly* (28 January 2009: 16, 3 March 2010: 25).
2 In 2009/10, ranking Israeli intelligence officials were quoted as stating that Hamas' arsenal significantly exceeded, in quality and quantity, its pre-war holdings, see online, available at: www.aviationweek.com/aw/generic/story_channel.

jsp?channel=defense&id=news/dti/2010/03/01/DT_03_01_2010_p41–204703.xml; www.haaretz.com/print-edition/news/analysis-israel-preparing-public-for-a-new-war-in-gaza-1.4839 (27 November 2010).
3 Online, available at: www.reuters.com/article/2008/12/30/idUSLU273804; www.jpost.com/Home/Article.aspx?id=159461 (27 November 2010).
4 UNSC (2006a: 10): Online, available at: www.un.org/ga/search/view_doc.asp?symbol=S/2010/565 (27 November 2010).
5 *Jane's Defence Weekly* (7 May 2008: 22); UNSC (2006b).
6 Rubin (2007).

References

Bermudez, Jr, Joseph S. (1999) *A History of Ballistic Missile Development in the DPRK*, Monterey, CA: Center for Nonproliferation Studies, Monterey Institute of International Studies. Online, available at: http://cns.miis.edu/opapers/op2/op2.pdf (25 March 2011)

Carus, W. Seth (1992) *Cruise Missile Proliferation in the 1990s*, Westport, CT: Praeger Security International.

Cordesman, Anthony H. (2002) *Peace and War: The Arab-Israeli Military Balance Enters the 21st Century*, Westport, CT: Praeger Security International.

Cordesman, Anthony H. (2006) *Arab-Israeli Forces in an Era of Asymmetric Wars*, Westport, CT: Praeger Security International.

Gormley, Dennis M. (2008) *Missile Contagion: Cruise Missile Proliferation and the Threat to International Security*, Westport, CT: Praeger Security International.

IISS/International Institute for Strategic Studies (2010a) *The Military Balance 2010*, London: Routledge.

IISS/International Institute for Strategic Studies (2010b) *Iran's Ballistic Missile Capabilities: A Net Assessment*, London: IISS.

IISS/International Institute for Strategic Studies (2010c) 'Hizbullah's Alledged Scuds Raise Storm Clouds over Lebanon', *Strategic Comments*, 10(20): 1–4.

INSS/Institute for National Security Studies (2011) *ME Military Forces Database*. Online, available at: http://www.inss.org.il/weapons.php?cat=250 (23 March 2011).

McGregor, Andrew (2006) 'Hezbollah's Rocket Strategy', *Terrorism Monitor*, 4(16): 2.

Moshe, Yaalon (2008) *Iran's Race for Regional Supremacy: Strategic Implications for the Middle East*, Jerusalem: Jerusalem Center for Public Affairs.

Rubin, Uzi (2006) 'Hizballah's Rocket Campaign against Northern Israel: A Preliminary Report', *JCPA Jerusalem Issue Briefs*, 6(10). Online, available at: www.jcpa.org/brief/brief006–10.htm (27 November 2010).

Rubin, Uzi (2007) 'The Rocket Campaign against Israel during the 2006 Lebanon War', *BESA Mideast Security and Policy Studies*, 71: 5–6. Online, available at: www.biu.ac.il/Besa/MSPS71.pdf (27 November 2010).

SIPRI/Stockholm International Peace Research Institute (2011) *Arms Transfers Database*. Online, available at: http://armstrade.sipri.org (23 March 2011).

UNSC/United Nations Security Council (2006a) 'Fourteenth Report of the Secretary-General on the Implementation of Security Council Resolution 1701', S/2010/565, New York: United Nations. Online, available at: www.un.org/ga/search/view_doc.asp?symbol=S/2010/565 (27 November 2010).

UNSC/United Nations Security Council (2006b) 'Report of the Secretary-General on the Implementation of Security Council Resolution 1701', S/2010/565, New York: United Nations. Online, available at: www.un.org/ga/search/view_doc.asp?symbol=S/2008/135 (27 November 2010).

Index

2006 War (Second Lebanon War) 43, 53, 77, 92, 203, 167, 173, 175–6

Abdullah, Crown Prince of Saudi Arabia 198
Additional Protocol (AP) 17, 107, 109, 113, 115, 118–19, 121–2, 126, 129, 284–6
Ahmadinejad, Mahmoud 69, 102, 115, 117–18, 285
airborne surveillance 155, 158, 163
Air Defence (AD) 40, 71, 77, 97, 134, 136, 140, 143, 226
Allon, Yigal 111
Al Qaida 190, 197, 223, 301
Anti-Ballistic Missile (ABM) 41, 220, 259–60, 263
Anti-Ship Cruise Missile (ASCM) 51, 77, 134, 136, 238–9, 247
Arab League 68, 91, 110, 120, 175, 197–8, 222, 226–8, 262, 270
Arab Peace Initiative (API) 18, 68, 103, 196, 217, 222, 230, 262, 300
Arms Control and Regional Security (ACRS) 2, 16, 27–37, 81, 106, 223–4, 226, 251–2, 260–1, 263–5, 270–4, 293–5, 298, 301
Army Tactical Missile System (ATACMS) 134, 136
Arrow (-2/-3) 48, 55, 57, 67, 101, 133, 238, 248, 267
Asad, Hafiz-al (Asad regime/Asad dynasty) 76–7, 82, 97, 168–9, 179–80, 182
Australia, New Zealand, and United States of America (ANZUS) 129, 144

Baker, James 29, 34
Ballistic Missile Defence 41, 48, 51–2
Barak, Ehud 167, 174, 176, 179, 183
Begin, Menachem 112
Begin doctrine 90
Ben-Gurion, David 93
biological weapons (B-weapons) 31, 93, 110, 114, 118, 236
Biological and Toxin Weapons Convention (BTWC) 110, 114, 118, 121, 284, 285
Blix Commission (WMD Commission) 4
Brandt, Willy 220, 222
Bush, George W. 76, 135, 202, 252, 260

Carter, Jimmy (Carter administration) 138, 200
Central Command (CENTCOM) 69, 137
chemical weapons (C-weapons) 11, 27, 57, 65, 75, 79, 80, 84, 98, 106–8, 110, 114, 155, 284–5
Chemical Weapons Convention (CWC) 57, 110, 114, 118, 284–5
Clinton, Bill (Clinton administration) 135, 138, 179
Clinton, Hillary 116
Comprehensive Test Ban Treaty (CTBT) 17, 106, 109, 113, 118–20, 156, 228, 284–5
Conference on Disarmament (CD) 113, 229, 242
Conference on Security and Co-operation in Europe (CSCE) 13, 19, 217–24
confidence-building measures (CBMs) 10, 19, 34, 83, 116, 120, 143, 159, 200, 218, 240–1, 290, 295
confidence- and security-building measures (CSBMs) 2, 14–16, 20,

29–30, 32–3, 35, 150, 152–3, 159, 163, 217–20, 222–6, 230, 251–3, 255, 259, 261, 264, 271, 277, 293–5, 301
Conventional Armed Forces in Europe (CFE) Treaty 151, 159, 258, 260, 263, 273
Cooperative Security Concept (CSC) 2, 13–14, 277, 281, 296, 300–1
crisis stability 1, 13, 56, 151, 160, 219, 253, 255–6, 258–9, 266–7, 269, 271–2, 288, 297–8
cruise missile 4–6, 17, 39–42, 45, 49, 51–2, 56–7, 77, 96, 102–3, 128, 130, 133–4, 136, 149, 157–8, 225, 235, 238–9, 241–3, 247, 265, 267, 278, 296
CSS (-2 (DF-3)/-6/-7/-8/-N-1/-N-2/N-8) 47, 49–50, 133–6, 142
Czempiel, Ernst-Otto 11, 15, 21

Dayan, Moshe 111
Delivery Vehicles (DVs) 2–4, 16–17, 40, 48, 65, 80, 106, 150, 151, 160–2, 257, 261, 265–6, 272, 279, 286, 288–9, 299

Erdogan, Recep Tayyip 183
Eshkol, Levy 94

First Gulf War (Iran–Iraq War) 114, 117, 182
Fissile Material Cut-off Treaty (FMCT) 107, 109–10, 113, 118–20, 284–6

Gaza War (Operation Cast Lead) 43, 53, 73, 91–2, 186, 191–5, 201–2, 204, 206, 208
Ghadr-1 (Shahab-3M) 80, 237
Golan Heights Agreement (Golan accord) 152, 224, 251, 295
Gorbachev, Mikhail 220, 222, 258, 259
Graduated Reciprocated Initiatives in Tension Reduction (GRIT) 209, 222, 224
ground-based surveillance 154–6, 289
Ground-Launched Cruise Missile (GLCM) 45, 57
Gulf Cooperation Council (GCC) 67–73, 81, 118, 135, 281, 282
Gulf wars 5–6, 20, 47, 82, 278
Gulf War 2003 (Third Gulf War) 245, 278

Hague Code of Conduct Against Ballistic Missile Proliferation (The Hague Code/HCOC) 6–7, 9, 19, 125, 234, 240–7, 287, 296, 297
Hamas 5, 7, 14, 18, 20, 32, 40, 42–4, 52–5, 66, 73–4, 91–2, 99, 108, 168, 172, 174, 179, 186–98, 200–9
Hezbollah 5, 7, 14, 18, 20, 32, 40, 42–3, 46, 51–6, 66, 73, 77, 90–2, 99, 102, 108, 133, 138, 167–83, 189, 229, 230–1, 239, 252, 265, 274, 278–80, 283, 290–3, 298
human inspections 154–7, 161–2, 288–9
Hussein, Saddam 27, 47–8, 78–9, 98–9, 108, 117, 137, 157, 168–9, 226

Intercontinental Ballistic Missile (ICBM) 257
International Atomic Energy Agency (IAEA) 33, 107–8, 110, 114–15, 117, 120–2, 127, 157, 284–5, 294
International Code of Conduct Against Ballistic Missile Proliferation (ICOC) 240, 242
Intermediate-Range Ballistic Missile (IRBM) 72, 80, 139
Intermediate-Range Nuclear Forces Treaty (INF) 20, 45, 143, 151, 153, 155, 157, 161, 259–60, 299
Intifada 33, 53, 90, 172, 182, 186–8, 208
Iron Dome 54–5, 174, 192, 280, 292
Islamic Revolutionary Guard Corps (IRGC) 56
Israel Defense Forces (IDF) 43–4, 54, 76, 90–1, 93–4, 97, 103
Israeli Air Force (IAF) 58, 77, 93, 97–9, 122, 173, 227
Israeli Army 188, 192, 204
Israeli–Egyptian Peace Treaty 73, 111, 152, 294
Israeli–Jordanian Peace Treaty 28, 37, 224
Israeli Navy 96

Jericho (-1/-2/-3) 51, 96, 103
Jihad 170, 201

Katyusha 43, 51–4, 58, 167, 174, 191, 229, 279, 292
Keohane, Robert O. 12, 15
Khamenei, Ali 115, 285
Khomeini, Ruhollah 114, 168–9, 182, 291

Land-Attack Cruise Missile (LACM) 40–2, 51–2, 57

Magic Wand (David's Sling) 55, 133
Man-Portable Air Defence System (MANPADS) 225
Medvedev, Dmitriy (Medvedev administration/government) 136, 139
Meir, Golda 95
Missile Defence (MD) 3–4, 6, 19, 39–42, 44, 48, 50–2, 54–7, 65–6, 71–3, 78, 101–2, 128, 131, 134–5, 137–8, 141–3, 168, 174, 180, 192, 207, 238, 251–2, 264–9, 272–3, 278–82, 286, 290–2, 297–8, 300
Missile Free Zone (MFZ) 1–3, 12, 14, 16–19, 65, 67, 81–2, 126–8, 130–1, 141–4, 149–50, 152–4, 157–63, 167, 174, 181, 186, 208, 217, 225, 230, 251, 252, 262, 264, 268–9, 272, 277, 282–3, 286–9, 292, 295, 297, 301
Missile Technology Control Regime (MTCR) 6–7, 9, 19, 45–6, 125, 140, 142–4, 234–6, 238–40, 242–8, 287, 296–7
Moravcsik, Andrew 11, 15, 21
Mubarak, Hosni 27–8, 74–5, 82, 108–10
Multiple Independently Targetable Re-entry Vehicle (MIRV) 96, 255, 257, 259, 273
Mutual and Balanced Force Reductions (MBFR) 258, 260

Nasrallah, Hasan 43, 171, 172
Nasser, Gamal Abdel 43, 93–4, 97, 107
National Technical Means (NTMs) 150–2, 154, 162–3
Netanyahu, Benjamin (Netanyahu government) 113, 179, 183, 194–5, 204
Nixon, Richard M. 95
Nixon doctrine 133
Nodong (-B) 79, 127, 237–8
North Atlantic Treaty Organization (NATO) 76, 103, 138, 220, 273
Nuclear Non-Proliferation Treaty (NPT) 2, 6, 27–8, 33, 36, 81, 95, 100, 106–15, 118–20, 127, 144, 163, 228, 252, 284–6, 294
Nuclear Weapon Free Zone (NWFZ) 3–4, 17–18, 27–8, 33, 36, 81, 108, 111–12, 120, 126–9, 149, 284, 286
Nuclear Weapons (N-weapons) 3–5, 7, 11, 28, 37, 48–9, 66, 73, 79, 84, 93, 95, 99, 100, 106–15, 118, 120, 128–30, 160, 228, 241, 243–4, 266, 280, 283–5

Obama, Barack (Obama administration) 3, 70, 76, 78, 128, 183, 195, 245
Olmert, Ehud (Olmert government) 175, 179, 204
Open Skies Treaty 155, 159–60, 165, 258, 264
Organization for Security and Co-operation in Europe (OSCE) 224, 268
Oslo Declaration (Oslo process/Oslo negotiations) 33, 91, 183, 188, 204

Palestine Liberation Organization (PLO) 52, 90–2, 96, 169, 182, 187, 193, 199, 201–2
Palestinian Authority (PA) 193–7, 201–3, 206
Patriot Advanced Capability (Pac) 20, 41, 48–9, 57, 72, 79, 101, 131, 134–5, 137, 267, 298
Peres, Shimon 94, 111, 179
Powell, Colin 131
Pre-Launch Notification (PLN) 241–2
Proliferation Security Initiative (PSI) 237, 239–40
Putin, Vladimir (Putin administration) 136, 139

Qassam (-1/-2/-3) 16, 42–3, 53–4, 174, 189, 191–2, 229, 279–80

Rabin, Yitzhak 33, 111, 138
Reagan, Ronald 20, 101, 164, 260
Regional Security Centre (RSC) 30, 33, 224, 270–1, 293, 301
Register of Conventional Arms in the Middle East (ROCAME) 227, 264, 295
Remotely Piloted Vehicles (RPV) 5, 225
Research and Development (R&D) 6, 84, 161, 288
Rice, Condoleezza 138, 239
Royal Saudi Air Force (RSAF) 71–2

Scud (-B/-C/-D) 6, 43, 47–50, 71, 74–5, 77, 79, 98–9, 125, 133–4, 235, 237–9, 245, 248, 278, 296
Second Gulf War (1991 Gulf War/ Operation Desert Storm) 28, 41, 46–8, 71, 79, 82, 90, 98, 101, 131, 137–8, 245, 267, 278, 298
Shahab (-1/-2/-3) 50, 57, 79–80, 83, 101, 157, 237–8

Short-Range Ballistic Missile (SRBM) 41, 50, 74–5, 101, 235
Sinai I (Separation of Forces Agreement) 152–3, 224, 294
Sinai II (Interim Agreement) 152–3, 224, 294
Skyguard 54–5, 192
space-based surveillance 154–5, 163, 289
Space Launch Vehicle (SLV) 51, 80, 149, 157–8, 162, 235, 241
SS (-1c/-18/-21/-26/-N-6-/R-27) 49–50, 57, 77, 133–4, 136, 237, 239, 273
Strategic Arms Limitation Talks (SALT I/II) 255, 260, 262
Strategic/arms race stability 1, 70, 151, 253, 255–6, 259, 261, 265–6, 268–9, 271–2, 291, 297
Strategic Arms Reduction Treaty (START) 159, 257, 260
Strategic Defense Initiative 101
Strategic Offensive Reductions Treaty (SORT) 257
Supreme National Security Council (SNSC) 115
Surface-to-Air Missile (SAM) 55, 71–2, 134, 167, 247
Surface-to-Surface Missile (SSM) 77, 96, 99, 102, 134, 135, 142
Synthetic Aperture Radar (SAR) 154–5

United Nations Emergency Force (UNEF) 152–3
United Nations General Assembly 5, 8, 20, 27, 108, 111, 126–7, 222, 225, 243–4
United Nations Institute for Disarmament Research (UNIDIR) 4, 20
United Nations Panel of Governmental Experts on Missiles 2, 4–5, 8–9, 20, 158, 243, 296
United Nations Register of Conventional Arms (UN Register, UNROCA) 5, 18, 19, 139, 217, 225–31, 265, 295
United Nations Security Council (UNSC) 5, 69, 78, 115–16, 121, 139, 157, 173, 176–7, 198, 208, 236, 238, 240, 247, 268, 285–6
Unmanned Aerial Vehicle (UAV) 4–6, 17, 39–40, 44–6, 51, 56–7, 157–8, 161, 235, 265, 267, 279, 288, 299
U.S. Sinai Support Mission 152

Waltz, Kenneth 9–10, 15
War of the Cities 46, 47, 57, 169, 278
weapons of mass destruction (WMD) 2–6, 17, 27–9, 31–2, 35–7, 40, 45, 48, 65–6, 79, 81, 98–9, 106–8, 110, 113–14, 118–22, 127, 144–5, 180, 226–8, 235–6, 239, 241, 252, 261–2, 265, 267, 274, 277, 280, 283, 285–6, 294, 298–300
Weapons of Mass Destruction Free Zone (WMDFZ) 2–3, 27–8, 33, 35–6, 68, 81, 106, 108–9, 118–19, 121, 126, 144, 149, 269, 272, 277, 284, 286, 294–5, 298, 300–1
Wendt, Alexander 11, 15

Yassin, Ahmad Sheikh 187, 201
Yemen Civil War 46, 278
Yom Kippur War 46, 91, 97, 278

For Product Safety Concerns and Information please contact our EU
representative GPSR@taylorandfrancis.com
Taylor & Francis Verlag GmbH, Kaufingerstraße 24, 80331 München, Germany

I ncontent.com/pod-product-compliance
P ent Group UK Ltd.
UK on Keynes, MK11 3LW, UK
47 3180926
00005B/1000
* 9 4 1 5 5 1 6 3 9 6 *